Geotechnics of Roads: Advanced Analysis and Modeling

T0199708

Geotechnics of Roads: Advanced Analysis and Modeling

Bernardo Caicedo

Department of Civil and Environmental Engineering
Los Andes University, Bogotá, Colombia

CRC Press
Taylor & Francis Group
Boca Raton London New York

CRC Press is an imprint of the
Taylor & Francis Group, an **informa** business

Library of Congress Cataloging-in-Publication Data
Names: Caicedo, Bernardo, author.
Title: Geotechnics of roads : fundamentals / Bernardo Caicedo (Department of Civil and Environmental Engineering, Los Andes University, Bogtâa, Colombia.
Description: Leiden, The Netherlands : CRC Press/Balkema, [2019] |
Includes bibliographical references and index.
Identifiers: LCCN 2018051953 (print) | LCCN 2018044343 (ebook) |
ISBN 9781138600577 (hardcover) | ISBN 9780429025914 (ebook)
Subjects: LCSH: Geotechnical engineering. | Roads–Design and construction. | Road materials. | Soil mechanics.
Classification: LCC TA705 .C27 2019 (ebook) | LCC TA705 (print) |
DDC 625.7/32–dc23
LC record available at https://lccn.loc.gov/2018051953

ISBN: 978-1-138-60057-7 (hbk Volume 1)
ISBN: 978-0-429-02591-4 (ebk Volume 1)
DOI: 10.1201/9780429025914
https://doi.org/10.1201/9780429025914

ISBN: 978-1-138-60058-4 (hbk Volume 2)
ISBN: 978-0-429-02592-1 (ebk Volume 2)
DOI: 10.1201/9780429025921
https://doi.org/10.1201/9780429025921

ISBN: 978-1-138-02956-9 (hbk set of 2 volumes)
ISBN: 978-1-315-22639-2 (ebk set of 2 volumes)

ISBN: 978-0-367-70778-1 (pbk)

Typeset in Times New Roman
by codeMantra

Dedication

To Gloria
Alejandro and Nicolás.

Contents

Author Biography

Bernardo Caicedo obtained his undergraduate degree in civil engineering at the Universidad del Cauca in Colombia in 1985. He received the Francisco José de Caldas Medal for his academic achievements, which is only awarded about once in every 10 years. He did his doctoral work at L'Ecole Centrale de Paris in 1991 and is working at the Laboratoire Central des Ponts et Chaussées in Paris.

He joined the "Universidad de los Andes, Uniandes" in Colombia in 1991. He has been involved in teaching, research, and administrative duties here for around 29 years now. During his first period at Uniandes, his research focused on bringing the laboratory to a high standard of international competitiveness. This effort has met with relatively good success. One example of this work was the design and construction, under his leadership, of two geotechnical centrifuges used for teaching and research. One of these machines is equipped with an environmental chamber including leading technologies to simulate climate in experiments using the centrifuge. Other major laboratory apparatus designed and constructed under his direction were a large hollow cylinder apparatus, a linear test track for physical modeling of pavements, and a 60-ton shaking table.

His research activities cover a broad spectrum of areas in geotechnical engineering. They include studies of soil dynamics, unsaturated soils, physical modeling of unsaturated soils (expansive and collapsing soils), study of the behavior of unbound granular materials for pavements, the development of new pavement design methods based on mechanistic concepts with climatic interactions, and the study of the mechanical behavior of multi-phase soils (unsaturated soils) including chemical (chemo-mechanics) and biological aspects. His efforts to develop laboratory facilities have allowed him to publish several documents: to date, he has published more than 200 documents including 2 books, 65 papers in indexed journals, 5 keynote lectures, 89 papers in reviewed international conferences, 12 in other international conferences, and 51 in local conferences. He is a member of the editorial panel of two international journals: *Transportation Geotechnics* (Elsevier) and *Acta Geotechnica*. In addition, he is a member of the board of the TC202 committee (Transportation Geotechnics) and vice chair of the TC 106 committee (Unsaturated Soils) of the International Society of Soil Mechanics and Geotechnical Engineering ISSMGE.

He has received several awards from the Institution of Civil Engineers (ICE): the Telford Premium in 2016, the Geotechnical Research Medal in 2018, and the Mokshagundam Visvesvaraya Award in 2020.

Acknowledgments

This book is the result of two years of work, and it was finalized during the pandemic of 2019–2020. For me, it is a pleasure to acknowledge the people who made it possible.

First of all, I would like to express my gratitude to my wife Gloria and my sons Alejandro and Nicolas. They encourage me in all moments and also provide me an environment of happiness that makes all the big projects possible.

A significant recognition is due to the University of Los Andes and its Department of Civil and Environmental Engineering which supported this work. Special gratitude goes to my colleagues from our Geomaterials and Infrastructure Systems research group: Silvia Caro, Laura Ibagón, Julieth Monroy, Mauricio Sánchez, Nicolás Estrada, Miguel Angel Cabrera, and to my colleagues and friends in the Department.

Research in geotechnical engineering has given me the pleasure of finding a community of friends and bright people; this book would not have been possible without the knowledge and friendship they have offered me.

I would express my gratitude to some graduate students who helped me in reviewing some of the examples and MATLAB® scripts: Carlos Vladimir Benavides, Lina María Pua, Juan Villacreses, Jorge Mario Lozano, and many others that I had the privilege to advise in the last 30 years.

List of mathematical symbols

The following table presents the list of the main mathematical symbols used throughout the book. However, it is important to remark that sometimes the same symbol has several meanings. For this reason, the reader must verify the context and the definition of the symbols presented below each equation.

ROMAN LETTERS

Symbol	Definition
A_i	Constant for the Burmister's method.
A_{σ_v}	Slope relating the Young's modulus and the vertical stress.
A_ω	Amplitude of the fundamental frequency of a vibratory drum compactor.
$A_{2\omega}$	Amplitude of the first harmonic of a vibratory drum compactor.
a	Radius of a circular loaded area, or
	half axis of a super elliptical contact area, or
	half contact width for the Hertz contact theory, or
	parameter of the water retention curve.
$a_{i,j}$	Decompaction coefficient of a granular mixture.
B_i	Constant for the Burmister's method.
b	Half axis of a super elliptical contact area, or
	intermediate principal stress parameter.
$b_{j,i}$	Wall coefficient of a granular mixture.
C_{AK}	Proportionality factor for defining the contact stress between a tire and the road.
C_c	Compression coefficient measured in oedometric compression tests.
C_i	Constant for the Burmister's method.
C_r	Recompression coefficient.
C_s	Coefficient of the dashpot representing the soil below a drum compactor.
C_θ	Specific water capacity.
CMV	Compaction meter value.
c	Dashpot constant.
c_H	Specific heat capacity.
c_{H_v}	Volumetric heat capacity of the soil.
c_h	Coefficient of horizontal consolidation.
c_s	Specific heat of the solid grains.

(*Continued*)

Symbol	Definition
c_u	Undrained shear strength in saturated state.
c_v	Coefficient of vertical consolidation.
c_w	Specific heat of water.
c_1	Damping coefficient of the tire for defining the truck-road interaction.
c_2	Damping coefficient of the suspension's shock absorber for defining the truck-road interaction.
D_{hg}	Time difference between the time of a site and the time zone's reference point.
D_i	Constant for the Burmister's solution.
d_n	Julian day.
d_x	Size of the sieve corresponding to x% by weight of those which passed through the sieve.
d_{60}	Grain size corresponding to 60 % by mass of those that passed through this dimension.
E	Young's modulus.
E^*	Equivalent Young's modulus.
E_r	Resilient Young's modulus.
E_{r-opt}	Resilient modulus at optimum moisture content
E_u	Undrained Young's Modulus
E_0	Correction factor for the direct radiation.
$E_{0\sigma_v}$	Young's resilient modulus for zero vertical stress.
e	Void's ratio, or eccentricity of the mass of a vibratory drum.
e_m	Microstructural void's ratio for the microstructural BBM.
e_0	Initial void's ratio for computing settlements due to consolidation.
$f(t)$	Vector that represents the road profile as a time signal.
G	Shear modulus.
G_a	Fitting coefficient of the Boyce's model.
G_r	Resilient shear modulus.
g	Acceleration of the gravity.
H	Normalized thickness of the road structure.
h	Angle between the center of the disc of the sun and the horizon, the solar zenith angle.
h_c	Convection coefficient
I	Irradiance
I_b	Direct irradiance.
I_d	diffuse irradiance.
I_0	Solar constant.
J_0	Bessel function of the first kind and order zero.
J_1	Bessel function of the first kind and order one.
K	Coefficient of volumetric compressibility, or packing coefficient of a granular mixture.
K_a	Fitting coefficient of the Boyce's model.
K_e	Kersten's number.
K_i	Coefficient that relates the packing state of grains i within a granular mixture.

(*Continued*)

Symbol	Definition
K_N	Coefficient relating the increase in density depending on the loading cycles.
K_r	Resilient coefficient of volumetric compressibility.
k_c	Parameter of the BBM model relating the increase in tensile strength due to an increase of suction.
k_H	Thermal conductivity.
k_{Hi}	Thermal conductivity of ice.
k_{Hs}	Thermal conductivity of the solid grains.
k_{Hw}	Thermal conductivity of water.
k_s	Slope of the unloading-reloading curve in the space relating matric suction to the degree of saturation.
k_s^e	Coefficient of the spring representing the soil below a drum compactor.
k_w	Hydraulic conductivity.
k_1	Fitting coefficient to model the resilient Young's modulus, or spring constant of a tire for defining the truck-road interaction.
k_2	Fitting coefficient to model the resilient Young's modulus, or spring constant of the suspension for defining the truck-road interaction.
k_3	Fitting coefficient to model the resilient Young's modulus.
$l_T(y)$	Length of the contact area between a tire and the road.
M	Slope of the critical state line in the $p\,q$ plane.
m	Mass, or parameter of the water retention curve, or $m = 1/\nu$.
m_d	Mass of a vibratory drum.
m_e	Eccentric mass of a vibratory drum.
m_f	Mass of the frame of a vibratory drum
m_W	Parameters of the water retention model that depends on void's ratio.
m_1	Mass of the axle for defining the truck-road interaction.
m_2	Mass of the body of the vehicle for defining the truck-road interaction.
$N(s)$	Specific volume in unsaturated state for a mean net stress of p^c.
$N(0)$	Specific volume in saturated state for a mean net stress of p^c.
n	Porosity, or exponent representing the rectangularity of a super elliptical contact area, or parameter of the water retention curve, or fitting coefficient of the Boyce's model.
n_c	Number of grain classes in a granular mixture.
n_{sm}	Smoothing parameter of the function representing the effective degree of saturation.
n_W	Parameters of the water retention model that depends on void's ratio.
P_{200}	Proportion of material that pass through the # 200 U.S. Standard Sieve.
PI	Plasticity Index.
p	Mean stress.
$\overline{p}$	Constitutive mean stress for the microstructural BBM.
p_a, p_{atm}	Atmospheric pressure.
p_c	Cyclic mean stress.
p^c	Reference stress for the BBM.
$\overline{p}_c$	Reference mean constitutive stress for the microstructural BBM.

(*Continued*)

Symbol	Definition
p_0	Maximum stress at for the Hertz contact theory.
p_0^*	Over consolidation mean stress in the saturated state.
q	Deviator stress, or
	uniform load over a circular loaded area.
q_c	Cyclic deviator stress.
q_{conv}	Heat flux due to convection.
q_{rad}	Heat flux due to radiation.
q_{sens}	Sensible heat flux.
q_{th}	Heat flux due to thermal emissions
R	Radial distance $R = \sqrt{x^2 + y^2 + z^2}$.
R_e	Reynolds number.
RMV	Resonant meter value.
r	Constant of the BBM relating maximum stiffness at infinite.
$\bar{r}$	Material parameter for the microstructural BBM.
S_r	Degree of saturation.
$\bar{S}_r$	Effective degree of saturation for the microstructural BBM.
S_{r-opt}	Degree of saturation at the optimum water content.
S_{r0}	Degree of saturation at the beginning of unloading.
$S_z(\Omega)$	Power Spectral Density, PSD, of the road's profile.
s	Matric suction.
s_{aev}	Suction corresponding to the air entry value.
s_b	Air entry suction.
s_{res}	Residual suction.
s_0	Suction at the beginning of unloading.
$\bar{s}_\lambda$	Material parameter for the microstructural BBM.
T	Temperature.
T_a	Air temperature.
T_d	Dew point temperature.
T_l	Local time in the zone of a particular site.
T_s	Temperature at the surface of the road.
T_{sv}	True solar time at a particular site.
T_{sky}	Hypothetical temperature above the surface of the road.
T_v	Time factor for vertical consolidation.
t	Time.
$U(t)$	Degree of consolidation achieved at time t.
$U_v(t)$	Degree of vertical consolidation.
U_{vr}	Degree of consolidation in the radial and vertical directions.
U_w	Relative humidity.
$\boldsymbol{u}$	Vector representing the displacements of the axle and the body of a vehicle.
u	Displacement towards the x axis.
u_a	Pore air pressure.
u_g	Function for defining the road's profile.
u_v	Vapor pressure.
u_{vs}	Saturation vapor pressure.
u_w	Pore water pressure.
$\bar{u}_{wc}$	Average excess pore water pressure.

(*Continued*)

Symbol	Definition
V_a	Air velocity.
v	Displacement towards the y axis, or
	Specific volume ($v = 1 + e$).
w	Gravimetric water content, or
	displacement towards the z axis, or
	hour angle indicating the position of the sun in a day, or
	waviness index for defining the road's profile.
$w_T(x)$	Width of the contact area between a tire and the road
x	Cartesian coordinate.
y	Cartesian coordinate.
y_i	Volumetric proportion of grains in a granular mixture.
Z_n	Amplitude of the nth harmonic for defining the road's profile.
z	Cartesian coordinate.
z_s	Soil displacement below a drum compactor.

GREEK LETTERS

Symbol	Definition
α	Mean absorptivity coefficient, or
	normalized radius of a circular loaded area.
α_K	Parameter describing the contact stress between a tire and the road.
β	Parameter of the BBM giving the shape of the function for the increase of stiffness due to an increase of suction, or
	angle to define the point where the stress is calculated using the Fröhlich's solution, or
	coefficient of the Boyce's model.
$\bar{\beta}$	Material parameter of the microstructural BBM.
β_i	Residual compacity of grains i in a granular mixture.
Γ	Angle indicating the position of the earth in its orbit for a particular day of the year.
γ	Unit weight, or
	virtual compacity of a granular mixture.
ΔT_l	Time difference between the local time and the standard time.
$\Delta\Omega$	Width of each frequency band for defining the road's profile.
δ	Solar declination.
δ_K	Parameter describing the contact stress between a tire and the road.
ϵ	Emissivity coefficient.
ϵ_v	Volumetric strain.
ϵ_1, ϵ_2	Geometrical variables to define a triangular load over a half space.
η	Empirical parameter for computing the thermal conductivity of soils.
θ	Volumetric water content, or
	bulk stress, or
	Angle between a normal vector on the surface of the road and a direct line joining the sun.
θ_i	Volumetric fraction of ice in a soil.

(*Continued*)

Symbol	Definition
θ_{res}	Residual volumetric water content.
θ_s	Volumetric fraction of the solid grains in a soil.
$\theta'(s)$	First derivative of the water retention curve.
θ_{sat}	Saturated volumetric water content.
θ_w	Volumetric fraction of water in a soil.
κ	Road's roughness coefficient for defining the road's profile, or slope of the overconsolidated domain, in logarithmic scale, of the compression line in the saturated state.
κ_H	Empirical parameter for computing the thermal conductivity of soils.
κ_s	Parameter of the BBM relating the change in specific volume due to an increase of matric suction.
λ	Normalized depth for the Burmister's solution, or slope of the virgin compression line in the saturated state.
$\bar{\lambda}$	Compressibility coefficient for the microstructural BBM.
$\lambda(0)$	Slope, in logarithmic scale, of the compression line in the saturated state.
$\lambda(s)$	Slope, in logarithmic scale, of the compression lines in the unsaturated state
$\bar{\lambda}(\bar{s})$	Coefficient of compressibility depending on the effective suction.
ν	Poisson's ratio.
ξ	Fröhlich's concentration factor.
ξ_m	Variable describing the microstructural state of a soil.
ρ	Radial distance in the cilindrical coordinate system.
ρ_c	Settlement due to primary consolidation.
ρ_d	Dry density.
ρ_H	Normalized radial distance used in Burmister's solution.
ρ_i	Immediate settlement.
ρ_w	Density of water.
σ	Total stress, or Stefan-Boltzmann constant.
σ'_c	Over consolidation effective stress for computing settlements due to consolidation.
σ_{net}	Net stress.
σ_{oct}	Octahedral stress.
σ_{u_g}	Standard deviation of the ISO road's profile.
$\sigma_{u'_g}$	Standard deviation of the first derivative of the ISO road's profile.
σ_v	Vertical stress.
$\sigma_{v_{N_C}}$	Vertical stress required to reach a specific density by applying N_C loading cycles.
$\sigma_{v_{N_C=1}}$	Vertical stress required to reach a particular density when applying one loading cycle.
$\sigma_x, \sigma_y, \sigma_z$	Normal stresses in the Cartesian coordinate system.
σ'_{z0}	Initial effective vertical stress for computing settlements due to consolidation.
$\sigma_1, \sigma_2, \sigma_3$	Principal stresses.
τ_{oct}	Octahedral shear stress.
Φ	Compacity of a granular mixture.
Φ_i	Partial volume of class i in a granular mixture.

(*Continued*)

Symbol	Definition
ϕ	Latitude on the earth of a particular site.
ϕ'	Friction angle in effective stresses.
ϕ_n	Phase angle of the nth harmonic for defining a road's profile.
ϕ_W	Parameters of the water retention model that depends on the void's ratio.
χ	Effective stress parameter.
χ_H	Empirical parameter for computing the thermal conductivity of soils.
ψ	Water potential.
ψ_W	Parameters of the water retention model that depends on the void's ratio.
Ω	Angular frequency.
Ω_L	Lower limit for the spatial reference frequency to define a road's profile.
Ω_n	Central frequency of the nth band to define a road's profile.
Ω_U	Upper limit of the spatial reference frequency to define a road's profile.
Ω_0	Spatial reference frequency to define a road's profile.
ω	Angular frequency, or angle to define the point where the stress is calculated using the Fröhlich's solution.

Introduction

This book develops 23 extended examples that cover most of the theoretical aspects presented in the book *Geotechnics of Roads, Fundamentals*, [10]. Moreover, for most examples, this book describes algorithms for solving complex problems and provides MATLAB scripts for their solution. Consequently, this book is a natural complement of the book *Geotechnics of Roads, Fundamentals*.

Although most of the theories required to solve the examples were described in detail in [10], each chapter in this book summarizes the set of equations required to solve the examples. This book has seven chapters as follows:

Chapter 1 of this book deals with the distribution of stresses and strains in road structures. It develops six examples that cover, first, the analysis of the stresses and displacements produced by vertical or horizontal loads in elastic half-spaces. Then, it describes the analysis of the tire–road interaction using Hertz's theory and the Fröhlich stress distribution. Besides, the chapter describes in detail the use of Burmister's method to calculate the stress distribution in structures of multilayered roads. Concerning elastodynamic solutions, this chapter presents an example that describes the calculation of the vehicle–road interaction that produces dynamic loads on the road. These loads could exceed the static load obtained by forgetting the dynamic interaction.

Chapter 2 deals with the unsaturated soil mechanics applied to road structures. It describes the methodology to assess the water retention curve using the empirical model proposed in the Mechanistic Empiric Pavement Design Gide (MEPDG). Then, it presents the method for calculating the unsaturated hydraulic conductivity based on the water retention curve. Moreover, regarding the flow of water in road structures, the chapter describes a simplified calculation of water infiltration. Then, it presents a more rigorous methodology based on the numerical calculation of water flow in unsaturated materials. Finally, regarding heat flow, the chapter presents an example describing the methodology for the numerical calculation of the evolution of temperature in road structures.

Chapter 3 analyzes soil compaction through two examples. The first one focuses on the compaction process interpreted as a hardening process of unsaturated soils. Therefore, this example uses the elastoplastic model Barcelona Basic Model (BBM) to analyze the evolution of irreversible volumetric strains within the soil produced by a tire compactor. The chapter also explains a linear packing model that allows assessing the density of compacted materials based on its grain size distribution, the second example of this chapter uses this methodology to compute the maximum density of Proctor's tests.

Chapter 4 develops three examples regarding the construction and the performance of embankments. The first example deals with the construction of embankments on soft soils; it explains in detail the use of the methodology of staged construction to design an embankment that has a proper safety factor and also analyzes the magnitude of settlement to estimate the correct height of the fill. The second and third examples analyze the collapse of embankments and link the compaction characteristics with the long-term deformation of embankments subject to wetting. For this purpose, these examples use the theory of unsaturated soil mechanics and the elastoplastic BBM. Such examples consider or not the effect of soil's microstructure.

Chapter 5 explains the essential subject of the mechanical behavior of road materials. Methodologies explained in this chapter are crucial to use mechanistic approaches that allow consideration of different load and climatic conditions that can affect roads. Two examples illustrate this effect; the first one explains the methodologies to adjust the laboratory measurements of the resilient Young's modulus to the different models that allow a mathematical description of its evolution regarding stresses, water content, or matric suction. The second example of this chapter uses the change of the resilient Young's modulus depending on the matric suction to assess the effect of the water content of the granular layer on the fatigue life of a low traffic road structure.

Chapter 6 is devoted to the climate effects on road structures. The chapter develops three examples. The first one explains the process to compute the evolution of the temperature in a road structure along a day of the year; the analysis includes several climatic variables such as the solar radiation, the air temperature, the wind velocity, the relative humidity, the latitude of the site, and the day of the year. Moreover, this chapter develops two other examples regarding the flow of water, one of them analyzes the effect of the cracks of the bituminous layer in the infiltration of water toward the road structure, and the second one analyses the capacity of a drainage layer to evacuate the infiltrations of water.

Finally, **Chapter 7** explains the theoretical tools that allow analyzing the performance of vibratory compactors and the procedure to examine their movement to apply the methodology of continuous compaction control. The chapter presents two examples; the first one considers the layer under compaction as an elastic material, while the second considers the elastoplastic response of the material undergoing compaction.

Chapter 1

Distribution of stresses and strains in roads

1.1 RELEVANT EQUATIONS

The performance of road structures results from the interaction between the stress produced by the external loads and the behavior of the different constituent materials. Also, since roads must undergo thousands to millions of load repetitions, their constituent materials must sustain loads without suffering irreversible strains. Therefore, they must remain in the elastic domain of behavior.

This chapter focuses on the calculation of stresses in road structures. First, it describes methodologies that allow calculating the stress distribution produced by vertical and horizontal loads in a homogeneous half-space. Then, the chapter describes a methodology for calculating road loads due to interaction with vehicles. Finally, the chapter describes Burmister's method, which is the usual method to calculate stresses and strains in road structures having multiple layers.

1.1.1 Boussinesq's solution

In 1878, Boussinesq proposed a solution that allows calculating the distribution of stresses in a half-space beneath a concentrated load [6]. The set of equations that permit to obtain the stress components resulting from a concentrated vertical load P located in the Cartesian coordinate system, as indicated in Figure 1.1, are

$$\sigma_x = \frac{3P}{2\pi}\left\{\frac{x^2 z}{R^5} - \frac{m-2}{3m}\left[-\frac{1}{R(R+z)} + \frac{(2R+z)x^2}{(R+z)^2 R^3} + \frac{z}{R^3}\right]\right\}, \tag{1.1}$$

$$\sigma_y = \frac{3P}{2\pi}\left\{\frac{y^2 z}{R^5} - \frac{m-2}{3m}\left[-\frac{1}{R(R+z)} + \frac{(2R+z)y^2}{(R+z)^2 R^3} + \frac{z}{R^3}\right]\right\}, \tag{1.2}$$

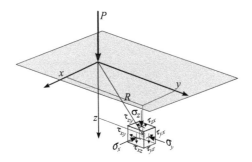

Figure 1.1 Geometric layout to describe the Boussinesq solution in Cartesian coordinates.

$$\sigma_z = \frac{3P}{2\pi}\frac{z^3}{R^5}, \tag{1.3}$$

$$\tau_{xy} = \frac{3P}{2\pi}\left\{\frac{xyz}{R^5} - \frac{m-2}{3m}\left[\frac{(2R+z)xy}{(R+z)^2 R^3}\right]\right\}, \tag{1.4}$$

$$\tau_{yz} = \frac{3P}{2\pi}\frac{yz^2}{R^5}, \tag{1.5}$$

$$\tau_{zx} = \frac{3P}{2\pi}\frac{xz^2}{R^5}, \tag{1.6}$$

where x, y, z are the Cartesian coordinates, $R = \sqrt{x^2 + y^2 + z^2}$, $m = 1/\nu$, and ν is Poisson's ratio.

Also, the expressions that provide the horizontal displacements u and v, and the vertical displacement w are

$$u = \frac{1+\nu}{2\pi E}\left[\frac{xz}{R^3} - \frac{(1-2\nu)x}{R(R+z)}\right]P, \tag{1.7}$$

$$v = \frac{1+\nu}{2\pi E}\left[\frac{yz}{R^3} - \frac{(1-2\nu)y}{R(R+z)}\right]P, \tag{1.8}$$

$$w = \frac{1+\nu}{2\pi E}\left[\frac{z^2}{R^3} + \frac{2(1-\nu)}{R}\right]P, \tag{1.9}$$

where E is Young's modulus.

1.1.2 Cerruti's solution

Shortly after Boussinesq, in 1882, Cerruti proposed a set of expressions for calculating the stress distribution in half-space beneath a horizontal load [16]. His solution provides the following expressions for the stresses' distribution produced by a concentrated horizontal load H located in the Cartesian plane as depicted in Figure 1.2:

$$\sigma_x = -\frac{Hx}{2\pi R^3}\left\{-\frac{3x^2}{R^2} + \frac{1-2\nu}{(R+z)^2}\left[R^2 - y^2 - \frac{2Ry^2}{R+z}\right]\right\}, \tag{1.10}$$

$$\sigma_y = -\frac{Hx}{2\pi R^3}\left\{-\frac{3y^2}{R^2} + \frac{1-2\nu}{(R+z)^2}\left[3R^2 - x^2 - \frac{2Rx^2}{R+z}\right]\right\}, \tag{1.11}$$

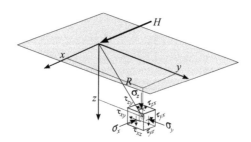

Figure 1.2 Geometric layout to describe the Cerruti solution in Cartesian coordinates.

$$\sigma_z = \frac{3Hxz^2}{2\pi R^5} \tag{1.12}$$

$$\tau_{xy} = -\frac{Hy}{2\pi R^3}\left\{-\frac{3x^2}{R^2}+\frac{1-2\nu}{(R+z)^2}\left[-R^2+x^2+\frac{2Rx^2}{R+z}\right]\right\}, \tag{1.13}$$

$$\tau_{yz} = \frac{3Hxyz}{2\pi R^5}, \tag{1.14}$$

$$\tau_{zx} = \frac{3Hx^2z}{2\pi R^5}, \tag{1.15}$$

Also, x, y, z are the Cartesian coordinates, and $R = \sqrt{x^2+y^2+z^2}$.

Moreover, the displacements u, v, and w provided by Cerruti's solution are

$$u = \frac{H}{4\pi GR}\left\{1+\frac{x^2}{R^2}+(1-2\nu)\left[\frac{R}{R+z}-\frac{x^2}{(R+z)^2}\right]\right\}, \tag{1.16}$$

$$v = \frac{H}{4\pi GR}\left\{\frac{xy}{R^2}-(1-2\nu)\frac{xy}{(R+z)^2}\right\}, \tag{1.17}$$

$$w = \frac{H}{4\pi GR}\left\{\frac{xz}{R^2}+(1-2\nu)\frac{x}{R+z}\right\}, \tag{1.18}$$

where G is the shear modulus, and ν is Poisson's ratio.

1.1.3   Fröhlich solution

Fröhlich [32] introduced a "concentration factor", ξ, into Boussinesq's equations. The Fröhlich solution is a rough alternative to account for the modification of the stress distribution due to plasticity. However, Fröhlich's solution is not exact because it respects the equilibrium conditions but without considering the equations for displacement's compatibility. Despite the approximation of the Fröhlich solution, it produces good agreement with the measures of field stresses, mainly when a layer of soil undergoes compaction [40].

Equation 1.19 gives Fröhlich's solution for the radial stress produced by a combination of vertical and horizontal point loads (P, H) [39,41].

$$\sigma_R = \frac{\xi P}{2\pi R^2}\cos^{\xi-2}\beta + (\xi-2)\frac{\xi H}{2\pi R^2}\cos\omega\sin\beta\cos^{\xi-3}\beta, \tag{1.19}$$

$$\sigma_\theta = 0. \tag{1.20}$$

In Equation 1.19, β is the angle formed by the vertical axis and a vector joining the point where the load is applied and the point where the stress is calculated. At the same time, ω is the angle between the horizontal load vector and a vertical plane including the vector mentioned above (in other words, the plane Ω represented in Figure 1.3).

In Cartesian coordinates, Equations 1.19 and 1.20 lead to the following stresses:

$$\sigma_x = \sigma_R\sin^2\beta\cos^2\omega = \sigma_R\frac{x^2}{R^2}, \tag{1.21}$$

$$\sigma_y = \sigma_R\sin^2\beta\sin^2\omega = \sigma_R\frac{y^2}{R^2}, \tag{1.22}$$

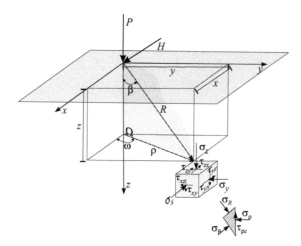

Figure 1.3 Geometric layout to describe the Fröhlich solution in Cartesian coordinates. (Adapted from Ref. [39].)

$$\sigma_z = \sigma_R \cos^2 \beta = \sigma_R \frac{z^2}{R^2} \tag{1.23}$$

$$\tau_{xy} = \sigma_R \sin^2 \beta \cos \omega \sin \omega = \sigma_R \frac{xy}{R^2}, \tag{1.24}$$

$$\tau_{xz} = \sigma_R \cos \beta \sin \beta \cos^2 \omega = \sigma_R \frac{xz}{R^2}, \tag{1.25}$$

$$\tau_{yz} = \sigma_R \cos \beta \sin \beta \sin^2 \omega = \sigma_R \frac{yz}{R^2}. \tag{1.26}$$

1.1.4   Tire–soil interaction

The contact area of the footprint below a tire can be approximated by a superellipse represented by Equation 1.27 as first suggested in Ref. [40]:

$$\left| \frac{x}{a} \right|^n + \left| \frac{y}{b} \right|^n = 1, \tag{1.27}$$

where a and b are the half axes of the super-ellipse, and n is an exponent representing its rectangularity.

Then, the footprint area is limited by Equation 1.28:

$$\Omega = \left\{ (x, y) \left(|x/a|^n + |y/b|^n \le 1 \right) \right\}. \tag{1.28}$$

Equation 1.29 provides a stress distribution function accounting for different tires, which was proposed in Ref. [40]. From this equation, the transverse and longitudinal distributions of vertical stress over the loaded area are

$$\sigma_z(x = 0, y) = C_{AK} \left(0.5 - \frac{y}{w_T(x)} \right) e^{-\delta_K (0.5 - y/w_T(x))} \quad \text{for} \quad 0 \le y \le \frac{w_T(x)}{2}, \quad \text{and}$$

$$\sigma_z(x, y) = \sigma_z(x = 0, y) \left[1 - \left(\frac{x}{l_T(y)/2} \right)^{\alpha_K} \right] \quad \text{for} \quad 0 \le x \le \frac{l_T(y)}{2}, \tag{1.29}$$

where $w_T(x)$ and $l_T(y)$ are the width and the length of the contact area, δ_K and α_K are parameters given in Ref. [40] that depend on the tire characteristics, and C_{AK} is a proportionality factor that accounts for the total load of the tire.

Another possibility allowing an approximate evaluation of the interaction tire–soil is to use of the Hertz contact theory. This procedure is explained in Example 3.

1.1.5   Road–vehicle interaction

Roads generally have a certain roughness of different wavelengths; at the same time, vehicles, which can be described as a dynamic system, interact with the roughness of the road. This interaction produces a dynamic load on the road whose magnitude depends on the speed of the vehicle, its dynamic characteristics, and the roughness of the road.

The 'quarter car model' permits to schematize the road–vehicle interactions. It describes the vehicle, with its tire and suspension, as a set of masses and springs, as shown in Figure 1.4.

The movement of the components of the quarter car is represented by the following second-order linear differential equation with two degrees of freedom [65]:

$$M\ddot{u} + C\dot{u} + Ku = f(t),\tag{1.30}$$

where $u = [u_1, u_2]^T$ is the vector representing the displacements of the axle and the body of the vehicle, positive in the downward direction, while the mass, damping, and stiffness matrices of the system are

$$M = \begin{bmatrix} m_1 & 0 \\ 0 & m_2 \end{bmatrix}, \quad C = \begin{bmatrix} c_1 + c_2 & -c_2 \\ -c_2 & c_2 \end{bmatrix}, \quad K = \begin{bmatrix} k_1 + k_2 & -k_2 \\ -k_2 & k_2 \end{bmatrix},\tag{1.31}$$

where k_1 is the spring constant of the tire, k_2 is the spring constant of the suspension, m_1 is the mass of the axle, m_2 is the mass of the body of the vehicle, c_1 is the damping of the tire, and c_2 is the constant of the suspension's shock absorber.

The road's profile, described by the function u_g, is positive in the downward direction. The vector that represents the road profile as a time signal $f(t)$ is

$$f(t) = \begin{bmatrix} k_1 u_g(t) + c_1 \dot{u}_g(t) + m_1 g \\ m_2 g \end{bmatrix}.\tag{1.32}$$

A methodology for the numerical solution of Equation 1.30 is described in Example 4.

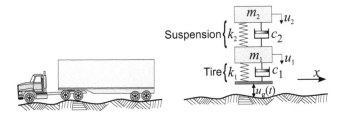

Figure 1.4 Quarter car model to analyze vehicle–road interaction.

1.1.5.1 Mathematical description of road profiles

Based on field data, the standard ISO 8608 [29] proposes to describe the profile of a road in the form of power spectral density (PSD). From Refs. [25,26], the spectral equation that describes the path profile of a single track is

$$S_z(\Omega) = \kappa \left(\frac{\Omega_0}{\Omega} \right)^w,$$

(1.33)

$$w = 2 \text{ for } \Omega_L \le \Omega \le \Omega_0, \text{ or}$$
$$w = 1.5 \text{ for } \Omega_o \le \Omega \le \Omega_U,$$

where $S_z(\Omega)$ is the PSD of the road profile, whose unit is 'm^3/cycle'; the frequencies $\Omega_0 = 1/2\pi$, $\Omega_L = 0.01$, and $\Omega_U = 10$, all in cycles/m, are the spatial reference frequency, and the respective lower and upper limits for frequency; κ is the road roughness coefficient in m^3/cycle and w is the waviness index.

In addition, ISO has proposed a road classification system having the seven classes of roughness shown in Table 1.1, and the geometric mean of the coefficient of road roughness κ can be assessed as $\kappa = 4^R \cdot 10^{-6}$.

A particular road profile, described by its power spectrum, can be obtained using the method of superposition of harmonics (SOH) proposed in Refs. [52,53]. The method divides the power spectrum into N frequency bands, each band corresponding to one harmonic. The superposition equation proposed in Ref. [25] is

$$u_g(x) = \sum_{n=1}^{N} Z_n \sin(2\pi \Omega_n x + \phi_n), \quad n = 1, 2, 3, \ldots, N,$$

(1.34)

where $u_g(x)$ is the elevation of road profile, x is the forward distance traveled by the vehicle, Z_n is the amplitude of the nth harmonic, $N = (\Omega_U - \Omega_L)/\Delta\Omega$ is the number of frequency bands into which the total PSD spectrum is divided, $\Delta\Omega$ is the width of each frequency band, and ϕ_n is the phase angle of the nth harmonic, which is assumed randomly using a uniform distribution in the interval $[0, 2\pi]$.

The discretization of the PSD leads to the following expression for the central frequency of the nth band:

$$\Omega_n = \omega_L + \frac{2n-1}{2} \Delta\Omega.$$

(1.35)

Table 1.1 ISO ranking of road profiles using a Roughness coefficient κ

Road Class	R	$\kappa \cdot 10^6$ m^3/Cycle	
		Range	Geometric Mean
A (very good)	1	<8	4
B (good)	2	8–32	16
C (average)	3	32–128	64
D (poor)	4	128–512	256
E (very poor)	5	512–2,048	1,024
F	6	2,048–8,192	4,096
G	7	8,192–32,768	16,384

This methodology requires using a minimum length of road for the simulation, which is $L \geq 1/\Omega_L$. However, a better description of low frequencies requires longer L, for example, $L = 3/\Omega_L$ [25].

Moreover, the amplitude Z_n of the nth harmonic is

$$Z_n = \sqrt{2S_z(\Omega_n)\Delta\Omega}, \tag{1.36}$$

where Ω_n is the central frequency of the nth band.

When assuming a constant waviness index for all frequencies (*i.e.*, $w = 2$), the standard deviation of the ISO road's profile σ_{u_g} and the standard deviation of its gradient $\sigma_{u'_g}$ are

$$\sigma_{u_g} = \sqrt{\kappa\Omega_0^2\left(\frac{1}{\Omega_L} - \frac{1}{\Omega_U}\right)}, \tag{1.37}$$

$$\sigma_{u'_g} = \sqrt{\kappa(\Omega_U - \Omega_L)}. \tag{1.38}$$

Example 4 describes the procedure carried out for the numerical computation of road profiles using the method of the SOH.

1.1.6 Burmister's method

Burmister's method allows calculating the stress distribution within a multilayer linear elastic system with a circularly loaded area (see Figure 1.5). The multilayer system is characterized by n layers having E_i and v_i as Young's moduli and Poisson's ratios. The depth of the bottom level of layer i is z_i for layers $i = 1$ to $i = n - 1$, while the depth of the layer n is infinite.

Moreover, the vertical uniformly distributed circular load can be characterized as $\sigma_z(\rho, 0) = q$ for $\rho \leq a$, and $\sigma_z(\rho, 0) = 0$ for $\rho > a$. Where a is the radius of the circular loaded area, q is the magnitude of the uniform load, and ρ and z are the radial and vertical distances in a cylindrical coordinate system whose symmetry axis crosses the center of the circular loaded area.

The Hankel transform allows describing σ_z as the following addition of Bessel functions to infinity:

$$q(\rho, 0) = q\alpha \int_0^\infty J_0\,(m\rho)\,J_1\,(m\alpha)\,dm, \tag{1.39}$$

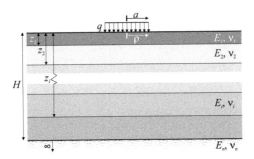

Figure 1.5 Schematic drawing of a road structure modeled as a multilayer system.

where m is the integration parameter and J_0 and J_1 are Bessel functions of the first kind with orders zero and one, respectively; α is the normalized radius of the loaded area defined as $\alpha = a/H$, and H is the thickness of the road structure.

Computing stresses and strains within the multilayer structure requires solving a system of $4n - 2$ equations with $4n - 2$ unknowns, which is built based on elementary matrices written for each interface of the structure.

The system of equations for the first layer is

$$
\begin{bmatrix} e^{-m\lambda_1} & 1 \\ e^{-m\lambda_1} & -1 \end{bmatrix} \begin{bmatrix} A_1 \\ B_1 \end{bmatrix} + \begin{bmatrix} -(1 - 2v_1)e^{-m\lambda_1} & 1 - 2v_1 \\ 2v_1 e^{-m\lambda_1} & 2v_1 \end{bmatrix} \begin{bmatrix} C_1 \\ D_1 \end{bmatrix} = \begin{bmatrix} 1 \\ 0 \end{bmatrix}.
$$

(1.40)

For the subsequent layers, the system of equations for the case when materials are bonded at the interface is

$$
\begin{bmatrix} 1 & F_i & -(1 - 2v_i - m\lambda_i) & (1 - 2v_i + m\lambda_i)F_i \\ 1 & -F_i & 2v_i + m\lambda_i & (2v_i - m\lambda_i)F_i \\ 1 & F_i & 1 + m\lambda_i & -(1 - m\lambda_i)F_i \\ 1 & -F_i & -(2 - 4v_i - m\lambda_i) & -(2 - 4v_i + m\lambda_i)F_i \end{bmatrix} \begin{bmatrix} A_i \\ B_i \\ C_i \\ D_i \end{bmatrix}
$$

$$
= \begin{bmatrix} F_{i+1} & 1 & -(1 - 2v_{i+1} - m\lambda_i)F_{i+1} & 1 - 2v_{i+1} + m\lambda_i \\ F_{i+1} & -1 & (2v_{i+1} + m\lambda_i)F_{i+1} & 2v_{i+1} - m\lambda_i \\ R_i F_{i+1} & R_i & (1 + m\lambda_i)R_i F_{i+1} & -(1 - m\lambda_i)R_i \\ R_i F_{i+1} & -R_i & -(2 - 4v_{i+1} - m\lambda_i)R_i F_{i+1} & -(2 - 4v_{i+1} + m\lambda_i)R_i \end{bmatrix}
$$

$$
\times \begin{bmatrix} A_{i+1} \\ B_{i+1} \\ C_{i+1} \\ D_{i+1} \end{bmatrix},
$$

(1.41)

while, when the layers are unbonded at their interface, the system of equations is

$$
\begin{bmatrix} 1 & F_i & -(1 - 2v_i - m\lambda_i) & (1 - 2v_i + m\lambda_i)F_i \\ 1 & -F_i & -(2 - 4v_i - m\lambda_i) & -(2 - 4v_i + m\lambda_i)F_i \\ 1 & -F_i & 2v_i + m\lambda_i & (2v_i - m\lambda_i)F_i \\ 0 & 0 & 0 & 0 \end{bmatrix} \begin{bmatrix} A_i \\ B_i \\ C_i \\ D_i \end{bmatrix}
$$

$$
= \begin{bmatrix} F_{i+1} & 1 & -(1 - 2v_{i+1} - m\lambda_i)F_{i+1} & 1 - 2v_{i+1} + m\lambda_i \\ R_i F_{i+1} & -R_i & -(2 - 4v_{i+1} - m\lambda_i)R_i F_{i+1} & -(2 - 4v_{i+1} + m\lambda_i)R_i \\ 0 & 0 & 0 & 0 \\ F_{i+1} & -1 & (2v_{i+1} + m\lambda_i)F_{i+1} & 2v_{i+1} - m\lambda_i \end{bmatrix}
$$

$$
\times \begin{bmatrix} A_{i+1} \\ B_{i+1} \\ C_{i+1} \\ D_{i+1} \end{bmatrix},
$$

(1.42)

where λ is the normalized depth $\lambda = z/H$, $F_i = e^{-m(\lambda_i - \lambda_{i-1})}$, and $R_i = \frac{E_i}{E_{i+1}} \frac{1 + v_{i+1}}{1 + v_i}$.

Stresses and displacements are calculated using Equations 1.43–1.48.

$$
\sigma_{z_i}(m) = -mJ_0(m\rho_H)\left\{[A_i - C_i(1 - 2v_i - m\lambda)]e^{-m(\lambda_i - \lambda)} \right.
$$

$$
\left. + [B_i + D_i(1 - 2v_i + m\lambda)]e^{-m(\lambda - \lambda_{i-1})}\right\},
$$

(1.43)

$$\sigma_{\rho_i}(m) = \left[mJ_0(m\rho_H) - \frac{J_1(m\rho_H)}{\rho_H} \right] \left\{ [A_i + C_i(1 + m\lambda)] e^{-m(\lambda_i - \lambda)} \right.$$
$$\left. + [B_i - D_i(1 - m\lambda)] e^{-m(\lambda - \lambda_{i-1})} \right\},$$
$$+ 2v_i m J_0(m\rho_H) \left[C_i e^{-m(\lambda_i - \lambda)} - D_i e^{-m(\lambda - \lambda_{i-1})} \right], \tag{1.44}$$

$$\sigma_{\theta_i}(m) = \frac{J_1(m\rho_H)}{\rho_H} \left\{ [A_i + C_i(1 + m\lambda)] e^{-m(\lambda_i - \lambda)} + [B_i - D_i(1 - m\lambda)] e^{-m(\lambda - \lambda_{i-1})} \right\},$$
$$+ 2v_i m J_0(m\rho_H) \left[C_i e^{-m(\lambda_i - \lambda)} - D_i e^{-m(\lambda - \lambda_{i-1})} \right], \tag{1.45}$$

$$\tau_{\rho z_i}(m) = m J_1(m\rho_H) \left\{ [A_i + C_i(2v_i + m\lambda)] e^{-m(\lambda_i - \lambda)} \right.$$
$$\left. - [B_i - D_i(2v_i - m\lambda)] e^{-m(\lambda - \lambda_{i-1})} \right\}, \tag{1.46}$$

$$u_i(m) = \frac{1 + v_i}{E_i} H J_1(m\rho_H) \left\{ [A_i + C_i(1 + m\lambda)] e^{-m(\lambda_i - \lambda)} \right.$$
$$\left. + [B_i - D_i(1 - m\lambda)] e^{-m(\lambda - \lambda_{i-1})} \right\}, \tag{1.47}$$

$$w_i(m) = -\frac{1 + v_i}{E_i} H J_0(m\rho_H) \left\{ [A_i - C_i(2 - 4v_i - m\lambda)] e^{-m(\lambda_i - \lambda)} \right.$$
$$\left. - [B_i + D_i(2 - 4v_i + m\lambda)] e^{-m(\lambda - \lambda_{i-1})} \right\}, \tag{1.48}$$

where ρ_H is a dimensionless variable accounting for the distance of the computational point, measured from the center of the load, and given by $\rho_H = \rho/H$.

The solution for the stresses and displacements requires using again the Hankel transform, which in a discrete form is as follows:

$$\sigma_{z_i} = q\alpha \sum_{k=1}^{n_k} \left(\frac{\sigma_{z_i}(m)}{m} J_1(m\alpha) \Delta m \right), \tag{1.49}$$

$$\sigma_{\rho_i} = q\alpha \sum_{k=1}^{n_k} \left(\frac{\sigma_{\rho_i}(m)}{m} J_1(m\alpha) \Delta m \right), \tag{1.50}$$

$$\sigma_{\theta_i} = q\alpha \sum_{k=1}^{n_k} \left(\frac{\sigma_{\theta_i}(m)}{m} J_1(m\alpha) \Delta m \right), \tag{1.51}$$

$$\tau_{\rho z_i} = q\alpha \sum_{k=1}^{n_k} \left(\frac{\tau_{\rho z_i}(m)}{m} J_1(m\alpha) \Delta m \right), \tag{1.52}$$

$$u_i = q\alpha \sum_{k=1}^{n_k} \left(\frac{u_i(m)}{m} J_1(m\alpha) \Delta m \right), \tag{1.53}$$

$$w_i = q\alpha \sum_{k=1}^{n_k} \left(\frac{w_i(m)}{m} J_1(m\alpha) \Delta m \right). \tag{1.54}$$

Examples 5 and 6 describe the procedure for computing the stresses in multilayer systems using Burmister's method.

1.2   EXAMPLE 1: CALCULATION OF THE STRESS DISTRIBUTION PRODUCED BY VERTICAL LOADS USING BOUSSINESQ'S SOLUTION

The purpose of this example is to use Boussinesq's solution for computing the stresses and displacements in an elastic homogeneous half-space produced by a tire that carries a load of 35 kN.

The geometry of the tire can be approximated by a superellipse whose width is 0.18 m, its length is 0.3 m, and its rectangularity coefficient is 4. The example assumes a uniform distribution of contact stress under the tire.

The elastic properties of the half-space are Young's modulus $E = 50\ MPa$ and Poisson's ratio $v = 0.3$.

Assuming the Mohr–Coulomb yield criterion with a friction angle of 35° evaluate, up to a depth of 0.5 m, the requirement of cohesion for avoiding the shear strength failure within the material.

Finally, analyze the effect of the friction angle regarding the requirement of cohesion of the road material.

Stresses produced by the tire in the elastic half-space can be obtained with the following procedure:

1   Discretize the loaded area into small rectangular areas and approximate the uniform stress within the discrete area using a set of concentrated point loads.
2   Superpose the stress produced by each concentrated load.
3   Analyze the requirement of cohesion depending on the angle of friction.

1.2.1   Loaded area and uniform stress

As described in Caicedo [10], the loaded area under a tire can be approximated using a superellipse represented by Equation 1.27 as

$$\left|\frac{x}{a}\right|^{n} + \left|\frac{y}{b}\right|^{n} = 1,$$

where x and y are the Cartesian coordinates, a and b are the half axes of the superellipse, and n is an exponent representing its rectangularity.

Then, the loaded area Ω is represented by the following inequality:

$$\Omega = \left\{(x, y)\left(|x/a|^{n} + |y/b|^{n} \leq 1\right)\right\}.$$

Using the data of the example which are $a = 0.09$ m, $b = 0.15$ m, and $n = 4$; Equation 1.28 becomes

$$\Omega = \left\{(x, y)\left(|x/0.09|^{4} + |y/0.15|^{4} \leq 1\right)\right\}. \qquad (1.55)$$

Figure 1.6 shows the footprint that results from Equation 1.55.

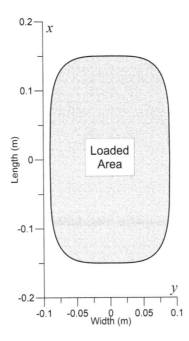

Figure 1.6 Footprint resulting from the superellipse given in Equation 1.55.

To discretize the uniform load, a rectangle of 0.3 m × 0.18 m, which circumscribes the superellipse, is divided into 48 × 48 rectangular areas. Therefore, the size of each rectangular area is $\Delta x \times \Delta y$, where $\Delta x = 0.3/48 = 0.00625$ m and $\Delta y = 0.18/48 = 0.00375$ m.

The Cartesian coordinates (x_P, y_P) of the center of each rectangular element (i, j) are

$$x_P(i, j) = -b + \frac{\Delta x}{2} + i\Delta x,$$

$$y_P(i, j) = -a + \frac{\Delta y}{2} + j\Delta y.$$

Among the 48 × 48 = 2,304 rectangular elements, only 2,152 elements are within the loaded area because these rectangles meet the inequality given in Equation 1.55: $\left(|x_P(i, j)/0.09|^4 + |y_P(i, j)/0.15|^4 \leq 1\right)$. Since the total load of 35 kN acts on these 2152 elements, the point load in each rectangle is 35/2, 152 = 0.01626 kN. So, the load matrix P becomes

$$P_{i,j} = 0 \quad \text{if } \left(|x_P(i, j)/0.09|^4 + |y_P(i, j)/0.15|^4 > 1\right),$$

$$P_{i,j} = 0.01626 \text{ kN} \quad \text{if } \left(|x_P(i, j)/0.09|^4 + |y_P(i, j)/0.15|^4 \leq 1\right).$$

Figure 1.7 shows the loaded area that results from the precedent analysis.

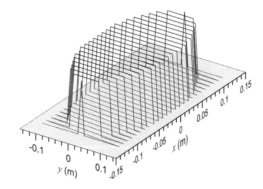

Figure 1.7 Distribution of vertical stress applied by the tire on the half-space.

1.2.2   Superposition of the stresses produced by each individual loaded area

Calculating the stresses requires creating a tridimensional grid within the half-space, the Cartesian coordinates of each calculation point are $(x_C(k, l, m), y_C(k, l, m), z_c(k, l, m))$. Therefore, as shown in Figure 1.8, the distances in each Cartesian direction between the loaded point and the point where the stress is calculated are

$$x = x_C(k, l, m) - x_P(i, j),$$
$$y = y_C(k, l, m) - y_P(i, j),$$
$$z = z_C(k, l, m), \quad \text{and}$$
$$R = \sqrt{x^2 + y^2 + z^2}. \tag{1.56}$$

Finally, the stresses in each point of the half-space, obtained by superposing Boussinesq's solution, given in Equations 1.1–1.6, become

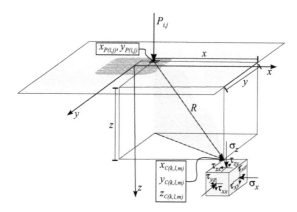

Figure 1.8 Location of the loading and computing points in the Cartesian plane.

$$\sigma_x(k,l,m) = \sum_{i=1}^{48}\sum_{j=1}^{48} \frac{3P_{i,j}}{2\pi}\left\{ \frac{x^2 z}{R^5} - \frac{1/0.3-2}{10}\left[-\frac{1}{R(R+z)} + \frac{(2R+z)x^2}{(R+z)^2 R^3} + \frac{z}{R^3} \right] \right\},$$

$$\sigma_y(k,l,m) = \sum_{i=1}^{48}\sum_{j=1}^{48} \frac{3P_{i,j}}{2\pi}\left\{ \frac{y^2 z}{R^5} - \frac{1/0.3-2}{10}\left[-\frac{1}{R(R+z)} + \frac{(2R+z)y^2}{(R+z)^2 R^3} + \frac{z}{R^3} \right] \right\},$$

$$\sigma_z(k,l,m) = \sum_{i=1}^{48}\sum_{j=1}^{48} \frac{3P_{i,j}}{2\pi}\frac{z^3}{R^5},$$

$$\tau_{xy}(k,l,m) = \sum_{i=1}^{48}\sum_{j=1}^{48} \frac{3P_{i,j}}{2\pi}\left\{ \frac{xyz}{R^5} - \frac{1/0.3-2}{10}\left[\frac{(2R+z)xy}{(R+z)^2 R^3} \right] \right\},$$

$$\tau_{yz}(k,l,m) = \sum_{i=1}^{48}\sum_{j=1}^{48} \frac{3P_{i,j}}{2\pi}\frac{yz^2}{R^5},$$

$$\tau_{zx}(k,l,m) = \sum_{i=1}^{48}\sum_{j=1}^{48} \frac{3P_{i,j}}{2\pi}\frac{xz^2}{R^5}.$$

Furthermore, from Equations 1.7–1.9, the displacements toward the Cartesian directions become

$$u(k,l,m) = \sum_{i=1}^{48}\sum_{j=1}^{48} \frac{1+0.3}{2\pi \cdot 50000}\left[\frac{xz}{R^3} - \frac{(1-2v)x}{R(R+z)} \right]P_{i,j},$$

$$v(k,l,m) = \sum_{i=1}^{48}\sum_{j=1}^{48} \frac{1+0.3}{2\pi \cdot 50000}\left[\frac{yz}{R^3} - \frac{(1-2v)y}{R(R+z)} \right]P_{i,j},$$

$$w(k,l,m) = \sum_{i=1}^{48}\sum_{j=1}^{48} \frac{1+0.3}{2\pi \cdot 50000}\left[\frac{z^2}{R^3} + \frac{2(1-v)}{R} \right]P_{i,j}.$$

Figure 1.9 shows the stresses along each plane of symmetry of the loaded area. Also, Figure 1.10 shows the displacements along the same planes. Stresses and displacements were computed with the MATLAB script provided with this book.

1.2.3 Requirements of Cohesion corresponding to the Mohr–Coulomb criterion

As shown in Figure 1.9, near the edges of the loaded area, the shear stress is high, and the vertical stress is low. Therefore, when assuming the Mohr–Coulomb criterion, the low value of vertical stress implies that the material can only remain below the yielding stress if the material has non-zero cohesion.

Estimating the required cohesion needs the evaluation of the principal stresses. This evaluation is possible after computing the components of the stress tensor depicted in Figure 1.9. The following equations permit to compute the principal stresses σ_1 and σ_3.

The principal stresses along the x axis are

$$\sigma_1 = \frac{\sigma_x + \sigma_z}{2} + \sqrt{\frac{1}{4}(\sigma_x - \sigma_z)^2 + \tau_{zx}^2}, \text{ and}$$

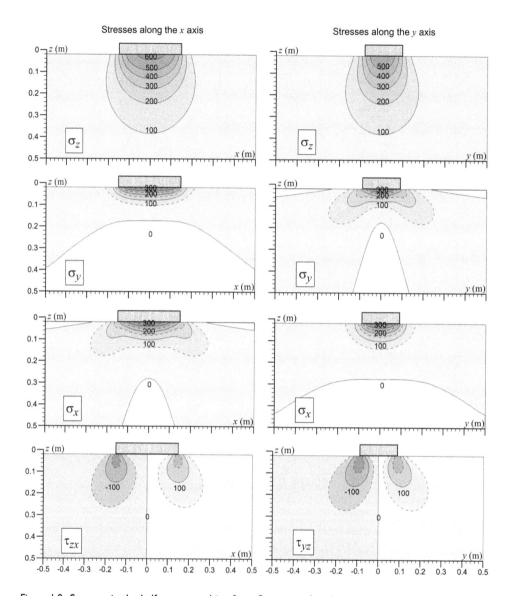

Figure 1.9 Stresses in the half-space resulting from Boussinesq's solution.

$$\sigma_3 = \frac{\sigma_x + \sigma_z}{2} - \sqrt{\frac{1}{4}\left(\sigma_x - \sigma_z\right)^2 + \tau_{zx}^2}.$$

And the principal stresses along the y axis are

$$\sigma_1 = \frac{\sigma_y + \sigma_z}{2} + \sqrt{\frac{1}{4}\left(\sigma_y - \sigma_z\right)^2 + \tau_{yz}^2}, \text{ and}$$

$$\sigma_3 = \frac{\sigma_y + \sigma_z}{2} - \sqrt{\frac{1}{4}\left(\sigma_y - \sigma_z\right)^2 + \tau_{yz}^2}.$$

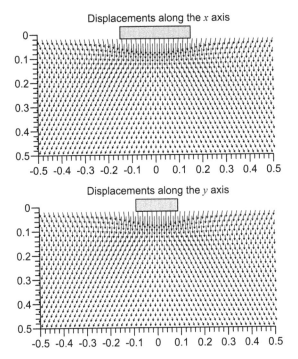

Figure 1.10 Displacements calculated using Boussinesq's solution.

After calculating the principal stresses, the geometrical construction shown in Figure 1.11 permits to calculate the cohesion required to remain below the yield limit given by the Mohr–Coulomb criterion.

From Figure 1.11, the sine of the friction angle, $\sin \phi$, is

$$\sin \phi = \frac{\frac{\sigma_1 - \sigma_3}{2}}{\frac{c}{\tan \phi} + \frac{\sigma_1 + \sigma_3}{2}}.$$

Therefore, the required cohesion, c, becomes

$$c = \frac{\sigma_1 - \sigma_3}{2 \cos \phi} - \frac{\sigma_1 + \sigma_3}{2} \tan \phi. \tag{1.57}$$

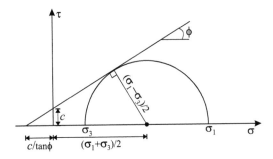

Figure 1.11 Required cohesion depending on the principal stresses σ_1 and σ_3.

Figure 1.12 shows the required cohesion along the axes x and y, evaluated using Equation 1.57.

The cohesion required at each depth corresponds to the maximum cohesion value evaluated for that level, and Equation 1.58 provides the mathematical expression for this condition. Alternatively, a graphical evaluation of the required cohesion is possible. In fact, for a given friction angle, it is possible to draw a straight line that is tangent to the envelope formed by all Mohr circles calculated at a particular depth.

$$c_{req}(z) = \max \left(\frac{\sigma_1(z) - \sigma_3(z)}{2 \cos \phi} - \frac{\sigma_1(z) + \sigma_3(z)}{2} \tan \phi \right). \tag{1.58}$$

Figure 1.13 shows the set of Mohr circles evaluated at depths 0.02, 0.1, 0.2, and 0.3 m, and these Mohr circles permit to outline the envelope made by all Mohr circles at each depth. This evaluation leads to Table 1.2, which shows the required cohesion along the x and y axes that result from assuming a friction angle of 35°.

The same procedure allows obtaining the required cohesion up to a depth of 0.5 m shown in Figure 1.14. This figure allows concluding that when considering the y axis, almost constant cohesion is required up to a depth of 0.15 m. Then, as the depth increases, the cohesion requirement decreases, but for a depth of 0.5 m, a certain amount of cohesion is still required.

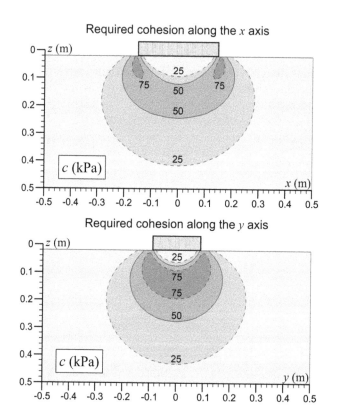

Figure 1.12 Required cohesion in different locations of the half-space along x and y axes.

Table 1.2 Required cohesion for a friction angle of $35°$

Depth (m)	$c_{req}(z)$ (kPa)	
	x Axis	y Axis
0.02	92.55	85.57
0.1	77.09	94.41
0.2	57.76	73.63
0.3	39.82	44.78

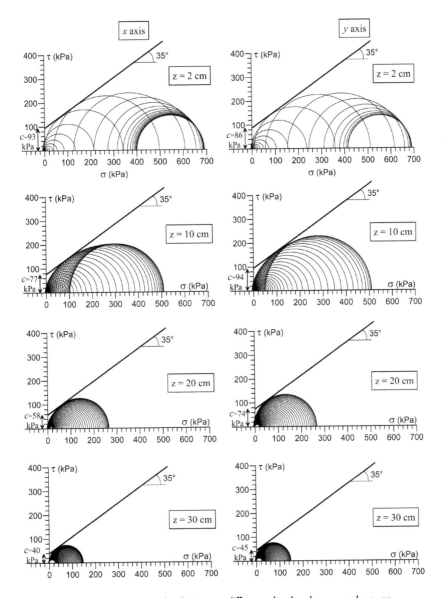

Figure 1.13 Mohr circles and required cohesion at different depths along x and y axes.

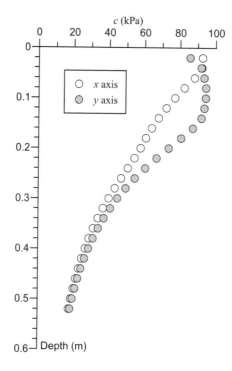

Figure 1.14 Required cohesion for different depths and a friction angle of $\phi = 35°$.

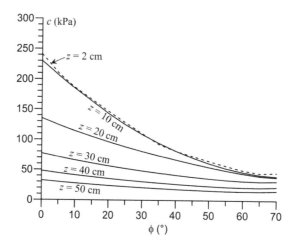

Figure 1.15 Effect of the friction angle on the required cohesion.

The cohesion requirement depends on the friction angle of the material; in fact, as the friction angle increases, the requirement of cohesion decreases. Figure 1.15 illustrates this dependency.

1.2.4   Concluding remarks

Although the analysis presented in this example uses the simplified assumptions of the Boussinesq solution (*i.e.,* homogeneous layer and isotropic linear elastic material), it permits to obtain essential conclusions that are useful for practical purposes:

- The road material below a tire can only remain into the elastic domain if the material has some cohesion.
- This cohesion could be the result of real cohesion created by bonding between particles made by hydraulic or asphaltic cement, or apparent cohesion resulting from the dilatancy in highly compacted materials or from capillary forces in unsaturated materials.
- Since the cohesion created by asphaltic bonding decreases as the temperature increases, when the layer heats, the material must have a high friction angle to sustain the shear stresses.
- The same case occurs for the capillary bonding. As the material wets, its apparent cohesion decreases, and therefore shear stresses must be sustained by the friction between particles.

1.3   EXAMPLE 2: USE OF CERRUTI'S SOLUTION TO CALCULATE THE STRESSES PRODUCED BY HORIZONTAL LOADS

Cerruti's solution allows calculating the stresses and displacements produced by horizontal loads in a homogeneous half-space. In this example, a tire whose geometry is similar to that of the previous example carries a horizontal load, which is uniformly distributed, and its magnitude is the 10% of the vertical load used in the previous example. Furthermore, the elastic properties of the half-space are the same as in Example 1.

The purpose of the example is to evaluate the stresses and displacements produced by the horizontal load and the cohesion requirement when assuming the Mohr–Coulomb criterion with a friction angle of 35°.

As in Example 1, stresses produced by the tire in the elastic half-space can be obtained with the following procedure:

1   Discretize the loaded area into small rectangular areas, applying in each discrete area a horizontal concentrated point load, and then superpose the stress produced by each one of those concentrated loads.

2   Analyze the requirement of cohesion for a friction angle of 35°.

The procedure for the first step is the same as in Example 1. However, the stress evaluation must be carried out using Cerruti's solution instead of Boussinesq's solution.

1.3.1   Stresses in the half-space

Figure 1.16 shows the point where the stress is computed whose Cartesian coordinates are $(x_C(k, l, m),\ y_C(k, l, m),\ \text{and}\ z_C(k, l, m))$, and the loading points located at $x_H(i, j)$ and $y_H(i, j)$. Therefore, the distances in each direction of the Cartesian plane between the point where the elementary load is applied, and the computing point are

$$x = x_C(k, l, m) - x_H(i, j),$$
$$y = y_C(k, l, m) - y_H(i, j),$$
$$z = z_C(k, l, m), \quad \text{and}$$
$$R = \sqrt{x^2 + y^2 + z^2}.$$

Equations 1.10–1.15 lead to the following superposition of stresses on the 48×48 sub-areas used in the Example 1 as

$$\sigma_x(k, l, m) = \sum_{i=1}^{48}\sum_{j=1}^{48} -\frac{H(i,j)x}{2\pi R^3}\left\{-\frac{3x^2}{R^2} + \frac{1-2v}{(R+z)^2}\left[R^2 - y^2 - \frac{2Ry^2}{R+z}\right]\right\},$$

$$\sigma_y(k, l, m) = \sum_{i=1}^{48}\sum_{j=1}^{48} -\frac{H(i,j)x}{2\pi R^3}\left\{-\frac{3y^2}{R^2} + \frac{1-2v}{(R+z)^2}\left[3R^2 - x^2 - \frac{2Rx^2}{R+z}\right]\right\}$$

$$\sigma_z(k, l, m) = \sum_{i=1}^{48}\sum_{j=1}^{48} \frac{3H(i,j)xz^2}{2\pi R^5},$$

$$\tau_{xy}(k, l, m) = \sum_{i=1}^{48}\sum_{j=1}^{48} -\frac{H(i,j)y}{2\pi R^3}\left\{-\frac{3x^2}{R^2} + \frac{1-2v}{(R+z)^2}\left[-R^2 + x^2 + \frac{2Rx^2}{R+z}\right]\right\},$$

$$\tau_{yz}(k, l, m) = \sum_{i=1}^{48}\sum_{j=1}^{48} \frac{3H(i,j)xyz}{2\pi R^5},$$

$$\tau_{zx}(k, l, m) = \sum_{i=1}^{48}\sum_{j=1}^{48} \frac{3H(i,j)x^2z}{2\pi R^5}.$$

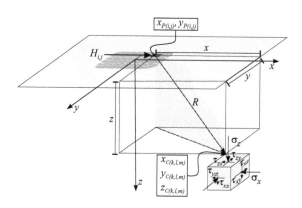

Figure 1.16 Horizontal load and computing points in the Cartesian plane.

Likewise, from Equations 1.16–1.18, displacements become

$$u(k, l, m) = \sum_{i=1}^{48} \sum_{j=1}^{48} \frac{H(i,j)}{4\pi\, GR} \left\{ 1 + \frac{x^2}{R^2} + (1-2v) \left[\frac{R}{R+z} - \frac{x^2}{(R+z)^2} \right] \right\},$$

$$v(k, l, m) = \sum_{i=1}^{48} \sum_{j=1}^{48} \frac{H(i,j)}{4\pi\, GR} \left\{ \frac{xy}{R^2} - (1-2v) \frac{xy}{(R+z)^2} \right\},$$

$$w(k, l, m) = \sum_{i=1}^{48} \sum_{j=1}^{48} \frac{H(i,j)}{4\pi\, GR} \left\{ \frac{xz}{R^2} + (1-2v) \frac{x}{R+z} \right\}.$$

Figures 1.17 and 1.18 show the stresses and displacements along the x axis of the loaded area. Note that stresses and displacements along the y axis are zero because it is the symmetry axis for the horizontal load. The MATLAB script used for computing the stresses and displacements of this example is provided with this book.

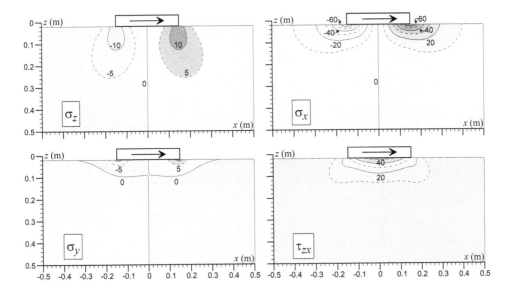

Figure 1.17 Stresses along the x axis computed using Cerruti's solution (in kPa).

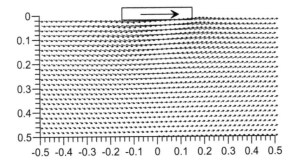

Figure 1.18 Displacements along the x axis computed using Cerruti's solution.

1.3.2 Requirements of cohesion for the Mohr–Coulomb criterion

The same procedure described in Example 1 allows evaluating the cohesion required for this example. Figure 1.19 indicates that the vertical stress σ_z increases in front of the horizontal load and decreases behind it. Therefore, because of the lack of confinement, the cohesion requirement for the Mohr–Coulomb criterion is greater behind the direction of the horizontal load, just as shown in Figure 1.19.

Similarly as in the previous example, the envelope made by the set of Mohr circles permits to assess the required cohesion for a friction angle of $35°$ (see Figure 1.20). Also, Table 1.3 shows the results of this evaluation.

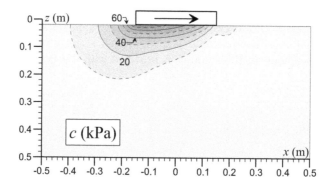

Figure 1.19 Requirement of cohesion along x axis for a friction angle of $35°$.

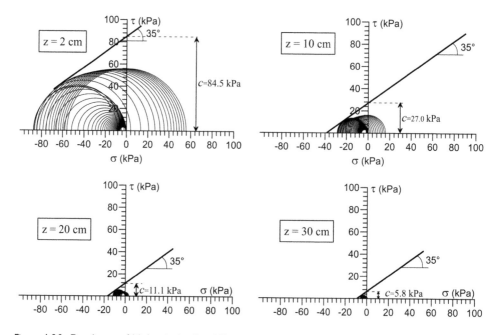

Figure 1.20 Envelopes of Mohr circles for different depths and requirement of cohesion for a friction angle of $35°$.

Table 1.3 Required cohesion for a friction angle of $35°$

Depth (m)	0.02	0.1	0.2	0.3
$c_{req}(z)$(kPa)	84.5	27.0	11.1	5.8

1.3.3 Concluding remarks

As in the example dealing with the Boussinesq solution, this example also uses the simplified assumptions of Cerruti's solution (*i.e.*, homogeneous layer and isotropic linear elastic material); nevertheless some relevant conclusions can be obtained:

- Bearing the stresses produced by horizontal loads while ensuring that the material remains into the elastic domain of behavior also requires a cohesive material.
- The cohesion requirement for a horizontal load whose magnitude is 10% of the vertical load is as high as the cohesion required for the vertical load. However, the cohesion required for horizontal loading decreases with depth faster than when applying vertical loads.
- Higher cohesion is required behind the direction of the horizontal load.
- Concerning displacements, the horizontal load creates horizontal displacements in the same direction of the load, as expected. Moreover, vertical displacements also appear: upward in the direction of the horizontal load and downward behind it.

1.4 EXAMPLE 3: TIRE–ROAD INTERACTION USING THE HERTZ THEORY AND THE FRÖHLICH STRESS DISTRIBUTION

The following example illustrates the procedure that allows a simplified computation of the tire–road interaction using the Hertz theory and the Fröhlich stress distribution.

In this example, the tire is approached as a homogeneous cylinder with two curvature radii whose geometric characteristics are: external radius of the tire 0.52 m, and radius of the tube of the tire 0.14 m.

Table 1.4 shows the results of a compression test of the tire that was carried out to assess its mechanical properties. The test consisted in applying incremental loads on the axle of the tire when it rests over a rigid concrete block. The elastic constants of that block are $E_c = 17$ GPa and $v_c = 0.25$.

For computing the tire–road interaction, let assume that the tire applies a load of 25 kN over a half-space having a Young's modulus of $E_R = 400$ MPa and a Poisson's ratio of $v_R = 0.3$. The distribution of vertical stresses in the half-space is computed using the Fröhlich stress distribution with a stress concentration factor of $\xi = 5$.

The procedure for evaluating the tire–road interaction requires the following steps:

1 Assess the elastic properties of the equivalent tire based on the results of the compression test.
2 Evaluate the contact stress applied by the tire on the road.

Table 1.4 Results of a compression test of a tire resting on a rigid concrete block

Compression Test	Load, F(kN)	Deflection, d (m)
	0	0
	1.93	0.0031
	4.01	0.0065
	5.71	0.0093
	7.19	0.0115
	8.98	0.0143
	10.46	0.0162
	11.86	0.0183
	14.16	0.0205
	16.98	0.0238
	20.25	0.0270
	22.99	0.0298
	25.96	0.0329
	28.55	0.0357
	31.22	0.0385
	33.22	0.0408
	35.30	0.0430

3   Discretize the loaded area into small rectangular areas with concentrated point loads, and superpose the stress produced by each concentrated load using the Fröhlich stress distribution.

1.4.1 Elastic properties of the equivalent tire

As described in Caicedo [10], the contact between an ellipsoid with two curvature radii R_1 and R_2 and a flat surface leads to an elliptical contact area whose semi-axes, a and b, are given by

$$a = \sqrt{R_1 d}, \quad \text{and} \quad b = \sqrt{R_2 d},$$

where d is the deflection, which is also known as the indentation, of the ellipsoid.

When considering the equation giving the area of an ellipse, the contact area becomes

$$A = \pi ab = \pi \tilde{R} d,$$

where $\tilde{R} = \sqrt{R_1 R_2}$ represents the Gaussian radius of curvature.

Furthermore, for the elliptical contact area resulting from a tire with curvature radii R_1 and R_2, the relationship between the loaded area and the load F is

$$ab = \left(\frac{3F\tilde{R}}{4E^*} \right)^{\frac{2}{3}} = \tilde{R} d. \tag{1.59}$$

In Equation 1.59, E^* is the equivalent Young's modulus given by

$$\frac{1}{E^*} = \frac{1 - v_T^2}{E_T} + \frac{1 - v_c^2}{E_c},$$

where E_T, E_c, v_T, and v_c are the elastic constants of the bodies in contact (*i.e.*, tire and concrete).

Considering that $a = \sqrt{R_1 d}$ and $b = \sqrt{R_2 d}$, then

$$d = \frac{a^2}{R_1} = \frac{b^2}{R_2}. \tag{1.60}$$

Therefore, the contact area can be found with the following equations:

$$b = a\sqrt{\frac{R_2}{R_1}}, \quad \text{and} \quad a^2\sqrt{\frac{R_2}{R_1}} = \left(\frac{3F\tilde{R}}{4E^*}\right)^{\frac{2}{3}}. \tag{1.61}$$

Since the tire compression test is performed on a rigid concrete block, the elastic constants of the second body are $E_c = 17$ GPa and $v_c = 0.25$. Therefore, the equivalent Young's modulus for the compression test is

$$E^* = \frac{1}{\frac{1-v_T^2}{E_T} + \frac{0.9375}{1.7 \cdot 10^7}}.$$

From Equation 1.61, the semi axis a of the ellipse is

$$a = \sqrt{\left(\frac{3F\tilde{R}}{4E^*}\right)^{\frac{2}{3}}\sqrt{\frac{R_1}{R_2}}}.$$

For the data of the example, $R_1 = 0.14$ m, $R_2 = 0.52$ m, the Gaussian radius of curvature is $\tilde{R} = \sqrt{0.52 \cdot 0.14} = 0.27$ m, and $\sqrt{R_1/R_2} = 0.519$, then the semi axis a becomes

$$a = 0.423F^{1/3}\left(\frac{1-v_T^2}{E_T} + \frac{0.9375}{1.7 \cdot 10^7}\right)^{1/3}, \quad \text{and} \quad b = \sqrt{\frac{0.52}{0.14}}a = 1.93a.$$

Finally, from Equation 1.60, the deflection of the tire is

$$d = \frac{a^2}{R_1} = 1.278F^{2/3}\left(\frac{1-v_T^2}{E_T} + \frac{0.9375}{1.7 \cdot 10^7}\right)^{2/3}. \tag{1.62}$$

The deflection computed using Equation 1.62 can be adjusted to the experimental results given in Table 1.4. Figure 1.21 shows the best adjustment of the experimental results obtained using an equivalent Young's modulus of the tire of $E_T = 5,700$ kPa and Poisson's ratio of $v_T = 0.31$. As shown in Figure 1.21, the experimental results have a more linear trend than the theoretical curve obtained using the Hertz theory because the shape of the tire differs from an ellipsoid with two curvature radii.

1.4.2 Contact stress applied by the tire on the road

The ellipsoid produces an elliptical loaded area whose maximum contact pressure is

$$p_0 = \left(\frac{6FE^{*2}}{\pi^3 \tilde{R}^2}\right)^{1/3}.$$

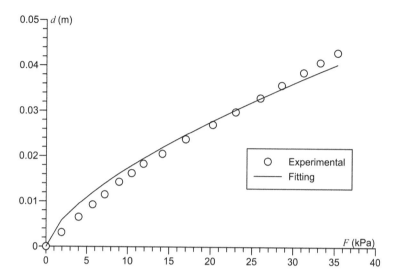

Figure 1.21 Fitting of the experimental results of the compression test of the tire.

However, when the tire rests on a road that has Young's modulus $E_R = 400$ MPa and Poisson's ratio $\nu_R = 0.3$, the equivalent Young's modulus becomes

$$E^* = \left(\frac{1 - \nu_T^2}{E_T} + \frac{1 - \nu_R^2}{E_R} \right)^{-1} = \left(\frac{1 - 0.31^2}{5,700} + \frac{1 - 0.3^2}{400,000} \right)^{-1} = 6,216.8 \text{ kPa.}$$

Therefore, when the tire has a load of 25 kN, the maximum pressure is

$$p_0 = \left(\frac{6 \cdot 25 \cdot 6,216.8^2}{\pi^3 \cdot 0.27^2} \right)^{1/3} = 1,371 \text{ kPa.}$$

And the semi-axes of the ellipsoid are

$$a = 0.423 \cdot 25^{1/3} \left(\frac{1}{6,216.8} \right)^{1/3} = 0.067 \text{ m, and } b = 1.93a = 0.13 \text{ m.}$$

Therefore, the following relationship gives the pressure distribution:

$$\sigma_z(x, y) = p_0 \sqrt{1 - \frac{x^2}{a^2} - \frac{y^2}{b^2}} = 1,371 \sqrt{1 - \frac{x^2}{0.067^2} - \frac{y^2}{0.13^2}}. \tag{1.63}$$

Figure 1.22 shows the distribution of vertical stresses on the road given by Equation 1.63.

1.4.3   Stresses in the half-space using the Fröhlich solution for stress distribution

As in the previous examples, the computation of the vertical stress requires creating a tridimensional grid within the half-space, the Cartesian coordinates of each computing

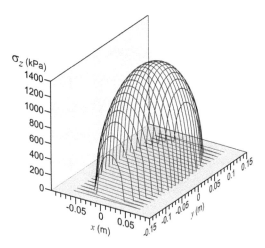

Figure 1.22 Pressure distribution on the road given by the Hertz theory.

point are $(x_C(k, l, m),\ y_C(k, l, m),\ z_c(k, l, m))$ and, as shown in Figure 1.23, distances in each direction of the Cartesian plane between the point where the elementary load is applied, and the computing point are

$$x = x_C(k, l, m) - x_P(i, j),$$

$$y = y_C(k, l, m) - y_P(i, j),$$

$$z = z_C(k, l, m), \quad \text{and,}$$

$$R = \sqrt{x^2 + y^2 + z^2},$$

$$\rho = \sqrt{x^2 + y^2},$$

$$\beta = \arctan \frac{\rho}{R}.$$

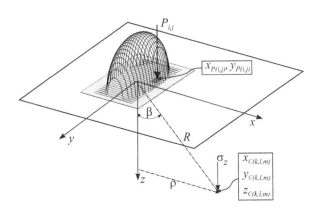

Figure 1.23 Hertzian load and computing points in the Cartesian plane for the Fröhlich solution.

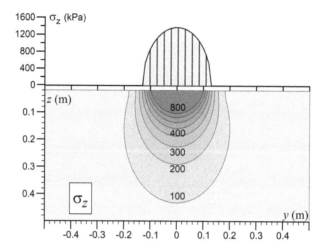

Figure 1.24 Vertical stresses computed using the Hertz contact theory and the Fröhlich solution for stress distribution.

Equations 1.19 and 1.23 permit computing the elementary vertical stress given by the Fröhlich solution as follows:

$$\Delta\sigma_R(k, l, m, i, j) = \frac{\xi\, P(i, j)}{2\pi\, R^2}\cos^{\xi-2}\beta,$$

$$\Delta\sigma_z(k, l, m, i, j) = \Delta\sigma_R(k, l, m)\frac{z^2}{R^2}.$$

Finally, discretizing the loaded area into 48×48 sub-areas, the superposition of the vertical stress becomes

$$\sigma_z(k, l, m) = \sum_{i=1}^{48}\sum_{j=1}^{48}\Delta\sigma_z(k, l, m, i, j).$$

Figure 1.24 shows the vertical stress along the y axis of the elliptical loaded area. This stress was computed using the MATLAB script provided with this book.

1.4.4   Concluding remarks

This example illustrates, in a simplified form, the interaction between an idealized tire and a homogeneous half-space. The example uses the Hertz contact theory; it is clear that this approach is only a rough approximation because it neglects the actual shape of a tire, which differs from a toroidal shape. More advanced theories regarding tire–soil interaction can be found in Caicedo [10]. Despite the simplified approach used in this example, it allows concluding the following points:

- Contact stresses on a road depend not only on the characteristics of the tire but also on the road's stiffness.

- The approach used in most of the mechanistic methods for pavement design, which consist in assuming the stress applied by the tire as a circular load with uniform contact stress is certainly a rough approximation of the actual contact pressure.
- The concentration factor used in the Fröhlich stress distribution modifies the shape of the distribution of the vertical stresses. This concentration factor is an indirect method to take in to account the plasticity of the material. This approach could be useful to analyze unpaved roads, for which the road material can reach the yield criterion.

1.5   EXAMPLE 4: CALCULATION OF THE VEHICLE–ROAD INTERACTION

The following example illustrates the methodology for calculating the force applied by a vehicle traveling at different velocities on roads of different roughness.

The vehicle is a two-axle truck, whose characteristics of the half-side of the rear axle (i.e., a quarter car) are

- *Spring constant of the tire $k_1 = 1.8 \cdot 10^6$ N/m,*
- *Spring constant of the suspension $k_2 = 1 \cdot 10^6$ N/m,*
- *Half mass of the axle $m_1 = 500$ kg,*
- *Half mass carried by the rear axle $m_2 = 4,500$ kg,*
- *Damping constant of the tire $c_1 = 2 \cdot 10^3$ Ns/m, and*
- *Damping constant of the suspension $c_2 = 1.5 \cdot 10^4$ Ns/m.*

Figure 1.25 represents the quarter car model for the data of the example.

In this example, the interaction is calculated for three different velocities of the truck (5, 10, and 20 m/s) and two types of rough roads:

1   *A flat road that has regular bumps located at a distance of 10 m between them, where each bump can be approximated by a sinusoidal function with an amplitude of 0.05 m and a wavelength of 2 m. The sinusoidal shape of each bump begins in the lower part of the sine function. Therefore, the total height of the bump is two times the amplitude of the function (i.e., 0.1 m).*

2   *Actual roads described by two power spectral densities, characterized by the following roughness coefficients: $\kappa = 4$ for a very good road, and $\kappa = 64$ for an average road.*

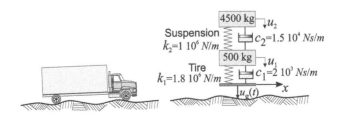

Figure 1.25 Quarter car model with the data of the example.

Solving this problem requires the following steps:

1 Discretize in time the second-order linear differential equation with two degrees of freedom, which represents the interaction between a quarter car and a road.
2 Describe the flat road with regular bumps and compute the interaction.
3 Describe the actual roads using the method of SOH and compute the interaction.

1.5.1   Discretization in time of the differential equation

As described in Section 1.1.5, the quarter car model that allows computing the vehicle–road interaction can be analyzed using the following second-order linear differential equation with two degrees of freedom:

$$M\ddot{u} + C\dot{u} + Ku = f(t). \tag{1.64}$$

For this example, the values of the mass, damping, and stiffness matrices of the system are

$$M = \begin{bmatrix} m_1 & 0 \\ 0 & m_2 \end{bmatrix} = \begin{bmatrix} 5,00 & 0 \\ 0 & 4,500 \end{bmatrix} \text{ kg,}$$

$$C = \begin{bmatrix} c_1 + c_2 & -c_2 \\ -c_2 & c_2 \end{bmatrix} = \begin{bmatrix} 17 & -15 \\ -15 & 15 \end{bmatrix} \cdot 10^3 \text{ Ns/m,}$$

$$K = \begin{bmatrix} k_1 + k_2 & -k_2 \\ -k_2 & k_2 \end{bmatrix} = \begin{bmatrix} 2.8 & -1 \\ -1 & 1 \end{bmatrix} \cdot 10^6 \text{ N/m.}$$

On the other hand, the finite difference method permits to discretize the derivatives in time of the vector of displacements u as

$$\ddot{u} = \frac{u^{t+\Delta t} - 2u^t + u^{t-\Delta t}}{\Delta t^2}, \quad \text{and} \quad \dot{u} = \frac{u^{t+\Delta t} - u^t}{\Delta t},$$

where Δt is the time step, and $u^{t+\Delta t}$, u^t, and $u^{t-\Delta t}$ represent the vectors of displacement in times $t + \Delta t$, t, and $t - \Delta t$, respectively.

The profile of the road can be defined as a function depending on time $u_g(t)$, or on the traveled distance x, $u_g(x)$. This second definition is the way to calculate the vehicle–road interaction because usually road profiles are described in geometrical terms. For this second definition, the displacement of the vehicle is $x = Vt$, where V is its velocity. Therefore, the discretization in time of the vector $f(t)$ becomes

$$f(t) = \begin{bmatrix} k_1 u_g(Vt) + c_1 \dot{u}_g(Vt) + m_1 g \\ m_2 g \end{bmatrix}$$

$$= \begin{bmatrix} 1.8 \cdot 10^6 u_g(Vt) + 2 \cdot 10^3 \frac{u_g(V(t+\Delta t)) - u_g(Vt)}{\Delta t} + 500 \cdot 9.81 \\ 4,500 \cdot 9.81 \end{bmatrix}. \tag{1.65}$$

Likewise, Equation 1.64, discretized in time, becomes

$$M \frac{u^{t+\Delta t} - 2u^t + u^{t-\Delta t}}{\Delta t^2} + C \frac{u^{t+\Delta t} - u^t}{\Delta t} + Ku^t = f(t)$$

which, rearranging the terms, becomes

$$u^{t+\Delta t}\left(\frac{M}{\Delta t^2}+\frac{C}{\Delta t}\right)=f(t)-M\frac{-2u^t+u^{t-\Delta t}}{\Delta t^2}+C\frac{u^t}{\Delta t}-Ku^t. \tag{1.66}$$

Equation 1.66 permits to obtain the following recursive equation that permits computing the displacement vector u in the time $t+\Delta t$ knowing the displacements u in times $t-\Delta t$ and t:

$$u^{t+\Delta t}=\left[\frac{1}{\Delta t^2}M+\frac{1}{\Delta t}C\right]^{-1}\left(f(t)+M\frac{2u^t-u^{t-\Delta t}}{\Delta t^2}+C\frac{u^t}{\Delta t}-Ku^t\right). \tag{1.67}$$

This equation is used to solve the road–vehicle interaction for different rough roads, which is developed in the following steps of the example.

1.5.2 Vehicle interaction in a bumpy road

In each time step, Δt, for which Equation 1.67 is solved, the vehicle advances a distance $\Delta x = V\Delta t$.

The number of points n_b for describing each bump is $n_b = \lambda_b/\Delta x$, and the amplitude of the sinusoidal function representing the bumps is 0.05 m. Then, an individual bump can be discretized into n_b points as

$$u_{\text{bump}}(j)=-0.05-0.05\sin\left(-\frac{\pi}{2}+\frac{2\pi j}{n_b}\right)\quad j=1,...,n_b.$$

Each bump can be placed at a distance of 10 m on the flat road, leading to the road's profile represented in Figure 1.26. Note that the positive direction of the vertical displacements is downwards.

After discretizing in time the vector $u_g(i)$, the discretization in time of the vector $f(t)$, given in Equation 1.65, is

$$f(i)=\left[\begin{array}{c}1.8\cdot10^6u_g(i)+2\cdot10^3\frac{u_g(i+1)-u_g(i)}{\Delta t}+500\cdot9.81\\4,500\cdot9.81\end{array}\right],$$

where each step i corresponds to a time lapse Δt and a distance Δx.

Finally, the movement of the masses, given by the vector u, is obtained by solving Equation 1.67 for the time step $i+1$, knowing the position of the masses in the times i and $i-1$ as follows:

$$u(i+1)=\left[\frac{1}{\Delta t^2}M+\frac{1}{\Delta t}C\right]^{-1}\left(f(i)+M\frac{2u(i)-u(i-1)}{\Delta t^2}+C\frac{u(i)}{\Delta t}-Ku(i)\right).$$

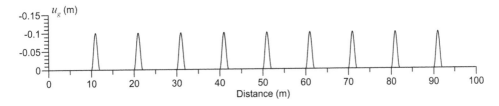

Figure 1.26 Profile of 100 m of the road that has regular bumps.

The MATLAB script provided with this book allows calculating the position of the masses as the vehicle advances along the bumpy road at different speeds.

Figures 1.27 and 1.28 show displacement of the quarter car computed for different speeds of the vehicle (remind that u_1 represents the movement of the axle of the truck, while u_2 represents the displacement of the body).

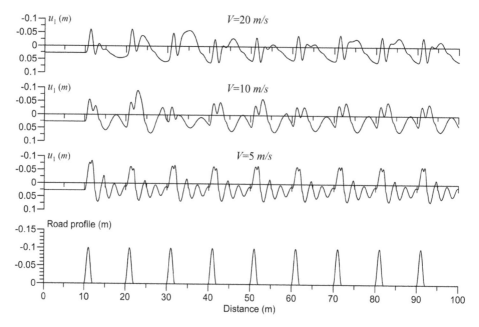

Figure 1.27 Movement of the mass m_1 for different velocities of the vehicle.

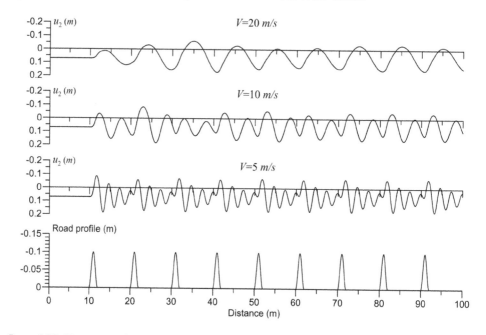

Figure 1.28 Movement of the mass m_2 for different velocities of the vehicle.

Results of the vehicle displacements, given by the body displacement u_2, are useful for analyzing the passenger comfort and the payload disturbance. However, regarding roads, the most important result is the computation of the dynamic load applied to its surface as the vehicle advances.

The dynamic load applied to the road results from the relative displacement between the truck axle and the road profile. This relative displacement produces compression in the spring, which has a constant k_1; moreover, the damper that has a constant c_1 reacts in proportion to the rate of compression. The compression C_o for the time step i and its rate of change are

$$C_o(i) = u_1(i) - u_g(i), \text{ and}$$

$$\dot{C}_o(i) = \frac{C_o(i) - C_o(i-1)}{\Delta t}.$$

However, it is essential to note that the reaction only exists when there is contact between the tire and the road (*i.e.*, when the compression is positive); otherwise, the reaction is zero because the tire loses its contact with the road. Therefore, the dynamic load, F, applied to the road is

$$F(i) = k_1 C_o(i) + c_1 \dot{C}_o(i) \quad \text{for } C_o(i) > 0,$$

$$F(i) = 0 \quad \text{for } C_o(i) \leq 0.$$

Furthermore, when the tire loses contact with the road, the constants k_1 and c_1 do not affect Equation 1.67. Therefore, the matrices $[C]$, $[K]$ and the vector $f(t)$ become

$$C = \begin{bmatrix} c_2 & -c_2 \\ -c_2 & c_2 \end{bmatrix}, \ K = \begin{bmatrix} k_2 & -k_2 \\ -k_2 & k_2 \end{bmatrix}, \ f(t) = \begin{bmatrix} m_1 g \\ m_2 g \end{bmatrix}.$$

And for the data of the example, these matrices are

$$C = \begin{bmatrix} 15 & -15 \\ -15 & 15 \end{bmatrix} \cdot 10^3 \text{ Ns/m,}$$

$$K = \begin{bmatrix} 1 & -1 \\ -1 & 1 \end{bmatrix} \cdot 10^6 \text{ N/m,}$$

$$f(t) = \begin{bmatrix} 500 \cdot 9.8 \\ 4,500 \cdot 9.8 \end{bmatrix} \text{ N.}$$

Figure 1.29 shows the dynamic load applied on the road for different velocities of the vehicle. For all velocities, before the first bump, the dynamic load equals the static load of $(4,500+500) \cdot 9.8 = 49$ kN; however, bumps amplify the dynamic load up to two times the static load. Also, in some places, the wheel loses contact, and therefore the dynamic load is zero.

Figure 1.30, drawn for a speed of 20 m/s, allows a more detailed analysis. In fact, the curve of dynamic load in this figure shows a significant increase when the wheel reaches the bumps, and this increase in the reaction of the road creates a vertical movement of the vehicle leading to a loose of contact. Afterward, the tire touches again the road producing a significant reaction and a subsequent loss of contact. This pattern appears for each bump. However, the magnitude of the loads and the distances of losing contact change because of the dynamic behavior of the system.

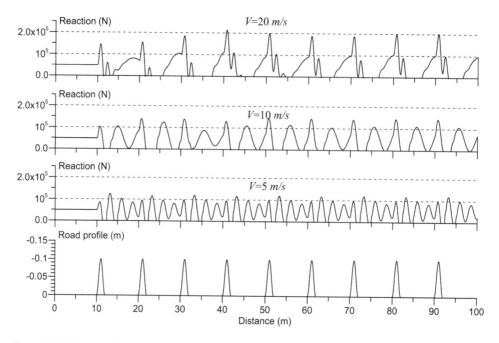

Figure 1.29 Dynamic load on the road for different velocities of the vehicle.

Finally, Figure 1.31 shows histograms of reaction forces for different velocities of the vehicle. It is noticeable how high can be the dynamic load in comparison with the static load, even when the vehicle advances at moderate speeds.

1.5.3   Vehicle interaction on actual roads

Obtaining the profile of actual roads is possible using the method of SOH explained in Section 1.1.5.1 of this book.

In this example, the power spectrum will be divided into 1,000 frequency bands, each band corresponding to one harmonic. Then, Equation 1.34 becomes

$$u_g(x) = \sum_{i=1}^{1,000} Z_i \sin(2\pi \Omega_i x + \phi_i),$$

where Z_i is the amplitude and ϕ_i the phase angle of the ith harmonic, which follows a random distribution into a uniform interval of $[0, 2\pi]$.

In addition, the lower and upper limits for the frequency are $\Omega_L = 0.01$ and $\Omega_U = 10$. Therefore, the width of each frequency band, $\Delta\Omega$, is $\Delta\Omega = (\Omega_U - \Omega_L)/1,000 = 0.00999$.

The discretization of the PSD leads to the following expression for the central frequency of the nth band:

$$\Omega_i = 0.01 + \frac{2i-1}{2} \cdot 0.00999.$$

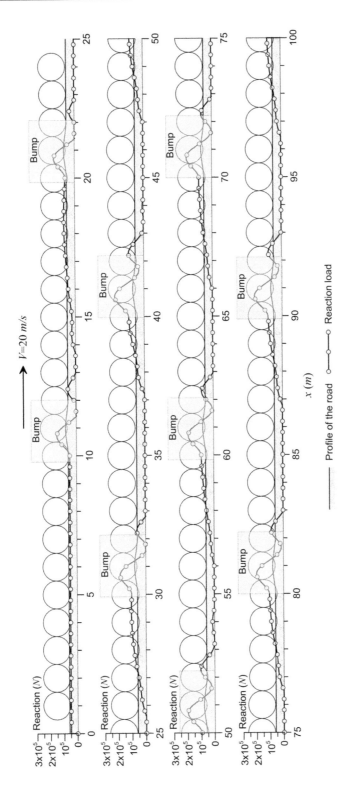

Figure 1.30 Wheel positions while advancing on a bumpy road at a velocity of 20 m/s.

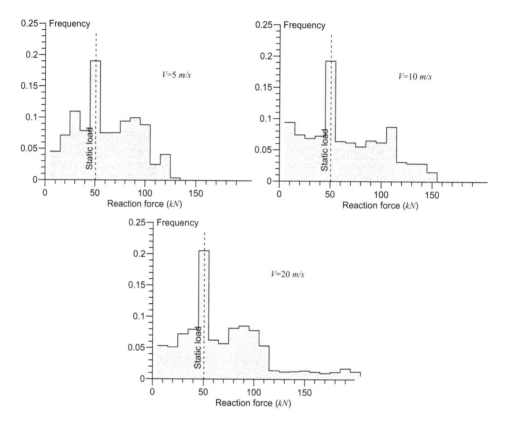

Figure 1.31 Histograms of the reaction force for different velocities of the vehicle.

Considering that $\Omega_0 = 1/2\pi$, Equation 1.33 becomes

$$S_z(\Omega_i) = \kappa \left(\frac{\pi}{2\Omega_i}\right)^w \text{ where}$$

$$w = 2 \text{ for } 0.01 \leq \Omega_i \leq \pi/2, \text{ or}$$
$$w = 1.5 \text{ for } \pi/2 \leq \Omega_i \leq 10.$$

Furthermore, from Equation 1.36, the amplitude Z_i of the ith harmonic is

$$Z_i = \sqrt{2S_z(\Omega_i) \cdot 0.00999}.$$

To include enough harmonics, the length of the road is $L = 3/\Omega_L = 3/0.01 = 300$ m. Then, the total length of the road L is divided into n_x segments with a constant length Δx. In this example, $\Delta x = 0.002$ m, and therefore, there are 150.000 points to calculate the road profile.

The algorithm for computing the actual profile of a road that has a coefficient of roughness κ is as follows:

1    $j = 1, ..., 150,000$
2    $i = 1, ..., 1,000$

3　$\Omega_i = 0.01 + \frac{2i-1}{2} \cdot 0.00999$

　　$w = 2$ for $0.01 \leq \Omega_i \leq \pi/2$, or $w = 1.5$ for $\pi/2 < \Omega_i \leq 10$

4　$S_z(\Omega_i) = \kappa \left(\frac{\pi}{2\Omega_i}\right)^w$

5　$Z_i = \sqrt{2S_z(\Omega_i) \cdot 0.00999}$

6　$\phi_i = rand[0, 2\pi]$

7　$u_g(j) = \sum_{i=1}^{1,000} Z_i \sin(2\pi \Omega_i j \Delta x + \phi_i)$

Figure 1.32 shows different profiles of the road segment of 300 m length when assuming different roughness coefficients, from a very good road with $\kappa = 4$ to a poor road with $\kappa = 256$. These profiles were obtained using the MATLAB script provided with this book.

　　Once obtained the road's profile, computing the interaction between the quarter car and the road follows the same procedure as in the case of a bumpy road. A MATLAB script that permits to calculate the interaction road–vehicle on actual roads is provided with this book. However, it is essential to keep in mind that the simulation of the actual profile of the road involves a random process. Therefore each repetition of the simulation produces different results.

　　Figures 1.33 and 1.34 show the dynamic loads applied on the road when the vehicle advances at different velocities on a very good road, $\kappa = 4 \cdot 10^{-6}$, and on an average road, $\kappa = 64 \cdot 10^{-6}$. From these figures, it is noticeably the increase of the dynamic load as the roughness of the road and the travel velocity increases.

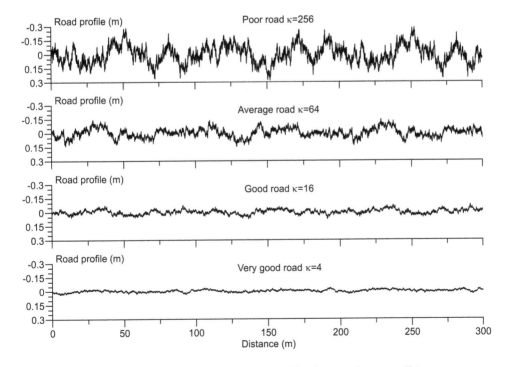

Figure 1.32 Profile of different quality roads characterized by their roughness coefficient κ.

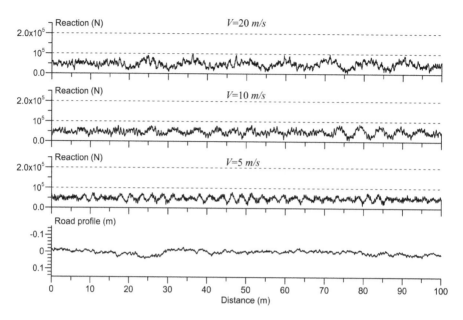

Figure 1.33 Dynamic load applied by the vehicle when traveling at different velocities on a very good road, characterized by a roughness coefficient of $\kappa = 4 \cdot 10^{-6}$.

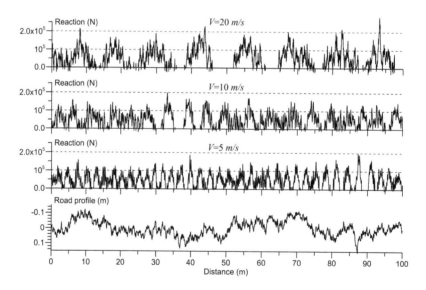

Figure 1.34 Dynamic load applied by the vehicle when traveling at different velocities on an average road, characterized by a roughness coefficient of $\kappa = 64 \cdot 10^{-6}$.

A more detailed analysis is possible in Figure 1.35, which presents the results when the vehicle circulates at the velocity of 20 m/s either on the good or the average road. This figure makes it evident the segments where the wheel loses its contact with the road, mainly when the vehicle circulates on the average road, while when it circulates on the good road, the wheel and the road are permanently in contact.

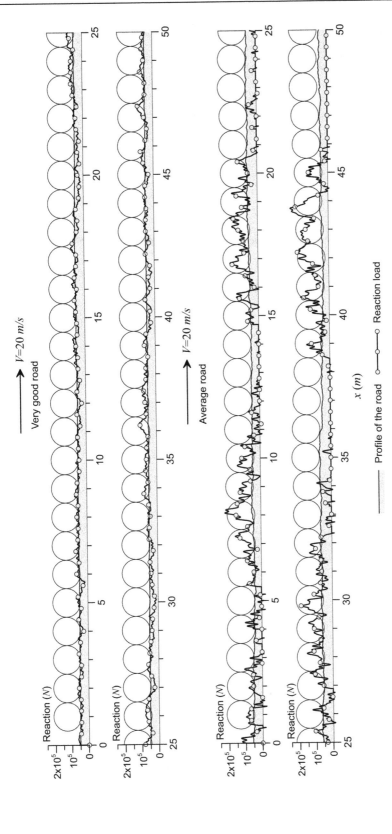

Figure 1.35 Position of the wheel when traveling at 20 m/s on roads of two different qualities.

Histograms presented in Figure 1.36 make evident the differences between static and dynamic loads. It is clear that when the vehicle circulates on a very good road, the histogram of dynamic loads is centered in the static load for all the velocities of the vehicle, and only a small proportion of dynamic loads are 50% higher than the static load. In contrast, when the vehicle circulates on an average road, the mean value of the dynamic loads is higher than the static load, and a significant proportion of dynamic loads exceed in more than 100% the static load.

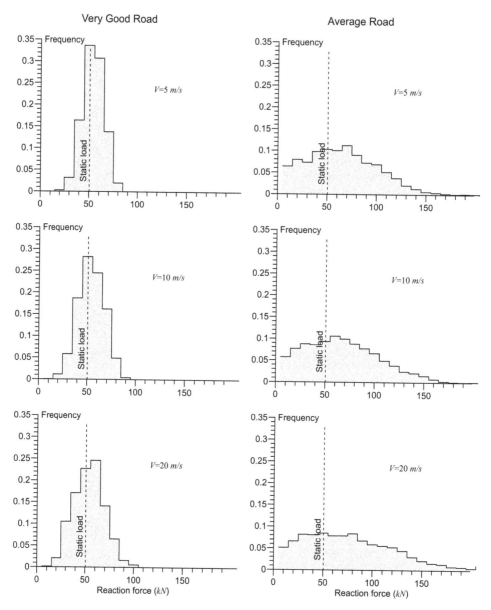

Figure 1.36 Histograms of dynamic load for different velocities of the vehicle and roads of very good and average quality.

1.5.4 Concluding remarks

The results of this example suggest that the procedure used in most mechanical pavement design methods that apply a constant static load to the pavement surface could be incorrect. In fact, the results of this example show that, even on good quality roads, the dynamic load could be 50 % higher than the static load. Also, the proportion of dynamic loads that exceed the static load grows as the road's roughness increases.

The methodology described in this example could be a tool to evaluate the evolution of real pavement damage. However, the use of the model requires an adequate characterization of the suspension components of the vehicle.

1.6 EXAMPLES 5: COMPUTATION OF STRESSES IN A THREE-LAYERED ROAD STRUCTURE USING BURMISTER'S METHOD

This example uses Burmister's method for computing the vertical and radial stresses (σ_z and σ_ρ) in a road structure that has three layers whose geometric and elastic characteristics are

- *Layer 1: Young's modulus $E_1 = 3,000$ MPa, Poisson's ratio $v_1 = 0.45$, thickness 0.3 m.*
- *Layer 2: Young's modulus $E_2 = 500$ MPa, Poisson's ratio $v_2 = 0.35$, thickness 0.7 m.*
- *Layer 3: Young's modulus $E_3 = 50$ MPa, Poisson's ratio $v_1 = 0.40$, thickness ∞.*

Layers are bonded between them.
The load is circular with radius $a = 0.1$ m, and it applies a uniform stress of $q = 0.67$ MPa. The stresses are computed at the interface of the layers and three different distances from the symmetry axis of the load ($\rho = 0.001$ m, $\rho = 0.1$ m, and $\rho = 0.2$ m).

The process for computing stresses using Burmister's method has two steps:

1   Approximate the applied load using a series of Bessel's functions.
2   Compute vertical and radial stresses using Burmister's method.

1.6.1 Approximation of the load using Bessel functions

The first stage of the problem requires to approximate the circular load that has the following characteristics: $q = 0.67$ MPa for $\rho \leq 0.1$ and $q = 0$ MPa for $\rho > 0.1$. The Hankel transform, presented in Section 1.1.6, permits such approximation by transforming the uniform circular load into in a discrete form by an addition of the following Bessel functions: $q(\rho, 0) = q\alpha \sum_{k=1}^{n_k} J_0(m\rho) J_1(m\alpha) \Delta m$. Since it is a discrete transformation, approaching the circular load requires a proper choice of the number of terms involved in the Bessel functions n_k and the integration increment Δm.

Figure 1.37 presents the results obtained with the MATLAB script provided with this book that permits to evaluate the circular load for different values of Δm and n_k.

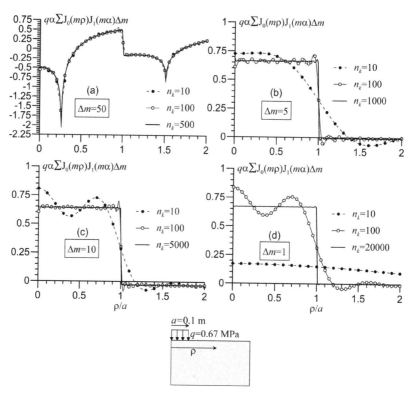

Figure 1.37 Approximation of a uniform circular load using the Hankel transform: (a) $\Delta m = 50$ $n_k = 500$, (b) $\Delta m = 5$ $n_k = 1,000$, (c) $\Delta m = 10$ $n_k = 5,000$, and (d) $\Delta m = 1$ $n_k = 20,000$.

It is clear that for high values of Δm, such as $\Delta m = 50$, whose results are shown in Figure 1.37a, the stress applied on the surface highly differs from the objective of a uniform circular load. On the other hand, as the integration increases, Δm decreases, and as the number of terms n_k increases, the approximation improves; it is the case in Figure 1.37d that uses $\Delta m = 1$ and $n_k = 20,000$. However, it is essential to take care of the number of terms, because a high value of n_k increases the computing cost.

1.6.2   Calculation of the vertical and radial stresses using Burmister's method

Computing stresses in the layered structure requires solving the system of equations made by Equation 1.40 for the surface of the structure and Equation 1.41 for each bonded interface. The dimensionless variables in these equations are $\lambda = z/H$, $F_i = e^{-m(\lambda_i - \lambda_{i-1})}$, and $R_i = \frac{E_i}{E_{i+1}} \frac{1+\nu_{i+1}}{1+\nu_i}$. Therefore, for the numerical data of the example, $H = 1$ m, $z_1 = 0.3$ m, $z_2 = 1.0$ m, $E_1 = 3,000$ MPa, $E_2 = 500$ MPa, $E_3 = 50$ MPa, $\nu_1 = 0.45$, $\nu_2 = 0.35$, and $\nu_3 = 0.40$, the values of the variables are

$$\lambda_1 = \frac{0.3}{1.0} = 0.30 \quad F_1 = e^{-0.3m} \quad R_1 = \frac{3,000(1+0.35)}{500(1+0.45)} = 5.59, \text{ and}$$

$$\lambda_2 = \frac{1.0}{1.0} = 1.0 \quad F_2 = e^{-0.7m} \quad R_2 = \frac{500(1+0.40)}{50(1+0.35)} = 10.37.$$

These values permit to build three matrices:

1   A matrix for the surface which is given in Equation 1.40 and becomes Equation 1.68 when using the numerical values of the example.
2   A matrix for the first interface (*i.e.*, between layers 1 and 2) given by Equation 1.41 which corresponds to Equation 1.70 when using the numerical data of the example.
3   The second interface (*i.e.*, between layer 2 and the subgrade) is also represented by Equation 1.41, but without the coefficients A_3 and C_3 because $F_3 = 0$. Then, Equation 1.72 represents the interface with the subgrade.

The equation for the surface, $z = 0$ is

$$\begin{bmatrix} e^{-0.3m} & 1 & -0.1e^{-0.3m} & 0.1 \\ e^{-0.3m} & -1 & 0.9e^{-0.3m} & 0.9 \end{bmatrix} \begin{bmatrix} A_1 \\ B_1 \\ C_1 \\ D_1 \end{bmatrix} = \begin{bmatrix} 1 \\ 0 \end{bmatrix}, \tag{1.68}$$

or in a shorter form,

$$[\Lambda_0] \begin{bmatrix} A_1 \\ B_1 \\ C_1 \\ D_1 \end{bmatrix} = \begin{bmatrix} 1 \\ 0 \end{bmatrix}. \tag{1.69}$$

The equation for the interface 1, located at $z = 0.3$ m, is

$$\begin{bmatrix} 1 & e^{-0.3m} & -(0.1-0.3m) & (0.1+0.3m)e^{-0.3m} \\ 1 & -e^{-0.3m} & 0.9+0.3m & (0.9-0.3m)e^{-0.3m} \\ 1 & e^{-0.3m} & 1+0.3m & -(1-0.3m)e^{-0.3m} \\ 1 & -e^{-0.3m} & -(0.2-0.3m) & -(0.2+0.3m)e^{-0.3m} \end{bmatrix} \begin{bmatrix} A_1 \\ B_1 \\ C_1 \\ D_1 \end{bmatrix}$$

$$- \begin{bmatrix} e^{-0.7m} & 1 & -(0.3-0.3m)e^{-0.7m} & (0.3+0.3m) \\ e^{-0.7m} & -1 & (0.7+0.3m)e^{-0.7m} & 0.7-0.3m \\ 5.87e^{-0.7m} & 5.87 & (1+0.3m)5.87e^{-0.7m} & -(1-0.3m)5.87 \\ 5.87e^{-0.7m} & -5.87 & -(0.6-0.3m)5.87e^{-0.7m} & -(0.6+0.3m)5.87 \end{bmatrix}$$

$$\times \begin{bmatrix} A_2 \\ B_2 \\ C_2 \\ D_2 \end{bmatrix} = \begin{bmatrix} 0 \\ 0 \\ 0 \\ 0 \end{bmatrix}, \tag{1.70}$$

or in a shorter form,

$$[\Lambda_1] \begin{bmatrix} A_1 \\ B_1 \\ C_1 \\ D_1 \end{bmatrix} - [\Gamma_1] \begin{bmatrix} A_2 \\ B_2 \\ C_2 \\ D_2 \end{bmatrix} = \begin{bmatrix} 0 \\ 0 \\ 0 \\ 0 \end{bmatrix}. \tag{1.71}$$

Finally, the equation for the interface with the subgrade, located at $z = 1$ m, is

$$
\begin{bmatrix}
1 & e^{-0.7m} & -(0.3-m) & (0.3+m)e^{-0.7m} \\
1 & -e^{-0.7m} & 0.7+m & (0.7-m)e^{-0.7m} \\
1 & e^{-0.7m} & 1+m & -(1-m)e^{-0.7m} \\
1 & -e^{-0.7m} & -(0.6-m) & -(0.6+m)e^{-0.7m}
\end{bmatrix}
\begin{bmatrix}
A_2 \\ B_2 \\ C_2 \\ D_2
\end{bmatrix}
$$

$$
-
\begin{bmatrix}
1 & 0.2+m \\
-1 & 0.8-m \\
10.36 & -(1-m)10.36 \\
-10.36 & -(0.4+m)10.36
\end{bmatrix}
\begin{bmatrix}
B_3 \\ D_3
\end{bmatrix}
=
\begin{bmatrix}
0 \\ 0 \\ 0 \\ 0
\end{bmatrix},
\tag{1.72}
$$

and also, in a shorter form, it becomes

$$
[\Lambda_2]
\begin{bmatrix}
A_2 \\ B_2 \\ C_2 \\ D_2
\end{bmatrix}
- [\Gamma_2]
\begin{bmatrix}
B_3 \\ D_3
\end{bmatrix}
=
\begin{bmatrix}
0 \\ 0 \\ 0 \\ 0
\end{bmatrix}.
\tag{1.73}
$$

Equations 1.69, 1.71, and 1.73 form a 10×10 matrix represented in Equation 1.74:

$$
\begin{bmatrix}
\Lambda_0 & 0 & 0 \\
\Lambda_1 & -\Gamma_1 & 0 \\
0 & \Lambda_2 & -\Gamma_2
\end{bmatrix}
\begin{bmatrix}
A_1 \\ B_1 \\ \hdashline C_1 \\ D_1 \\ A_2 \\ B_2 \\ \hdashline C_2 \\ D_2 \\ B_3 \\ D_3
\end{bmatrix}
=
\begin{bmatrix}
1 \\ 0 \\ \hdashline 0 \\ 0 \\ 0 \\ 0 \\ \hdashline 0 \\ 0 \\ 0 \\ 0
\end{bmatrix}
\tag{1.74}
$$

Equation 1.74 must be solved n_k times for each value of m involved in the Bessel functions and the Hankel transform. These solutions lead to n_k values of the constants $A_{1,2}$, $B_{1,2,3}$, $C_{1,2}$, and $D_{1,2,3}$. For each set of constants A, B, C, D corresponding to a value of m, the increments of stresses, $\Delta\sigma_z$ and $\Delta\sigma_\rho$, are computed using Equations 1.43 and 1.44.

Considering the numerical values of the dimensionless variables of each layer, Equations 1.43 and 1.44 become

Layer 1

$$
\Delta\sigma_z(m)^{\text{Top}} = -mJ_0(m\rho_H)\{[A_1 - C_1(1-2\nu_1)]F_1 + B_1 + D_1(1-2\nu_1)\}
$$

$$
= -mJ_0(m\rho)\left\{[A_1 - 0.1C_1]e^{-0.3m} + B_1 + 0.1D_1\right\}
$$

$$
\Delta\sigma_\rho(m)^{\text{Top}} = [mJ_0(m\rho_H) - J_1(m\rho_H)/\rho_H][(A_1 + C_1)F_1 + B_1 - D_1]
$$

$$
+ 2\nu_1 mJ_0(m\rho_H)(C_1 F_1 - D_1)
$$

$$
= [mJ_0(m\rho) - J_1(m\rho)/\rho]\left[(A_1 + C_1)e^{-0.3m} + B_1 - D_1\right]
$$

$$
+ 0.9mJ_0(m\rho)\left(C_1 e^{-0.3m} - D_1\right)
$$

$$\Delta\sigma_z(m)^{\text{Bottom}} = -mJ_0(m\rho_H)\{[A_1 - C_1(1 - 2v_1 - m\lambda_1)]$$
$$+ [B_1 + D_1(1 - 2v_1 + m\lambda_1)]F_1\}$$
$$= -mJ_0(m\rho)\left\{[A_1 - C_1(0.1 - 0.3m)]\right.$$
$$\left.+ [B_1 + D_1(0.1 + 0.3m)]e^{-0.3m}\right\}$$

$$\Delta\sigma_\rho(m)^{\text{Bottom}} = [mJ_0(m\rho_H) - J_1(m\rho_H)/\rho_H]\{[A_1 + C_1(1 + m\lambda_1)]$$
$$+ [B_1 - D_1(1 - m\lambda_1)]F_1\} + 2 * v_1mJ_0(m\rho_H)(C_1 - D_1F_1)$$
$$= [mJ_0(m\rho) - J_1(m\rho)/\rho]\left\{[A_1 + C_1(1 + 0.3m)]\right.$$
$$\left.+ [B_1 - D_1(1 - 0.3m)]e^{-0.3m}\right\} + 0.9mJ_0(m\rho)\left(C_1 - D_1e^{-0.3m}\right)$$

Layer 2

$$\Delta\sigma_z(m)^{\text{Top}} = -mJ_0(m\rho_H)\{[A_2 - C_2(1 - 2v_2 - m\lambda_1)]F_2$$
$$+ [B_2 + D_2(1 - 2v_2 + m\lambda_1)]\}$$
$$= -mJ_0(m\rho)\left\{[A_2 - C_2(0.3 - 0.3m)]e^{-0.7m}\right.$$
$$\left.+ [B_2 + D_2(0.3 + 0.3m)]\right\}$$

$$\Delta\sigma_\rho(m)^{\text{Top}} = [mJ_0(m\rho_H) - J_1(m\rho_H)/\rho_H]\{[A_2 + C_2(1 + m\lambda_1)]F_2$$
$$+ (B_2 - D_2(1 - m\lambda_1)]\} + 2v_2mJ_0(m\rho_H)(C_2F_2 - D_2)$$
$$= [mJ_0(m\rho) - J_1(m\rho)/\rho]\left\{[A_2 + C_2(1 + 0.3m)]e^{-0.7m}\right.$$
$$\left.+ (B_2 - D_2(1 - 0.3m)]\right\} + 0.7mJ_0(m\rho)\left(C_2e^{-0.7m} - D_2\right)$$

$$\Delta\sigma_z(m)^{\text{Bottom}} = -mJ_0(m\rho_H)\{[A_2 - C_2(1 - 2v_2 - m\lambda_2)]$$
$$+ [B_2 + D_2(1 - 2v_2 + m\lambda_2)]F_2\}$$
$$= -mJ_0(m\rho)\left\{[A_2 - C_2(0.7 - m)] + [B_2 + D_2(0.7 + m)]e^{-0.7m}\right\}$$

$$\Delta\sigma_\rho(m)^{\text{Bottom}} = [mJ_0(m\rho_H) - J_1(m\rho_H)/\rho_H]\{[A_2 + C_2(1 + m\lambda_2)]$$
$$+ [B_2 - D_2(1 - m\lambda_2)]F_2\} + 2v_2mJ_0(m\rho_H)(C_2 - D_2F_2)$$
$$= [mJ_0(m\rho) - J_1(m\rho)/\rho]\left\{[A_2 + C_2(1 + m)]\right.$$
$$\left.+ \left[B_2 - D_2(1 - m)]e^{-0.7m}\right\} + 0.7mJ_0(m\rho)\left(C_2 - D_2e^{-0.7m}\right)\right.$$

Subgrade

$$\Delta\sigma_z(m)^{\text{Top}} = -mJ_0(m\rho_H)\{[B_3 + D_3(1 - 2v_3 + m\lambda_2)]\}$$
$$= -mJ_0(m\rho)\{[B_3 + D_3(0.2 + m)]\}$$

$$\Delta\sigma_\rho(m)^{\text{Top}} = [mJ_0(m\rho_H) - J_1(m\rho_H)/\rho_H]\{[B_3 - D_3(1 - m\lambda_2)]\}$$
$$+ 2v_3mJ_0(m\rho_H)(-D_3)$$
$$= [mJ_0(m\rho) - J_1(m\rho)/\rho]\{[B_3 - D_3(1 - m)]\} - 0.8mJ_0(m\rho)D_3\Delta$$

Finally, vertical and radial stresses are computed using Equations 1.49 and 1.50. For the numerical values of the example, $\alpha = 0.1/1 = 0.1$ and $q = 0.67$ MPa. Adopting $n_k = 20,000$ and $\Delta m = 1$, Equations 1.49 and 1.50 become

$$\sigma_z(\rho) = 0.067 \sum_{m=1}^{20,000} \left(\frac{\Delta\sigma_z(m)}{m} J_1(0.1m) \right) \text{ MPa},$$

$$\sigma_\rho(\rho) = 0.067 \sum_{m=1}^{20,000} \left(\frac{\Delta\sigma_\rho(m)}{m} J_1(0.1m) \right) \text{ MPa}.$$

Figure 1.38 presents the algorithm for computing stresses in a multilayer system. Also, a MATLAB script is provided with this book. This script is restricted to three layers, but Example 6 leads to a more general script.

Tables 1.5 and 1.6 present the results obtained using the MATLAB script, and these results agree very well with the results of commercial software based on Burmister's method.

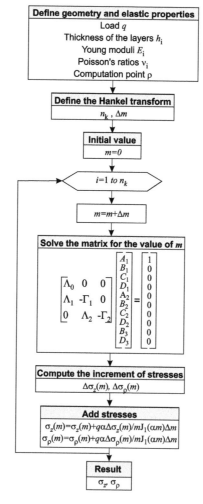

Figure 1.38 Algorithm for stress computation in a multilayer structure using Burmister's method.

Table 1.5 Vertical stress σ_z in MPa in the contact of the layers

σ_z (MPa)	ρ (m)		
z (m)	0.001	0.1	0.2
0.0	−0.669	−0.335	0.000
0.3	−0.043	−0.037	−0.024
0.3	−0.043	−0.037	−0.024
1.0	−0.002	−0.002	−0.002
1.0	−0.002	−0.002	−0.002

Table 1.6 Radial stress σ_ρ in MPa in the contact of the layers

σ_ρ (MPa)	ρ (m)		
z (m)	0.001	0.1	0.2
0.0	−0.745	−0.401	−0.069
0.3	0.164	0.125	0.058
0.3	0.005	0.002	−0.003
1.0	0.013	0.013	0.012
1.0	0.0002	0.0002	0.0002

1.6.3 Concluding remarks

Burmister's method is a useful method that has a low computational cost and produces accurate results regarding stresses, strains, and displacements of multilayer systems. Nevertheless, it is essential to keep in mind the following limitations of the method:

- The method is approximate; its level of accuracy depends on a proper choice of the discretization of the Hankel transform.
- The method assumes that the materials of all layers are homogeneous, isotropic, and linear elastic. It is clear that regarding road materials, some of these assumptions are too restrictive.
- The method is essentially static, using Young's moduli that depend on the excitation frequency could overcome this limitation. However, even with this option, the method remains static because it neglects the inertial term in the equilibrium equation (*i.e.*, the product of the mass and the acceleration).

1.7 EXAMPLE 6: TRIDIMENSIONAL DISTRIBUTION OF STRESSES PRODUCED BY MOVING WHEEL LOADS

The purpose of the following example is to calculate the evolution of the mean and deviator stresses (p, q) as well as the angle of rotation of the principal stresses produced by a moving wheel load. Figure 1.39 represents the idealized road structure of the example. This structure is exposed to a double wheel loading represented by two circular loaded areas with radius a = 0.125 m, with separation between centers of 3a and applying a vertical stress q = 0.67 MPa.

Stresses are calculated using Burmister's method at levels A to I of the interme-diate layer shown in Figure 1.39 (this layer represents a granular layer). The depths z of those levels are (0.1, 0.225, 0.35, 0.475, 0.6, 0.725, 0.85, 0.975, and 1.1 m), and stresses must be analyzed along the middle axis between circular loads and in the forward direction of the movement.

The geometric and elastic characteristics of the three layers of the road structure are

- **Layer 1:** *Young's modulus $E_1 = 4,000$ MPa, Poisson's ratio $v_1 = 0.45$, thickness 0.1 m.*

- **Layer 2:** *Young's modulus $E_2 = 400$ MPa, Poisson's ratio $v_2 = 0.35$, thickness 1.0 m.*

- **Layer 3:** *Young's modulus $E_3 = 50$ MPa, Poisson's ratio $v_1 = 0.40$, thickness ∞.*

The following steps are necessary for evaluating the stresses in the three axes of the Cartesian plane and then evaluating the principal stresses and their rotation:

1   Compute the stresses produced by one load in a cylindrical coordinate system.

2   Transform the stresses into a Cartesian coordinate system and superpose the stresses produced by each circular load.

3   Compute the principal stresses, the mean and deviator stresses (p, q), and the angle of rotation of the principal stresses α_{yz}.

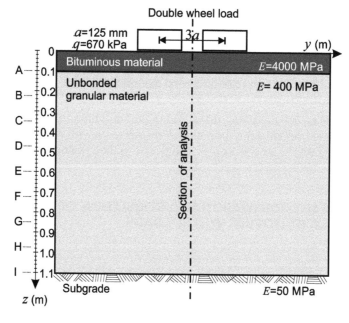

Figure 1.39 Layout of the three-layer road structure whose intermediate layer is divided into eight sub-layers.

1.7.1 Stresses produced by a circular load in a cylindrical coordinate system

Burmister's method described in Example 5 permits to evaluate the distribution of stresses below the road structure. Although the road's structure in this example has three layers, as in the previous example, to compute the stresses at the depths A to I of the granular layer, it is convenient to divide this layer into eight sublayers as shown in Figure 1.40.

After dividing the granular layer into eight sublayers, the whole structure turns into a structure with ten layers leading to a system of $4n-2 = 38$ equations with 38 unknowns.

The dimensionless variables in these equations are $\lambda = z_i/H$, $F_i = e^{-m(\lambda_i - \lambda_{i-1})}$, and $R_i = \frac{E_i}{E_{i+1}} \frac{1+\nu_{i+1}}{1+\nu_i}$, and the numerical data of the example are $H = 1.1$ m, $z_1 = 0.1$ m, $z_2 = 0.225$ m, $z_3 = 0.35$ m, $z_4 = 0.475$ m, $z_5 = 0.6$ m, $z_6 = 0.725$ m, $z_7 = 0.85$ m, $z_8 = 0.975$ m, $z_9 = 1.1$ m, $E_1 = 4,000$ MPa, $E_2, ..., E_9 = 400$ MPa, $E_{10} = 50$ MPa, $\nu_1 = 0.45$, $\nu_2, ..., \nu_9 = 0.35$, and $\nu_{10} = 0.40$.

These numerical values lead to the following dimensionless constants λ_i, F_i and R_i:

$$\lambda_1 = \frac{0.1}{1.1} = 0.0909 \quad F_1 = e^{-0.909m} \quad R_1 = \frac{4,000(1+0.35)}{400(1+0.45)} = 9.31,$$

$$\lambda_2 = \frac{0.225}{1.1} = 0.2045 \quad F_2 = e^{-0.1136m} \quad R_2 = \frac{400(1+0.35)}{400(1+0.35)} = 1.0,$$

$$\lambda_3 = \frac{0.35}{1.1} = 0.3181 \quad F_3 = e^{-0.1136m} \quad R_3 = \frac{400(1+0.35)}{400(1+0.35)} = 1.0,$$

$$\lambda_4 = \frac{0.475}{1.1} = 0.4318 \quad F_4 = e^{-0.1136m} \quad R_4 = \frac{400(1+0.35)}{400(1+0.35)} = 1.0,$$

$$\lambda_5 = \frac{0.6}{1.1} = 0.5454 \quad F_5 = e^{-0.1136m} \quad R_5 = \frac{400(1+0.35)}{400(1+0.35)} = 1.0,$$

$$\lambda_6 = \frac{0.725}{1.1} = 0.6591 \quad F_6 = e^{-0.1136m} \quad R_6 = \frac{400(1+0.35)}{400(1+0.35)} = 1.0,$$

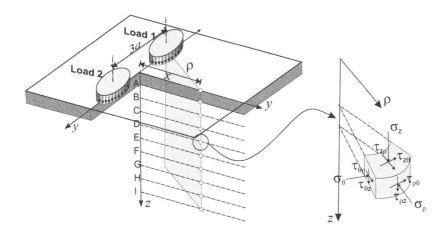

Figure 1.40 Computational points for the stresses in cylindrical coordinates.

$$\lambda_7 = \frac{0.85}{1.1} = 0.7727 \quad F_7 = e^{-0.1136m} \quad R_7 = \frac{400(1+0.35)}{400(1+0.35)} = 1.0,$$

$$\lambda_8 = \frac{0.975}{1.1} = 0.8863 \quad F_8 = e^{-0.1136m} \quad R_8 = \frac{400(1+0.35)}{400(1+0.35)} = 1.0,$$

$$\lambda_9 = \frac{1.1}{1.1} = 1.0 \quad F_9 = e^{-0.1136m} \quad R_9 = \frac{400(1+0.40)}{50(1+0.35)} = 8.30.$$

Equations for each interface of the road structure lead to the following matrices.

For the surface, the value of the constants are $z = 0$, $\lambda_1 = 0.0909$ and $v_1 = 0.45$; therefore, Equation 1.40 leads to

$$\begin{bmatrix} e^{-0.0909m} & 1 & -0.1e^{-0.0909m} & 0.1 \\ e^{-0.0909m} & -1 & 0.9e^{-0.0909m} & 0.9 \end{bmatrix} \begin{bmatrix} A_1 \\ B_1 \\ C_1 \\ D_1 \end{bmatrix} = \begin{bmatrix} 1 \\ 0 \end{bmatrix}. \tag{1.75}$$

For interfaces 1–8, Equation 1.41 leads to

Interface 1 ($F_1 = e^{-0.0909m}$, $v_1 = 0.45$, $\lambda_1 = 0.0909$, $F_2 = e^{-0.1136m}$, $v_2 = 0.35$, $R_1 = 9.31$)

$$\begin{bmatrix} 1 & e^{-0.0909m} & -(0.1 - 0.0909m) & (0.1 + 0.0909m)e^{-0.0909m} \\ 1 & -e^{-0.0909m} & 0.9 + 0.0909m & (0.9 - 0.0909m)e^{-0.0909m} \\ 1 & e^{-0.0909m} & 1 + 0.0909m & -(1 - 0.0909m)e^{-0.0909m} \\ 1 & -e^{-0.0909m} & -(0.2 - 0.0909m) & -(0.2 + 0.0909m)e^{-0.0909m} \end{bmatrix} \begin{bmatrix} A_1 \\ B_1 \\ C_1 \\ D_1 \end{bmatrix}$$

$$= \begin{bmatrix} e^{-0.1136m} & 1 & -(0.3 - 0.0909m)e^{-0.1136m} & 0.3 + 0.0909m \\ e^{-0.1136m} & -1 & (0.7 + 0.0909m)e^{-0.1136m} & 0.7 - 0.0909m \\ 9.31e^{-0.1136m} & 9.31 & (1 + 0.0909m)9.31e^{-0.1136m} & -(1 - 0.0909m)9.31 \\ 9.31e^{-0.1136m} & -9.31 & -(0.6 - 0.0909m)9.31e^{-0.1136m} & -(0.6 + 0.0909m)9.31 \end{bmatrix}$$

$$\begin{bmatrix} A_2 \\ B_2 \\ C_2 \\ D_2 \end{bmatrix}. \tag{1.76}$$

Interface 2 ($F_2 = e^{-0.1136m}$, $v_2 = 0.35$, $\lambda_2 = 0.2045$, $F_3 = e^{-0.1136m}$, $v_3 = 0.35$, $R_2 = 1$)

$$\begin{bmatrix} 1 & e^{-0.1136m} & -(0.3 - 0.2045m) & (0.3 + 0.2045m)e^{-0.1136m} \\ 1 & -e^{-0.1136m} & 0.7 + 0.2045m & (0.3 - 0.2045m)e^{-0.1136m} \\ 1 & e^{-0.1136m} & 1 + 0.2045m & -(1 - 0.2045m)e^{-0.1136m} \\ 1 & -e^{-0.1136m} & -(0.6 - 0.2045m) & -(0.6 + 0.2045m)e^{-0.1136m} \end{bmatrix} \begin{bmatrix} A_2 \\ B_2 \\ C_2 \\ D_2 \end{bmatrix}$$

$$= \begin{bmatrix} e^{-0.1136m} & 1 & -(0.3 - 0.2045m)e^{-0.1136m} & 0.3 + 0.2045m \\ e^{-0.1136m} & -1 & (0.7 + 0.2045m)e^{-0.1136m} & 0.7 - 0.2045m \\ e^{-0.1136m} & 1 & (1 + 0.2045m)e^{-0.1136m} & -(1 - 0.2045m) \\ e^{-0.1136m} & -1 & -(0.6 - 0.2045m)e^{-0.1136m} & -(0.6 + 0.2045m) \end{bmatrix} \begin{bmatrix} A_3 \\ B_3 \\ C_3 \\ D_3 \end{bmatrix}. \tag{1.77}$$

Interface 3 ($F_3 = e^{-0.1136m}$, $v_3 = 0.35$, $\lambda_3 = 0.3181$, $F_4 = e^{-0.1136m}$, $v_4 = 0.35$, $R_3 = 1$)

$$\begin{bmatrix} 1 & e^{-0.1136m} & -(0.3 - 0.3181m) & (0.3 + 0.3181m)e^{-0.1136m} \\ 1 & -e^{-0.1136m} & 0.7 + 0.0.3181m & (0.3 - 0.3181m)e^{-0.1136m} \\ 1 & e^{-0.1136m} & 1 + 0.3181m & -(1 - 0.3181m)e^{-0.1136m} \\ 1 & -e^{-0.1136m} & -(0.6 - 0.3181m) & -(0.6 + 0.3181m)e^{-0.1136m} \end{bmatrix} \begin{bmatrix} A_3 \\ B_3 \\ C_3 \\ D_3 \end{bmatrix}$$

$$
= \begin{bmatrix}
e^{-0.1136m} & 1 & -(0.3-0.3181m)e^{-0.1136m} & 0.3+0.3181m \\
e^{-0.1136m} & -1 & (0.7+0.3181m)e^{-0.1136m} & 0.7-0.3181m \\
e^{-0.1136m} & 1 & (1+0.3181m)e^{-0.1136m} & -(1-0.3181m) \\
e^{-0.1136m} & -1 & -(0.6-0.3181m)e^{-0.1136m} & -(0.6+0.3181m)
\end{bmatrix}
\begin{bmatrix} A_4 \\ B_4 \\ C_4 \\ D_4 \end{bmatrix}.
$$

(1.78)

Interface 4 ($F_4 = e^{-0.1136m}$, $v_4 = 0.35$, $\lambda_4 = 0.4318$, $F_5 = e^{-0.1136m}$, $v_5 = 0.35$, $R_4 = 1$)

$$
\begin{bmatrix}
1 & e^{-0.1136m} & -(0.3-0.4318m) & (0.3+0.4318m)e^{-0.1136m} \\
1 & -e^{-0.1136m} & 0.7+0.4318m & (0.3-0.4318m)e^{-0.1136m} \\
1 & e^{-0.1136m} & 1+0.4318m & -(1-0.4318m)e^{-0.1136m} \\
1 & -e^{-0.1136m} & -(0.6-0.4318m) & -(0.6+0.4318m)e^{-0.1136m}
\end{bmatrix}
\begin{bmatrix} A_4 \\ B_4 \\ C_4 \\ D_4 \end{bmatrix}
$$

$$
= \begin{bmatrix}
e^{-0.1136m} & 1 & -(0.3-0.4318m)e^{-0.1136m} & 0.3+0.4318m \\
e^{-0.1136m} & -1 & (0.7+0.4318m)e^{-0.1136m} & 0.7-0.4318m \\
e^{-0.1136m} & 1 & (1+0.4318m)e^{-0.1136m} & -(1-0.4318m) \\
e^{-0.1136m} & -1 & -(0.6-0.4318m)e^{-0.1136m} & -(0.6+0.4318m)
\end{bmatrix}
\begin{bmatrix} A_5 \\ B_5 \\ C_5 \\ D_5 \end{bmatrix}.
$$

(1.79)

Interface 5 ($F_5 = e^{-0.1136m}$, $v_5 = 0.35$, $\lambda_5 = 0.5454$, $F_6 = e^{-0.1136m}$, $v_6 = 0.35$, $R_5 = 1$)

$$
\begin{bmatrix}
1 & e^{-0.1136m} & -(0.3-0.545m) & (0.3+0.545m)e^{-0.1136m} \\
1 & -e^{-0.1136m} & 0.7+0.545m & (0.3-0.545m)e^{-0.1136m} \\
1 & e^{-0.1136m} & 1+0.545m & -(1-0.545m)e^{-0.1136m} \\
1 & -e^{-0.1136m} & -(0.6-0.545m) & -(0.6+0.545m)e^{-0.1136m}
\end{bmatrix}
\begin{bmatrix} A_5 \\ B_5 \\ C_5 \\ D_5 \end{bmatrix}
$$

$$
= \begin{bmatrix}
e^{-0.1136m} & 1 & -(0.3-0.545m)e^{-0.1136m} & 0.3+0.545m \\
e^{-0.1136m} & -1 & (0.7+0.545m)e^{-0.1136m} & 0.7-0.545m \\
e^{-0.1136m} & 1 & (1+0.545m)e^{-0.1136m} & -(1-0.545m) \\
e^{-0.1136m} & -1 & -(0.6-0.545m)e^{-0.1136m} & -(0.6+0.545m)
\end{bmatrix}
\begin{bmatrix} A_6 \\ B_6 \\ C_6 \\ D_6 \end{bmatrix}.
$$

(1.80)

Interface 6 ($F_6 = e^{-0.1136m}$, $v_6 = 0.35$, $\lambda_6 = 0.6591$, $F_7 = e^{-0.1136m}$, $v_7 = 0.35$, $R_6 = 1$)

$$
\begin{bmatrix}
1 & e^{-0.1136m} & -(0.3-0.6591m) & (0.3+0.6591m)e^{-0.1136m} \\
1 & -e^{-0.1136m} & 0.7+0.6591m & (0.3-0.6591m)e^{-0.1136m} \\
1 & e^{-0.1136m} & 1+0.6591m & -(1-0.6591m)e^{-0.1136m} \\
1 & -e^{-0.1136m} & -(0.6-0.6591m) & -(0.6+0.6591m)e^{-0.1136m}
\end{bmatrix}
\begin{bmatrix} A_6 \\ B_6 \\ C_6 \\ D_6 \end{bmatrix}
$$

$$
= \begin{bmatrix}
e^{-0.1136m} & 1 & -(0.3-0.6591m)e^{-0.1136m} & 0.3+0.6591m \\
e^{-0.1136m} & -1 & (0.7+0.6591m)e^{-0.1136m} & 0.7-0.6591m \\
e^{-0.1136m} & 1 & (1+0.6591m)e^{-0.1136m} & -(1-0.6591m) \\
e^{-0.1136m} & -1 & -(0.6-0.6591m)e^{-0.1136m} & -(0.6+0.6591m)
\end{bmatrix}
\begin{bmatrix} A_7 \\ B_7 \\ C_7 \\ D_7 \end{bmatrix}.
$$

(1.81)

Interface 7 ($F_7 = e^{-0.1136m}$, $v_7 = 0.35$, $\lambda_7 = 0.7727$, $F_8 = e^{-0.1136m}$, $v_8 = 0.35$, $R_8 = 1$)

$$
\begin{bmatrix}
1 & e^{-0.1136m} & -(0.3-0.7727m) & (0.3+0.7727m)e^{-0.1136m} \\
1 & -e^{-0.1136m} & 0.7+0.7727m & (0.3-0.7727m)e^{-0.1136m} \\
1 & e^{-0.1136m} & 1+0.7727m & -(1-0.7727m)e^{-0.1136m} \\
1 & -e^{-0.1136m} & -(0.6-0.7727m) & -(0.6+0.7727m)e^{-0.1136m}
\end{bmatrix}
\begin{bmatrix} A_7 \\ B_7 \\ C_7 \\ D_7 \end{bmatrix}
$$

$$= \begin{bmatrix} e^{-0.1136m} & 1 & -(0.3 - 0.7727m)e^{-0.1136m} & 0.3 + 0.7727m \\ e^{-0.1136m} & -1 & (0.7 + 0.7727m)e^{-0.1136m} & 0.7 - 0.7727m \\ e^{-0.1136m} & 1 & (1 + 0.7727m)e^{-0.1136m} & -(1 - 0.7727m) \\ e^{-0.1136m} & -1 & -(0.6 - 0.7727m)e^{-0.1136m} & -(0.6 + 0.7727m) \end{bmatrix} \begin{bmatrix} A_8 \\ B_8 \\ C_8 \\ D_8 \end{bmatrix}.$$

(1.82)

Interface 8 ($F_8 = e^{-0.1136m}$, $v_8 = 0.35$, $\lambda_8 = 0.8863$, $F_9 = e^{-0.1136m}$, $v_9 = 0.35$, $R_8 = 1$)

$$\begin{bmatrix} 1 & e^{-0.1136m} & -(0.3 - 0.8863m) & (0.3 + 0.8863m)e^{-0.1136m} \\ 1 & -e^{-0.1136m} & 0.7 + 0.8863m & (0.3 - 0.88637m)e^{-0.1136m} \\ 1 & e^{-0.1136m} & 1 + 0.8863m & -(1 - 0.8863m)e^{-0.1136m} \\ 1 & -e^{-0.1136m} & -(0.6 - 0.8863m) & -(0.6 + 0.8863m)e^{-0.1136m} \end{bmatrix} \begin{bmatrix} A_8 \\ B_8 \\ C_8 \\ D_8 \end{bmatrix}$$

$$= \begin{bmatrix} e^{-0.1136m} & 1 & -(0.3 - 0.8863m)e^{-0.1136m} & 0.3 + 0.8863m \\ e^{-0.1136m} & -1 & (0.7 + 0.8863m)e^{-0.1136m} & 0.7 - 0.8863m \\ e^{-0.1136m} & 1 & (1 + 0.8863m)e^{-0.1136m} & -(1 - 0.8863m) \\ e^{-0.1136m} & -1 & -(0.6 - 0.8863m)e^{-0.1136m} & -(0.6 + 0.8863m) \end{bmatrix} \begin{bmatrix} A_9 \\ B_9 \\ C_9 \\ D_9 \end{bmatrix}.$$

(1.83)

For interface 9, which is the interface with the subgrade, the values of the dimensionless variables are $F_9 = e^{-0.1136m}$, $v_9 = 0.35$, $\lambda_9 = 1.0$, $F_{10} = 0$, $v_{10} = 0.4$, and $R_9 = 8.30$. For this layer, Equation 1.41 is also useful but, as $F_{10} = 0$, unknowns A_{10} and C_{10} do not affect the equation; therefore, this layer has only two unknowns B_{10} and D_{10}. As a result, the equation for the last interface becomes

$$\begin{bmatrix} 1 & e^{-0.1136m} & -(0.3 - m) & (0.3 + m)e^{-0.1136m} \\ 1 & -e^{-0.1136m} & 0.7 + m & (0.7 - m)e^{-0.1136m} \\ 1 & e^{-0.1136m} & 1 + m & -(1 - m)e^{-0.1136m} \\ 1 & -e^{-0.1136m} & -(0.6 - m) & -(0.6 + m)e^{-0.1136m} \end{bmatrix} \begin{bmatrix} A_9 \\ B_9 \\ C_9 \\ D_9 \end{bmatrix}$$

$$= \begin{bmatrix} 1 & 0.2 + m \\ -1 & 0.8 - m \\ 8.30 & -(1 - m)8.30 \\ -8.30 & -(0.4 + m)8.30 \end{bmatrix} \begin{bmatrix} B_{10} \\ D_{10} \end{bmatrix}.$$

(1.84)

In a shorter form, Equations 1.75–1.84 are

$$[\Lambda_0] \begin{bmatrix} A_1 \\ B_1 \\ C_1 \\ D_1 \end{bmatrix} = \begin{bmatrix} 1 \\ 0 \end{bmatrix},$$

(1.85)

$$[\Lambda_1] \begin{bmatrix} A_1 \\ B_1 \\ C_1 \\ D_1 \end{bmatrix} - [\Gamma_1] \begin{bmatrix} A_2 \\ B_2 \\ C_2 \\ D_2 \end{bmatrix} = \begin{bmatrix} 0 \\ 0 \\ 0 \\ 0 \end{bmatrix},$$

(1.86)

$$[\Lambda_2] \begin{bmatrix} A_2 \\ B_2 \\ C_2 \\ D_2 \end{bmatrix} - [\Gamma_2] \begin{bmatrix} A_3 \\ B_3 \\ C_3 \\ D_3 \end{bmatrix} = \begin{bmatrix} 0 \\ 0 \\ 0 \\ 0 \end{bmatrix},$$

(1.87)

$$[\Lambda_3] \begin{bmatrix} A_3 \\ B_3 \\ C_3 \\ D_3 \end{bmatrix} - [\Gamma_3] \begin{bmatrix} A_4 \\ B_4 \\ C_4 \\ D_4 \end{bmatrix} = \begin{bmatrix} 0 \\ 0 \\ 0 \\ 0 \end{bmatrix}, \tag{1.88}$$

$$[\Lambda_4] \begin{bmatrix} A_4 \\ B_4 \\ C_4 \\ D_4 \end{bmatrix} - [\Gamma_4] \begin{bmatrix} A_5 \\ B_5 \\ C_5 \\ D_5 \end{bmatrix} = \begin{bmatrix} 0 \\ 0 \\ 0 \\ 0 \end{bmatrix}, \tag{1.89}$$

$$[\Lambda_5] \begin{bmatrix} A_5 \\ B_5 \\ C_5 \\ D_5 \end{bmatrix} - [\Gamma_5] \begin{bmatrix} A_6 \\ B_6 \\ C_6 \\ D_6 \end{bmatrix} = \begin{bmatrix} 0 \\ 0 \\ 0 \\ 0 \end{bmatrix}, \tag{1.90}$$

$$[\Lambda_6] \begin{bmatrix} A_6 \\ B_6 \\ C_6 \\ D_6 \end{bmatrix} - [\Gamma_6] \begin{bmatrix} A_7 \\ B_7 \\ C_7 \\ D_7 \end{bmatrix} = \begin{bmatrix} 0 \\ 0 \\ 0 \\ 0 \end{bmatrix}, \tag{1.91}$$

$$[\Lambda_7] \begin{bmatrix} A_7 \\ B_7 \\ C_7 \\ D_7 \end{bmatrix} - [\Gamma_7] \begin{bmatrix} A_8 \\ B_8 \\ C_8 \\ D_8 \end{bmatrix} = \begin{bmatrix} 0 \\ 0 \\ 0 \\ 0 \end{bmatrix}, \tag{1.92}$$

$$[\Lambda_8] \begin{bmatrix} A_8 \\ B_8 \\ C_8 \\ D_8 \end{bmatrix} - [\Gamma_8] \begin{bmatrix} A_9 \\ B_9 \\ C_9 \\ D_9 \end{bmatrix} = \begin{bmatrix} 0 \\ 0 \\ 0 \\ 0 \end{bmatrix}, \tag{1.93}$$

$$[\Lambda_9] \begin{bmatrix} A_9 \\ B_9 \\ C_9 \\ D_9 \end{bmatrix} - [\Gamma_9] \begin{bmatrix} B_{10} \\ D_{10} \end{bmatrix} = \begin{bmatrix} 0 \\ 0 \\ 0 \\ 0 \end{bmatrix}. \tag{1.94}$$

Equations 1.85–1.94 form the following system of 38 equations with 38 unknowns:

$$\begin{bmatrix} \Lambda_0 & 0 & 0 & 0 & 0 & 0 & 0 & 0 & 0 & 0 \\ \Lambda_1 & -\Gamma_1 & 0 & 0 & 0 & 0 & 0 & 0 & 0 & 0 \\ 0 & \Lambda_2 & -\Gamma_2 & 0 & 0 & 0 & 0 & 0 & 0 & 0 \\ 0 & 0 & \Lambda_3 & -\Gamma_3 & 0 & 0 & 0 & 0 & 0 & 0 \\ 0 & 0 & 0 & \Lambda_4 & -\Gamma_4 & 0 & 0 & 0 & 0 & 0 \\ 0 & 0 & 0 & 0 & \Lambda_5 & -\Gamma_5 & 0 & 0 & 0 & 0 \\ 0 & 0 & 0 & 0 & 0 & \Lambda_6 & -\Gamma_6 & 0 & 0 & 0 \\ 0 & 0 & 0 & 0 & 0 & 0 & \Lambda_7 & -\Gamma_7 & 0 & 0 \\ 0 & 0 & 0 & 0 & 0 & 0 & 0 & \Lambda_8 & -\Gamma_8 & 0 \\ 0 & 0 & 0 & 0 & 0 & 0 & 0 & 0 & \Lambda_9 & -\Gamma_9 \end{bmatrix}_{38 \times 38} \begin{bmatrix} A_1 \\ B_1 \\ C_1 \\ D_1 \\ . \\ . \\ . \\ A_9 \\ B_9 \\ C_9 \\ D_9 \\ B_{10} \\ D_{10} \end{bmatrix}_{38 \times 1} = \begin{bmatrix} 1 \\ 0 \\ 0 \\ . \\ . \\ . \\ . \\ . \\ . \\ 0 \end{bmatrix}_{38 \times 1}.$$

Stresses in the interface i can be computed at the bottom of the layer i or the top of layer $i + 1$.

When considering the bottom of the layer i, the depth of the interface i is z_i, and then $\lambda = \lambda_i$, $e^{-m(\lambda_i-\lambda)} = 1$ and $F_i = e^{-m(\lambda-\lambda_{i-1})}$. As a result, Equations 1.43–1.46 for the bottom of the layer i become

$$\Delta\sigma_{z_i}(m) = -mJ_0(m\rho_H)\{[A_i - C_i(1 - 2\nu_i - m\lambda)] + [B_i + D_i(1 - 2\nu_i + m\lambda)]F_i\},$$

(1.95)

$$\Delta\sigma_{\rho_i}(m) = \left[mJ_0(m\rho_H) - \frac{J_1(m\rho_H)}{\rho_H}\right]\{[A_i + C_i(1 + m\lambda)] + [B_i - D_i(1 - m\lambda)]F_i\}$$
$$+ 2\nu_i mJ_0(m\rho_H)[C_i - D_iF_i],$$

(1.96)

$$\Delta\sigma_{\theta_i}(m) = \frac{J_1(m\rho_H)}{\rho_H}\{[A_i + C_i(1 + m\lambda)] + [B_i - D_i(1 - m\lambda)]F_i\}$$
$$+ 2\nu_i mJ_0(m\rho_H)[C_i - D_iF_i],$$

(1.97)

$$\Delta\tau_{\rho z_i}(m) = mJ_1(m\rho_H)\{[A_i + C_i(2\nu_i + m\lambda)] - [B_i - D_i(2\nu_i - m\lambda)]F_i\}.$$

(1.98)

Likewise, stresses at the interface i can be computed at the top of the layer $i+1$. For this level, also $\lambda = \lambda_i$, and therefore $e^{-m(\lambda_{i+1}-\lambda_i)} = F_{i+1}$, and $e^{-m(\lambda-\lambda_i)} = 1$. Then, Equations 1.43–1.46 for the top of the layer $i+1$ become

$$\Delta\sigma_{z_i}(m) = -mJ_0(m\rho_H)\{[A_{i+1} - C_{i+1}(1 - 2\nu_{i+1} - m\lambda)]F_{i+1}$$
$$+ [B_{i+1} + D_{i+1}(1 - 2\nu_{i+1} + m\lambda)]\},$$

(1.99)

$$\Delta\sigma_{\rho_i}(m) = \left[mJ_0(m\rho_H) - \frac{J_1(m\rho_H)}{\rho_H}\right]\{[A_{i+1} + C_{i+1}(1 + m\lambda)]F_{i+1}$$
$$+ [B_{i+1} - D_{i+1}(1 - m\lambda)]\} + 2\nu_{i+1}mJ_0(m\rho_H)[C_{i+1}F_{i+1} - D_{i+1}],$$

(1.100)

$$\Delta\sigma_{\theta_i}(m) = \frac{J_1(m\rho_H)}{\rho_H}\{[A_{i+1} + C_{i+1}(1 + m\lambda)]F_{i+1} + [B_{i+1} - D_{i+1}(1 - m\lambda)]\}$$
$$+ 2\nu_{i+1}mJ_0(m\rho_H)[C_{i+1}F_{i+1} - D_{i+1}],$$

(1.101)

$$\Delta\tau_{\rho z_i}(m) = mJ_1(m\rho_H)\{[A_{i+1} + C_{i+1}(2\nu_{i+1} + m\lambda)]F_{i+1}$$
$$- [B_{i+1} - D_{i+1}(2\nu_{i+1} - m\lambda)]\}.$$

(1.102)

For interface 1, the elastic properties of the first and second layers are different, which leads to different radial and tangential stresses ($\sigma_{\rho_1}^{top} \neq \sigma_{\rho_1}^{bottom}$ and $\sigma_{\theta_1}^{top} \neq \sigma_{\theta_1}^{bottom}$). Therefore, stresses in the granular layer must be calculated at the top of layer 2 where $F_2 = e^{-0.1136m}$, $\nu_2 = 0.35$, and $\lambda_1 = 0.0909$. Equations 1.99–1.102 result in

$$\Delta\sigma_{z_i}(m) = -mJ_0(m\rho_H)\{[A_2 - C_2(0.3 - 0.0909m)]e^{-0.1136m}$$
$$+ [B_2 + D_2(0.3 + 0.0909m)]\},$$

$$\Delta\sigma_{\rho_i}(m) = \left[mJ_0(m\rho_H) - \frac{J_1(m\rho_H)}{\rho_H}\right]\{[A_2 + C_2(1 + 0.0909m)]e^{-0.1136m}$$
$$+ [B_2 - D_2(1 - 0.0909m)]\} + 0.7mJ_0(m\rho_H)\left[C_2e^{-0.1136m} - D_2\right],$$

$$\Delta\sigma_{\theta_i}(m) = \frac{J_1(m\rho_H)}{\rho_H}\{[A_2 + C_2(1 + 0.0909m)]F_2 + [B_2 - D_2(1 - 0.0909m)]\}$$
$$+ 0.7mJ_0(m\rho_H)\left[C_2e^{-0.1136m} - D_2\right],$$

$$\Delta\tau_{\rho z_i}(m) = mJ_1(m\rho_H)\left\{[A_2 + C_2(0.7 + 0.0909m)]e^{-0.1136m}\right.$$
$$\left. - [B_2 - D_2(0.7 - 0.0909m)]\right\},$$

For bonded interfaces with equal elastic constants in the upper and lower layers, the condition of continuity of strains due to the bonded interface also implies continuity in stresses. Therefore, stresses at interfaces 2–9 can be calculated indistinctly at the top or the bottom of the corresponding layers. At the bottom of the layers, stresses can be calculated using Equations 1.95–1.98. For these levels, the values of F_i are also equal $(F_2 = \ldots = F_9 = e^{-0.1136m})$ as well as their Poisson's ratios $(\nu_2 = \ldots = \nu_9 = 0.35)$. Values of λ are $\lambda_2 = 0.2045$, $\lambda_3 = 0.3181$, $\lambda_4 = 0.4318$, $\lambda_5 = 0.5454$, $\lambda_6 = 0.6591$, $\lambda_7 = 0.7727$, $\lambda_8 = 0.8863$, and $\lambda_9 = 1.0$. Then, the stresses for layer i when $3 \le i \le 9$ become

$$\Delta\sigma_{z_i}(m) = -mJ_0(m\rho_H)\left\{[A_i - C_i(0.3 - m\lambda_i)] + [B_i + D_i(0.3 + m\lambda_i)]e^{-0.1136m}\right\},$$

$$\Delta\sigma_{\rho_i}(m) = \left[mJ_0(m\rho_H) - \frac{J_1(m\rho_H)}{\rho_H}\right]\left\{[A_i + C_i(1 + m\lambda_i)]\right.$$
$$\left. + [B_i - D_i(1 - m\lambda_i)]e^{-0.1136m}\right\} + 0.7mJ_0(m\rho_H)\left[C_i - D_i e^{-0.1136m}\right],$$

$$\Delta\sigma_{\theta_i}(m) = \frac{J_1(m\rho_H)}{\rho_H}\left\{[A_i + C_i(1 + m\lambda_i)] + [B_i - D_i(1 - m\lambda_i)]e^{-0.1136m}\right\}$$
$$+ 0.7mJ_0(m\rho_H)\left[C_i - D_i e^{-0.1136m}\right],$$

$$\Delta\tau_{\rho z_i}(m) = mJ_1(m\rho_H)\left\{[A_i + C_i(0.7 + m\lambda_i)] - [B_i - D_i(0.7 - m\lambda_i)]e^{-0.1136m}\right\}.$$

Analyzing the evolution of stresses produced by a moving wheel loading requires computing the stresses along the symmetry axis between the circular loads at different distances x. Considering that the distance between the center of the circular loads is $3a$, the radial distance ρ, corresponding to the advancement of the load along the symmetry axis x, is $\rho = \sqrt{x^2 + (1.5a)^2}$. And the dimensionless variable ρ_H becomes $\rho_H = \sqrt{x^2 + (1.5a)^2}/1.1$.

Using the proper set of equations, the algorithm described in Figure 1.38 allows computing the stresses at different depths and computing points.

1.7.2 Transformation of stresses from cylindrical into Cartesian coordinates

Burmister's method allows calculating the stresses of a multilayer structure in a cylindrical coordinates system. However, when there are multiple loads, the superposition of stresses is only possible in Cartesian coordinates. Therefore, it is necessary to transform the stresses from cylindrical into Cartesian coordinates (see Figure 1.41). The following rotation matrix allows such transformation:

$$\begin{bmatrix} \sigma_x \\ \sigma_y \\ \sigma_z \\ \tau_{yz} \\ \tau_{xz} \\ \tau_{xy} \end{bmatrix} = \begin{bmatrix} \cos^2\theta & \sin^2\theta & 0 & 0 & 0 & -\sin 2\theta \\ \sin^2\theta & \cos^2\theta & 0 & 0 & 0 & \sin 2\theta \\ 0 & 0 & 1 & 0 & 0 & 0 \\ 0 & 0 & 0 & \cos\theta & \sin\theta & 0 \\ 0 & 0 & 0 & -\sin\theta & \cos\theta & 0 \\ 0.5\sin 2\theta & -0.5\sin 2\theta & 0 & 0 & 0 & \cos 2\theta \end{bmatrix} \begin{bmatrix} \sigma_\rho \\ \sigma_\theta \\ \sigma_z \\ \tau_{\theta z} \\ \tau_{\rho z} \\ \tau_{\rho\theta} \end{bmatrix},$$

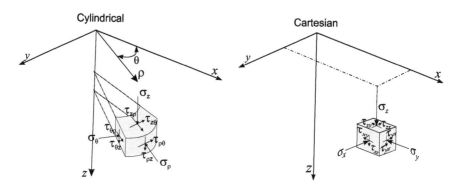

Figure 1.41 Stresses in cylindrical and Cartesian coordinates.

where the angle θ is the angle between the Cartesian and cylindrical axes, as shown in Figure 1.41.

Due to symmetry, the shear stresses in the plane $\theta\rho$ and θz are zero, leading to the following transformation matrix:

$$
\begin{bmatrix} \sigma_x \\ \sigma_y \\ \sigma_z \\ \tau_{yz} \\ \tau_{xz} \\ \tau_{xy} \end{bmatrix} = \begin{bmatrix} \cos^2\theta & \sin^2\theta & 0 & 0 \\ \sin^2\theta & \cos^2\theta & 0 & 0 \\ 0 & 0 & 1 & 0 \\ 0 & 0 & 0 & \sin\theta \\ 0 & 0 & 0 & \cos\theta \\ 0.5\sin 2\theta & -0.5\sin 2\theta & 0 & 0 \end{bmatrix} \begin{bmatrix} \sigma_\rho \\ \sigma_\theta \\ \sigma_z \\ \tau_{\rho z} \end{bmatrix}.
$$

Once the stresses are in the same Cartesian coordinate system, the stresses produced by multiple loads can be superposed. Tables 1.7–1.10 show the stresses σ_x, σ_y, and σ_z as well as the shear stress τ_{xz} that result from the superposition of the pair of loads of this example. The sign convention of these stresses is the useful convention used in geotechnical engineering (compression stress is positive, while traction is negative).

This book also provides a MATLAB script that uses Burmister's method for multiple layers and loads.

Table 1.7 Horizontal stress σ_x in MPa

Level					x (m)				
	0.000	0.125	0.250	0.375	0.500	0.750	1.000	1.500	2.000
A	0.027	0.044	0.045	0.03	0.017	0.006	0.003	0.001	0
B	0.002	0.013	0.026	0.026	0.02	0.009	0.004	0.001	0
C	−0.005	0.001	0.01	0.015	0.015	0.009	0.004	0.001	0
D	−0.008	−0.005	0.001	0.006	0.008	0.006	0.003	0.001	0.001
E	−0.011	−0.009	−0.005	−0.001	0.001	0.002	0.001	0.001	0.001
F	−0.014	−0.013	−0.01	−0.007	−0.005	−0.002	−0.001	0	0.001
G	−0.018	−0.018	−0.016	−0.013	−0.01	−0.005	−0.002	0	0.002
H	−0.025	−0.024	−0.022	−0.019	−0.015	−0.009	−0.004	0.001	0.002
I	−0.036	−0.035	−0.031	−0.026	−0.021	−0.011	−0.005	0.001	0.003

Table 1.8 Horizontal stress σ_y in MPa

Level	x (m)								
	0.000	0.125	0.250	0.375	0.500	0.750	1.000	1.500	2.000
A	0.101	0.07	0.031	0.013	0.006	0.003	0.002	0.001	0.001
B	0.034	0.027	0.015	0.007	0.003	0.001	0	0	0
C	0.009	0.007	0.004	0.001	−0.001	−0.001	−0.001	−0.001	0
D	−0.001	−0.002	−0.003	−0.003	−0.004	−0.004	−0.003	−0.002	−0.001
E	−0.007	−0.007	−0.007	−0.007	−0.007	−0.006	−0.005	−0.003	−0.002
F	−0.012	−0.012	−0.011	−0.011	−0.01	−0.008	−0.007	−0.004	−0.002
G	−0.017	−0.017	−0.016	−0.015	−0.014	−0.011	−0.009	−0.005	−0.003
H	−0.024	−0.024	−0.023	−0.021	−0.019	−0.015	−0.011	−0.006	−0.003
I	−0.035	−0.034	−0.032	−0.029	−0.026	−0.019	−0.014	−0.008	−0.004

Table 1.9 Vertical stress σ_z in MPa

Level	x (m)								
	0.000	0.125	0.250	0.375	0.500	0.750	1.000	1.500	2.000
A	0.179	0.123	0.051	0.015	0.002	−0.002	−0.001	−0.001	−0.001
B	0.144	0.114	0.062	0.026	0.008	−0.001	−0.001	−0.001	−0.001
C	0.1	0.086	0.056	0.029	0.013	0.001	−0.001	0	0
D	0.067	0.06	0.044	0.027	0.014	0.003	0	0	0
E	0.045	0.041	0.032	0.022	0.013	0.004	0.001	0	0
F	0.029	0.027	0.022	0.016	0.011	0.004	0.001	0	0
G	0.018	0.017	0.015	0.011	0.008	0.004	0.002	0.001	0
H	0.01	0.01	0.009	0.008	0.006	0.004	0.002	0.001	0
I	0.007	0.007	0.006	0.006	0.005	0.003	0.002	0.001	0

Table 1.10 Shear stress τ_{xz} in MPa

Level	x (m)								
	0.000	0.125	0.250	0.375	0.500	0.750	1.000	1.500	2.000
A	0.000	−0.052	−0.044	−0.027	−0.016	−0.008	−0.006	−0.004	−0.003
B	0.000	−0.039	−0.045	−0.034	−0.022	−0.011	−0.007	−0.005	−0.003
C	0.000	−0.024	−0.034	−0.031	−0.024	−0.013	−0.008	−0.005	−0.003
D	0.000	−0.015	−0.024	−0.025	−0.022	−0.014	−0.009	−0.005	−0.003
E	0.000	−0.01	−0.017	−0.02	−0.019	−0.014	−0.009	−0.005	−0.003
F	0.000	−0.007	−0.012	−0.015	−0.015	−0.012	−0.009	−0.004	−0.003
G	0.000	−0.005	−0.009	−0.011	−0.012	−0.01	−0.007	−0.004	−0.002
H	0.000	−0.003	−0.005	−0.007	−0.007	−0.007	−0.005	−0.003	−0.002
I	0.000	−0.001	−0.001	−0.002	−0.002	−0.002	−0.002	−0.001	−0.001

1.7.3 Principal stresses, rotation, and invariants p and q

As the stresses were computed along the symmetry plane between the pair of loads, shear stresses τ_{yx} and τ_{yz} are null; it means that the stress σ_y is one of the principal stresses. The other principal stresses act on the plane zx and can be computed based on the stresses σ_x, σ_z, and τ_{xz} using the following equations:

$$\sigma_1 = \frac{\sigma_x + \sigma_z}{2} + \sqrt{\frac{1}{4}(\sigma_x - \sigma_z)^2 + \tau_{xz}^2}, \text{ and}$$

$$\sigma_3 = \frac{\sigma_x + \sigma_z}{2} - \sqrt{\frac{1}{4}(\sigma_x - \sigma_z)^2 + \tau_{xz}^2}.$$

Tables 1.11 and 1.12 show the principal stresses σ_1 and σ_3 computed using the results shown in Tables 1.7, 1.9, and 1.10.

This example analyzes the pair of wheels' movement by calculating the stresses at different distances x from its central axis. The movement of the wheels imposes variable vertical, horizontal, and shear stresses whose combined effects produce continuous rotation of the principal stresses, as shown in Figure 1.42. Equation 1.103 permits computing the angle α_{xz} that describes the rotation of the principal stresses:

Table 1.11 Principal stress σ_1 in MPa

Level	\multicolumn{9}{c}{x (m)}								
	0.000	0.125	0.250	0.375	0.500	0.750	1.000	1.500	2.000
A	0.179	0.149	0.092	0.051	0.027	0.011	0.007	0.004	0.003
B	0.144	0.127	0.092	0.060	0.037	0.016	0.009	0.005	0.003
C	0.100	0.092	0.074	0.054	0.038	0.019	0.010	0.006	0.003
D	0.067	0.063	0.055	0.044	0.033	0.019	0.011	0.006	0.004
E	0.045	0.043	0.039	0.034	0.027	0.017	0.010	0.006	0.004
F	0.029	0.028	0.026	0.023	0.020	0.013	0.009	0.004	0.004
G	0.018	0.018	0.017	0.015	0.014	0.010	0.007	0.005	0.003
H	0.010	0.010	0.010	0.010	0.008	0.007	0.005	0.004	0.003
I	0.007	0.007	0.006	0.006	0.005	0.003	0.003	0.002	0.003

Table 1.12 Principal stress σ_3 in MPa

Level	\multicolumn{9}{c}{x (m)}								
	0.000	0.125	0.250	0.375	0.500	0.750	1.000	1.500	2.000
A	0.027	0.018	0.004	−0.006	−0.008	−0.007	−0.005	−0.004	−0.004
B	0.002	0.000	−0.004	−0.008	−0.009	−0.008	−0.006	−0.005	−0.004
C	−0.005	−0.005	−0.008	−0.010	−0.010	−0.009	−0.007	−0.005	−0.003
D	−0.008	−0.008	−0.010	−0.011	−0.011	−0.010	−0.008	−0.005	−0.003
E	−0.011	−0.011	−0.012	−0.013	−0.013	−0.011	−0.008	−0.005	−0.003
F	−0.014	−0.014	−0.014	−0.014	−0.014	−0.011	−0.009	−0.004	−0.003
G	−0.018	−0.019	−0.018	−0.017	−0.016	−0.011	−0.007	−0.004	−0.001
H	−0.025	−0.024	−0.023	−0.021	−0.017	−0.012	−0.007	−0.002	−0.001
I	−0.036	−0.035	−0.031	−0.026	−0.021	−0.011	−0.006	0.000	0.000

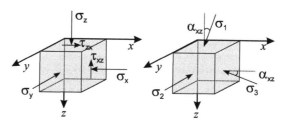

Figure 1.42 Principal stresses and rotation angle in the x, z plane.

$$\alpha_{xz} = \frac{1}{2} \arctan \frac{2\tau_{zy}}{\sigma_z - \sigma_y}. \tag{1.103}$$

Figure 1.43, from Ref. [13], illustrates how the stresses rotate depending on the distance x from the middle axis of the pair of loads and the depth z.

After computing the principal stresses, Equations 1.104 and 1.105 permit to obtain the magnitude of the invariants p and q in a tridimensional system of stresses. In the case of this example, the stress σ_y becomes the intermediate stress σ_2:

$$p = \frac{\sigma_1 + \sigma_2 + \sigma_3}{3}, \text{ and} \tag{1.104}$$

$$q = \sqrt{\frac{1}{2}\left[(\sigma_1 - \sigma_2)^2 + (\sigma_2 - \sigma_3)^2 + (\sigma_3 - \sigma_1)^2\right]}. \tag{1.105}$$

Tables 1.13 and 1.14 present the results of the mean stress p and the deviator stress q computed based on the principal stresses given in Table 1.8 for $\sigma_2 = \sigma_y$, and Tables 1.11 and 1.12 for σ_1 and σ_3. Also, Figure 1.44 shows the stress paths in the $p - q$ plane computed at different levels of the granular layer (*i.e.*, points A to I). These results show that the stress path at each level can be approximated to straight lines. However, it is important to note that for deeper levels, tension stresses appear.

Finally, the variation of the magnitude of the stress path and its direction can be characterized by the mean stress p, the deviator stress q, and the angle α_{xz}, as shown in Figure 1.45. Results show that the stress paths in the plane (p, q, α) follow an inclined bell curve with a slightly curved projection in the $p - q$ plane.

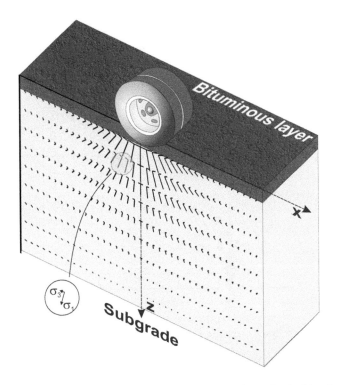

Figure 1.43 Direction of the principal stresses on the symmetry plane of a double wheel load.

Table 1.13 Mean stress p in MPa

Level	x (m)								
	0.000	0.125	0.250	0.375	0.500	0.750	1.000	1.500	2.000
A	0.102	0.079	0.042	0.019	0.008	0.002	0.001	0.000	0.000
B	0.060	0.051	0.034	0.020	0.010	0.003	0.001	0.000	0.000
C	0.035	0.031	0.023	0.015	0.009	0.003	0.001	0.000	0.000
D	0.019	0.018	0.014	0.010	0.006	0.002	0.000	0.000	0.000
E	0.009	0.008	0.007	0.005	0.002	0.000	−0.001	−0.001	0.000
F	0.001	0.001	0.000	−0.001	−0.001	−0.002	−0.002	−0.001	0.000
G	−0.006	−0.006	−0.006	−0.006	−0.005	−0.004	−0.003	−0.001	0.000
H	−0.013	−0.013	−0.012	−0.011	−0.009	−0.007	−0.004	−0.001	0.000
I	−0.021	−0.021	−0.019	−0.016	−0.014	−0.009	−0.006	−0.002	0.000

Table 1.14 Deviator stress q in MPa

Level	x (m)								
	0.000	0.125	0.250	0.375	0.500	0.750	1.000	1.500	2.000
A	0.132	0.114	0.078	0.049	0.031	0.016	0.011	0.007	0.005
B	0.129	0.116	0.089	0.062	0.041	0.021	0.013	0.009	0.005
C	0.099	0.092	0.077	0.059	0.044	0.024	0.015	0.009	0.005
D	0.072	0.069	0.061	0.051	0.041	0.026	0.016	0.009	0.005
E	0.054	0.052	0.048	0.044	0.037	0.026	0.017	0.009	0.006
F	0.042	0.041	0.039	0.036	0.032	0.023	0.017	0.008	0.006
G	0.036	0.036	0.035	0.031	0.029	0.022	0.015	0.009	0.006
H	0.035	0.034	0.033	0.031	0.026	0.021	0.014	0.009	0.006
I	0.043	0.042	0.038	0.034	0.029	0.020	0.014	0.009	0.006

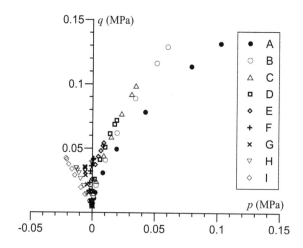

Figure 1.44 Stress paths in the $p - q$ plane produced by the moving wheels.

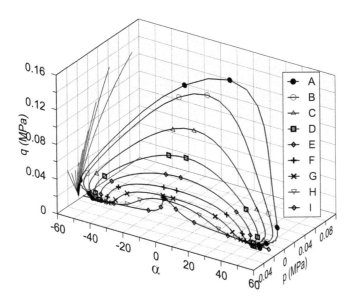

Figure 1.45 Stress path in the plane $p - q$ and rotation angle α computed in Ref. [13].

1.7.4 Concluding remarks

Results of this example permit to deliver the following concluding remarks:

- For each depth in the granular layer, the stress paths in the plane $p - q$ follow a line with a slight curvature that can be approximated to a straight line.
- The $p - q$ paths also indicate that the granular layer requires some cohesion to remain in the elastic domain. Even the deeper layers require some tensile strength.
- When the material cannot withstand some tensile stress, it enters into the plastic domain of behavior and, therefore, the Burmister's method is no longer valid.
- As the vehicle wheel moves, the principal stresses experience a significant rotation. This rotation is an important issue when stress paths are simulated using a triaxial apparatus because, in this device, the principal stresses remain in the same direction throughout the test. More advanced test devices, such as the hollow cylinder apparatus used in Ref. [13], can more accurately reproduce the stresses produced by a moving wheel.

Chapter 2

Unsaturated soil mechanics applied to road materials

2.1 RELEVANT EQUATIONS

The theory of unsaturated soil mechanics allows us to calculate the evolution of some state variables, such as the pore water pressure and the temperature of the materials within the structure of a road. Also, the mechanics of unsaturated soil allow the mechanical behavior of road materials to be calculated based on such state variables. Despite the broad applicability of unsaturated soil mechanics in road engineering, this chapter focuses only on heat and water transport mechanisms, because the mechanical response of road materials is explored in other chapters of this book.

Focusing on the phenomena of mass and heat transport, the main functions and variables regarding water transport are the water retention curve and hydraulic conductivity, while heat capacity and thermal conductivity are the variables required for heat transport.

2.1.1 Water retention curve

Several equations can describe the water retention curve of road materials. However, the Mechanistic Empiric Pavement Design Guide (MEPDG) proposes an empirical adjustment of the parameters involved in the model proposed by Fredlund and Xing [30]. The following equations give this model:

$$S_r = C(s)\frac{1}{\{\ln[\exp(1) + (s/a)^n]\}^m}, \quad C(s) = 1 - \frac{\ln(1 + s/s_{res})}{\ln(1 + 10^6/s_{res})}, \tag{2.1}$$

where S_r is the degree of saturation, s is the suction, s_{res} is the residual suction, and a, m, and n are parameters of the model.

The set of correlations to obtain the model parameters was proposed in Ref. [47]. Soils have been divided into those with plasticity indexes greater than zero or equal to zero. Those materials have different properties that are represented by the product $P_{200}PI$ or by the size d_{60}.

Parameters for soils with $PI > 0$ are

$$a = 0.00364(P_{200}PI)^{3.35} + 4P_{200}PI + 11, \tag{2.2}$$

$$m = 0.0514(P_{200}PI)^{0.465} + 0.5, \tag{2.3}$$

$$n = m(-2.313(P_{200}PI)^{0.14} + 5), \text{ and} \tag{2.4}$$

$$s_{res} = a32.44e^{0.0186P_{200}PI}, \tag{2.5}$$

were P_{200} is the proportion of material that pass through the # 200 U.S. Standard Sieve, expressed as a decimal, and PI is the plasticity index as a percentage (%).

For soils with plasticity index equal to zero, the parameters of the model are

$$a = 0.8627d_{60}^{-0.751}, \tag{2.6}$$

$$m = 0.1772\ln(d_{60}) + 0.7734, \tag{2.7}$$

$$n = 7.5, \text{ and} \tag{2.8}$$

$$s_{res} = \frac{a}{d_{60} + 9.7e^{-4}}. \tag{2.9}$$

Remember that d_{60} is the grain size, obtained from the grain size distribution, corresponding to 60% by mass of those that passed through this dimension, expressed in mm.

2.1.2   Assessment of the hydraulic conductivity based on the water retention curve

Similar to the water retention curve, there are several methods for evaluating the hydraulic conductivity; some of them allow estimating the hydraulic conductivity based on the water retention curve. This section describes the method proposed in Ref. [31], which is based on the water retention curve proposed by Fredlund and Xing [30].

According to Ref. [31], the hydraulic conductivity is assessed by evaluating the following equation:

$$k_{rw}(s) = \frac{\int_{\ln(s)}^{b} \frac{\theta(e^y) - \theta(s)}{e^y} \theta'(e^y) dy}{\int_{\ln(s_{aev})}^{b} \frac{\theta(e^y) - \theta_{sat}}{e^y} \theta'(e^y) dy}, \tag{2.10}$$

where $\theta(s)$ is the water retention curve defined in terms of the volumetric water content θ, $\theta'(s)$ is the first derivative of the water retention curve, s_{aev} is the suction corresponding to the air entry value, and θ_{sat} is the saturated volumetric water content.

The procedure for the numerical integration of Equation 2.10 is as follows:

1   Establish the integration limits a and b as

$$a = \ln s_{aev}, \text{ and } b = \ln(10^6).$$

2   Divide the interval $[a, b]$ into N sub-intervals of size Δy as

$$a = y_1 < y_2 <, ..., y_N < y_{N+1} = b, \quad \Delta y = \frac{b - a}{N}.$$

3   Evaluate the denominator of Equation 2.10 as

$$\int_{\ln(s_{aev})}^{b} \frac{\theta(e^y) - \theta_{sat}}{e^y} \theta'(e^y) dy \approx \Delta y \sum_{i=1}^{N} \frac{\theta(e^{\bar{y}_i}) - \theta_{sat}}{e^{\bar{y}_i}} \theta'(e^{\bar{y}_i}),$$

where $\bar{y}_i$ is the midpoint of the interval $[y_i, y_{i+1}]$, and θ' is the derivative of the water retention curve proposed in Ref. [30] which is given by

$$\theta'(s) = C'(s) \frac{\theta_{sat}}{\{\ln[e + (s/a)^n]\}^m} - C(s) \frac{\theta_{sat}}{\{\ln[e + (s/a)^n]\}^{m+1}} \frac{mn\left(\frac{s}{a}\right)^{n-1}}{a[e + (s/a)^n]}, \tag{2.11}$$

$$C'(s) = \frac{-1}{(s_{res} + s)\ln\left(1 + \frac{10^6}{s_{res}}\right)}.$$

(2.12)

4   Evaluate the numerator of Equation 2.10 as

$$\int_{\ln(s)}^{b} \frac{\theta(e^y) - \theta(s)}{e^y}\theta'(e^y)dy \approx \Delta y \sum_{i=j}^{N} \frac{\theta(e^{\bar{y}_i}) - \theta(s)}{e^{\bar{y}_i}}\theta'(e^{\bar{y}_i}).$$

5   Finally, the relative water permeability becomes

$$k_{rw}(s) = \frac{\sum_{i=j}^{N} \frac{\theta(e^{\bar{y}_i}) - \theta(s)}{e^{\bar{y}_i}}\theta'(e^{\bar{y}_i})}{\sum_{i=1}^{N} \frac{\theta(e^{\bar{y}_i}) - \theta_{sat}}{e^{\bar{y}_i}}\theta'(e^{\bar{y}_i})} \quad j \to \theta(s) \geq \theta(e^{\bar{y}_i}).$$

(2.13)

2.1.3   Flow of water in unsaturated materials

This section summarizes the equations that allow the calculation of water flow in unsaturated soils that do not undergo volumetric changes.

Calculating the changes in water content requires solving the differential equation representing the conservation law for the mass of water, and its solution requires considering some phenomenological relationships and boundary conditions.

The conservation equation for the mass of water, neglecting the flow of water in the vapor phase, is

$$\frac{\partial \theta_w}{\partial t} = \nabla \cdot (\nabla k_w \psi),$$

where θ_w is the volumetric water content, k_w is the unsaturated hydraulic conductivity, ψ is the water potential, and t is the time.

Relating the evolution of volumetric water content with its water retention curve, and considering a non-compressible material, the continuity equation for the flow of liquid water becomes

$$C_\theta \frac{\partial u_w}{\partial t} = \nabla \cdot \left[k_w(S_r)\nabla\left(z + \frac{u_w}{\gamma_w}\right)\right],$$

(2.14)

where $C_\theta = n\frac{\partial S_r}{\partial u_w}$,

C_θ is known as the specific water capacity.

Equation 2.14 is a nonlinear parabolic differential equation whose solution can only be reached by using numerical methods. The solution of this equation, using the explicit finite difference method (FDM), is described in detail in Example 10.

2.1.4   Thermal properties of unsaturated materials

The fundamental thermal properties for evaluating the heat transport in unsaturated soils are the thermal conductivity and the heat capacity.

Table 2.1 Empirical parameters to obtain the thermal conductivity

Particle Type	k_{Hdry} Parameters	
	χ_H	η
Gravels and crushed sand	1.70	1.80
Fine-grained soils and natural sands	0.75	1.20
Peat	0.30	0.87

Soil Type	κ_H Parameters	
	Unfrozen	Frozen
Well-graded gravels and coarse sands	4.60	1.70
Medium and fine sands	3.55	0.95
Silts and clays	1.90	0.85
Peat	0.60	0.25

Source: From Ref. [19].

The book *Geotechnics of Roads Fundamentals* [10] describes several methods for estimating the thermal conductivity, and this section summarizes only the method proposed by Côté and Konrad [19]. The set of equations of such method is

$$k_{Hsat} = k_{Hs}^{\theta_s}\, k_{Hw}^{\theta_w}\, k_{Hi}^{\theta_i} \quad \text{Saturated thermal conductivity,} \qquad (2.15)$$

where k_{Hs}, k_{Hw}, and k_{Hi}; and θ_s, θ_w, and θ_i are the thermal conductivities and the volumetric fractions of solids, water, and ice, respectively.

The dry thermal conductivity is

$$k_{Hdry} = \chi_H \cdot 10^{-\eta n}, \qquad (2.16)$$

where χ_H is a dimensional empirical parameter, given in [W/(mK)], η is another empirical parameter, and n is the porosity. Table 2.1 shows the values of the empirical parameters, χ_H, and η, depending on the type of soil.

Also, Kersten's number K_e, expressed as a function of the degree of saturation, is

$$K_e = \frac{\kappa_H S_r}{1 + (\kappa_H - 1)S_r}, \qquad (2.17)$$

where κ_H is also an empirical parameter.

Figure 2.1 describes the methodology for assessing the thermal conductivity of unsaturated soils.

Also, Table 2.2 presents the specific heat capacity c_H of various materials that form soils. Then, the heat capacity c_H of soils is given by the sum of the heat capacities of its different constituents [24], and it can be evaluated using the following equation:

$$c_H = \frac{c_s + w c_w}{1 + w}, \qquad (2.18)$$

where c_s and c_w are the specific heats, in J/kg °C, of particles of dry soil and water, respectively, while w is the gravimetric water content.

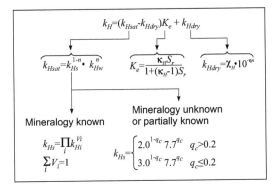

Figure 2.1 Schematic description of the method to estimate the thermal conductivity of soils [19,20,71].

Table 2.2 Specific heat capacity of common components in soils

Material	Heat Capacity (kJ/kg°C)	Material	Heat Capacity (kJ/kg°C)
Air 10°C	1.00	Orthoclase feldspar	0.79
Water 25°C	4,2	Quartz	0.79
Water vapor 1 atm 400 K	1.9	Basalt	0.84
Ice 0°C	2.04	Clay minerals	0.9
Augite	0.81	Granite	0.8
Hornblende	0.82	Limestone	0.91
Mica	0.86	Sandstone	0.92
		Shale	0.71

Source: From Refs. [67,71].

2.1.5 Heat flow in unsaturated materials

In the same way than the flow of water, the conservation equation for heat, neglecting the change in phases between liquid water and vapor or ice, is

$$c_{H_v} \frac{\partial T}{\partial t} = \nabla \cdot (\nabla k_H T), \tag{2.19}$$

where c_{H_v} is the volumetric heat capacity of the material, defined as $c_{H_v} = \rho c_H$, k_H is its thermal conductivity, and T is the temperature.

Equation 2.19 can be solved using the FDM as described in detail in Example 11.

2.2 EXAMPLE 7: ASSESSMENT OF THE WATER RETENTION CURVE USING THE EMPIRICAL MODEL PROPOSED IN THE MECHANISTIC EMPIRIC PAVEMENT DESIGN GUIDE (MEPDG)

This example describes the use of the empirical method proposed in the MEPDG for assessing the water retention curve. Two types of materials are considered here: one is a plastic soil, while the other is nonplastic. The characteristics of these two soils are

- *Plastic soil:*

  - *passing through the # 200 U.S. Standard Sieve $P_{200} = 0.3$, and plasticity index $PI = 15\%$.*

- *Nonplastic soil:*

  - *grain size corresponding to passing 60% by mass, $d_{60} = 0.4\,mm$.*

As described in Section 2.1.1, the Fredlund and Xing model, proposed in Ref. [30], is the model suggested in the MEPDG [5]. The following equations give the water retention curve for this model:

$$S_r = C(s) \frac{1}{\{\ln\left[\exp(1) + (s/a)^n\right]\}^m},$$

$$C(s) = 1 - \frac{\ln(1 + s/s_{\text{res}})}{\ln(1 + 10^6/s_{\text{res}})}.$$

Moreover, the set of empirical relationships represented in Equations 2.2–2.9 were proposed in Ref. [47] as parameters to be used in this model.

For the plastic soil, the empirical relationships depend on the product of the proportion of passing the # 200 sieve multiplied by the plasticity index. For this example, such a product is $P_{200}PI = 4.5$. Therefore, the parameters of the model are

$$a = 0.00364 \cdot 4.5^{3.35} + 4 \cdot 4.5 + 11 = 29.5615,$$

$$m = 0.0514 \cdot 4.5^{0.465} + 0.5 = 0.6034,$$

$$n = m(-2.313 \cdot 4.5^{0.14} + 5) = 0.6034 \cdot 2.1448 = 1.2943,$$

$$s_{\text{res}} = a32.44e^{0.0186 \cdot 4.5} = 29.5615 \cdot 32.44 \cdot 1.0873 = 1042.6967\ \text{kPa}.$$

In contrast, for the nonplastic soil, parameters depend on the d_{60}, which is 0.4 mm in this example, and therefore,

$$a = 0.8627 \cdot 0.4^{-0.751} = 1.7168,$$

$$m = 0.1772\ln(0.4) + 0.7734 = 0.6110,$$

$$n = 7.5,$$

$$s_{\text{res}} = \frac{a}{0.4 + 9.7e^{-4}} = \frac{1.7168}{0.5777} = 2.9719\ \text{kPa}.$$

As a result, the equation for the water retention curve for each soil type is

- for the plastic soil,

$$S_r = \left[1 - \frac{\ln(1 + s/1042.6967)}{\ln(1 + 10^6/1042.6967)}\right] \frac{1}{\{\ln\left[exp(1) + (s/29.5615)^{1.2943}\right]\}^{0.6034}},\ \text{and}$$

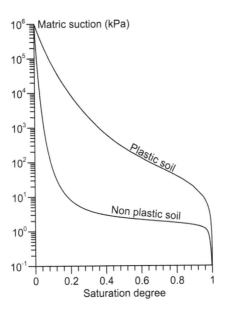

Figure 2.2 Water retention curves obtained using the empirical relationships proposed in the MEPDG.

- for the nonplastic soil,

$$S_r = \left[1 - \frac{\ln(1 + s/2.9719)}{\ln(1 + 10^6/2.9719)}\right] \frac{1}{\left\{\ln\left[\exp(1) + (s/1.7168)^{7.5}\right]\right\}^{0.6110}}.$$

The above equations allow drawing the water retention curve for each soil as shown in Figure 2.2.

2.3 EXAMPLE 8: METHOD FOR CALCULATING THE UNSATURATED HYDRAULIC CONDUCTIVITY BASED ON THE WATER RETENTION CURVE

This example describes the use of the model proposed in Ref. [31] for calculating the unsaturated hydraulic conductivity based on the water retention curve.

The example uses the experimental results presented by Brooks and Corey [9], which are summarized in Table 2.3. In fact, the water retention curve of this example corresponds to a silty loam denoted as GE3 in Ref. [9], whose fitting parameters, corresponding to the Fredlund and Xing model, are a = 8.34, n = 9.9, m = 0.44, and s_{res} = 30. Also, the analysis of the water retention curve shows that the suction pressure corresponding to the air entry value is s_{aev} = 6.8 kPa. Moreover, the porosity of the material is n = 0.48.

Also, Table 2.4 presents the experimental results of the unsaturated hydraulic conductivity, which are provided here to compare the experimental results with the results of the model.

Table 2.3 Experimental data of the water retention curve of a silty loam

s (kPa)	S_r	s (kPa)	S_r	s (kPa)	S_r	s (kPa)	S_r
0.6	1.000	7.2	0.947	11.8	0.553	18.8	0.358
1.3	0.989	8.5	0.883	13.5	0.479	23.7	0.305
3.5	0.982	8.5	0.762	15.4	0.436	27.9	0.294
5.5	0.961	10.1	0.642	17.6	0.401	30.9	0.266

Source: Presented in Ref. [9].

Table 2.4 Experimental data of the unsaturated hydraulic conductivity of a silty loam

s (kPa)	0.5	1.2	3.5	5.8	6.6	7.7	8.6	9.6	10.7	11.9
K_{wr}	0.960	0.975	1.000	0.960	0.898	0.887	0.577	0.207	0.096	0.045

Source: Presented in Ref. [9].

Section 2.1.2 describes a method that allows calculating the unsaturated hydraulic conductivity based on the water retention curve. The method requires the following steps:

1 Divide the range of suction within which the water retention curve is analyzed in sub-intervals, and establish the limits of integration of such discrete sub-intervals.
2 Calculate the volumetric water content and the derivative of the water content with respect to suction for the midpoint of each sub-interval.
3 Calculate the denominator of Equation 2.13.
4 Calculate the volumetric water content for any value of matric suction, and evaluate the numerator of Equation 2.13.

2.3.1   Limits of integration and sub-intervals

The limits for the discrete integration (in terms of the matric suction) are on the bottom side, the suction of the air entry value; and, on the higher side, 10^6 kPa, which is the maximum thermodynamically achievable suction pressure. Therefore, when 12 sub-intervals are used, the matric suctions in the middle of each one, which is s_i, become

$$y_1 = \ln(s_{aev}) = \ln(6.8) = 1.917,$$

$$y_{13} = \ln(10^6) = 13.816,$$

$$\Delta y = \frac{13.816 - 1.917}{12} = 0.9915,$$

$$y_{i+1} = y_i + \Delta y,$$

$$\bar{y}_i = \frac{1}{2}(y_{i+1} + y_i), \text{ and then}$$

$$s_i = e^{\bar{y}_i}.$$

Column 4 of Table 2.5 presents the results of s_i, which is the matric suction of the middle of each sub-interval.

2.3.2 Volumetric water content and derivative with respect to suction

The second step consists of evaluating Equation 2.1, for calculating the volumetric water content for each suction pressure s_i as follows

$$\theta(s_i) = 0.48 \left[1 - \frac{\ln(1 + s_i/30)}{\ln(1 + 10^6/30)}\right] \frac{1}{\left\{\ln\left[e + (s_i/8.34)^{9.9}\right]\right\}^{0.44}}, \quad \text{becoming}$$

$$\theta(s_i) = 0.48 \frac{1 - 0.096 \ln(1 + s_i/30)}{\left\{\ln\left[e + (s_i/8.34)^{9.9}\right]\right\}^{0.44}}.$$

Then, from Equations 2.11 and 2.12, the derivative of the volumetric water content is

$$\theta'(s_i) = C'(s_i) \frac{0.48}{\left\{\ln\left[e + (s_i/8.34)^{9.9}\right]\right\}^{0.44}}$$

$$- C(s_i) \frac{0.48}{\left\{\ln\left[e + (s_i/8.34)^{9.9}\right]\right\}^{1.44}} \frac{0.44 \cdot 9.9 \left(\frac{s_i}{8.34}\right)^{8.9}}{8.34 \left[e + (s_i/8.34)^{9.9}\right]},$$

where

$$C(s_i) = 1 - \frac{\ln(1 + s_i/30)}{\ln\left(1 + \frac{10^6}{30}\right)} = 1 - 0.096 \ln(1 + s_i/30), \quad \text{which is}$$

$$C'(s_i) = \frac{-1}{(30 + s_i) \ln\left(1 + \frac{10^6}{30}\right)} = -0.096/(30 + s_i).$$

Finally, the derivative of the volumetric water content is

$$\theta'(s_i) = -\frac{0.0461}{(30 + s_i) \left\{\ln\left[e + (s_i/8.34)^{9.9}\right]\right\}^{0.44}}$$

$$- \frac{0.251 \left[1 - 0.096 \ln(1 + s_i/30)\right]}{\left\{\ln\left[e + (s_i/8.34)^{9.9}\right]\right\}^{1.44} \left[e + (s_i/8.34)^{9.9}\right]} \left(\frac{s_i}{8.34}\right)^{8.9}.$$

Columns 5 and 6 of Table 2.5 show the results of the volumetric water content and its derivative for the midpoint of each sub-interval.

2.3.3 Denominator of Equation 2.13

The next step consists in evaluating the denominator of Equation 2.13, denoted here as $\Sigma 1$, which is

$$\Sigma 1 = \sum_{i=1}^{N} \frac{\theta(e^{\bar{y}_i}) - \theta_{\text{sat}}}{e^{\bar{y}_i}} \theta'(e^{\bar{y}_i}).$$

This summation is evaluated in column 7 of Table 2.5 and the result is

$$\Sigma 1 = \sum_{i=1}^{N} \frac{\theta(s_i) - \theta_{\text{sat}}}{s_i} \theta'(s_i) = 5.91 \cdot 10^{-4}.$$

Table 2.5 Computation of the unsaturated hydraulic conductivity using the model of Fredlund et al. [31]

							s (kPa)	6	8	10	12
							θ(s)	0.469	0.430	0.332	0.262
i	y_i	$\bar{y}_i$	$s_i = e^{\bar{y}_i}$	$\theta(s_i)$	$\theta'(s_i)$	$\Sigma 1$		$\Sigma 2$	$\Sigma 2$	$\Sigma 2$	$\Sigma 2$
1	1.917	2.413	11.16	0.286	$-3.27\cdot10^{-2}$	$5.68\cdot10^{-4}$		$5.35\cdot10^{-4}$	$4.22\cdot10^{-4}$	$1.35\cdot10^{-4}$	
2	2.908	3.404	30.09	0.146	$-1.92\cdot10^{-3}$	$2.13\cdot10^{-5}$		$2.06\cdot10^{-5}$	$1.81\cdot10^{-5}$	$1.18\cdot10^{-5}$	$7.37\cdot10^{-6}$
3	3.900	4.396	81.11	0.107	$-3.60\cdot10^{-4}$	$1.66\cdot10^{-6}$		$1.61\cdot10^{-6}$	$1.43\cdot10^{-6}$	$9.99\cdot10^{-7}$	$6.89\cdot10^{-7}$
4	4.892	5.387	218.62	0.083	$-9.12\cdot10^{-5}$	$1.66\cdot10^{-7}$		$1.61\cdot10^{-7}$	$1.45\cdot10^{-7}$	$1.04\cdot10^{-7}$	$7.48\cdot10^{-8}$
5	5.883	6.379	589.28	0.066	$-2.59\cdot10^{-5}$	$1.82\cdot10^{-8}$		$1.77\cdot10^{-8}$	$1.60\cdot10^{-8}$	$1.17\cdot10^{-8}$	$8.62\cdot10^{-9}$
6	6.875	7.370	1588.34	0.052	$-7.76\cdot10^{-6}$	$2.09\cdot10^{-9}$		$2.03\cdot10^{-9}$	$1.85\cdot10^{-9}$	$1.37\cdot10^{-9}$	$1.03\cdot10^{-9}$
7	7.866	8.362	4281.21	0.041	$-2.42\cdot10^{-6}$	$2.48\cdot10^{-10}$		$2.41\cdot10^{-10}$	$2.20\cdot10^{-10}$	$1.64\cdot10^{-10}$	$1.25\cdot10^{-10}$
8	8.858	9.354	11539.60	0.031	$-7.74\cdot10^{-7}$	$3.01\cdot10^{-11}$		$2.93\cdot10^{-11}$	$2.67\cdot10^{-11}$	$2.02\cdot10^{-11}$	$1.55\cdot10^{-11}$
9	9.849	10.345	31103.92	0.023	$-2.53\cdot10^{-7}$	$3.72\cdot10^{-12}$		$3.63\cdot10^{-12}$	$3.32\cdot10^{-12}$	$2.52\cdot10^{-12}$	$1.95\cdot10^{-12}$
10	10.841	11.337	83837.70	0.016	$-8.44\cdot10^{-8}$	$4.67\cdot10^{-13}$		$4.56\cdot10^{-13}$	$4.17\cdot10^{-13}$	$3.18\cdot10^{-13}$	$2.48\cdot10^{-13}$
11	11.832	12.328	225976.67	0.009	$-2.85\cdot10^{-8}$	$5.94\cdot10^{-14}$		$5.79\cdot10^{-14}$	$5.31\cdot10^{-14}$	$4.07\cdot10^{-14}$	$3.19\cdot10^{-14}$
12	12.824	13.320	609098.98	0.003	$-9.72\cdot10^{-9}$	$7.61\cdot10^{-15}$		$7.43\cdot10^{-15}$	$6.82\cdot10^{-15}$	$5.25\cdot10^{-15}$	$4.13\cdot10^{-15}$
13	13.816				Σ	$5.91\cdot10^{-4}$		$5.58\cdot10^{-4}$	$4.42\cdot10^{-4}$	$1.48\cdot10^{-4}$	$8.15\cdot10^{-6}$
						k_{wr}		0.9428	0.7473	0.2503	0.0138

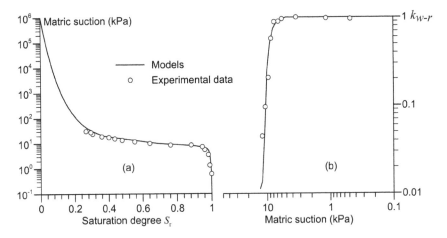

Figure 2.3 Comparisons between experimental data and models for the water retention curve and the relative unsaturated hydraulic conductivity presented in Refs. [30,31].

2.3.4 Numerator of Equation 2.13

The numerator of Equation 2.13, denoted here as $\Sigma 2$, is

$$\Sigma 2 = \sum_{i=j}^{N} \frac{\theta(s_i) - \theta(s)}{s_i} \theta'(s_i) \quad j \to \theta(s) \geq \theta(s_i).$$

As observed, there is a different value of $\Sigma 2$ for each value of $\theta(s)$, evidencing that $\Sigma 2$ depends on the suction pressure. Therefore, values of $\Sigma 2$ are evaluated in columns 8–11 of Table 2.5 for the following values of suction pressure: $s = 6, \ 8, \ 10, \ 12 \ \text{kPa}$.

Finally, the unsaturated relative hydraulic conductivity, for each matric suction, results from the ratio $\Sigma 2 / \Sigma 1$.

Figure 2.3 presents a comparison between the experimental data and the models presented in Refs. [30,31] for the water retention curve and the unsaturated relative hydraulic conductivity.

Figure 2.3 shows an excellent agreement between the experimental measures and the models. However, regarding the unsaturated relative hydraulic conductivity, the model was validated only for materials having low values of matric suction. Also, for a better agreement, Fredlund et al. [31] recommend multiplying Equation 2.13 by Θ_n^q, as in Mualem's equation, to take into consideration the tortuosity on the porous space.

2.4 EXAMPLE 9: SIMPLIFIED CALCULATION OF WATER INFILTRATION

The calculation of water infiltration in soils must consider the evolution of variables such as suction, water content, and unsaturated hydraulic conductivity, which evolve in a nonlinear way. In a simplified form, the algebraic integration of the equations involved in water infiltration is possible when considering a retention curve defined as follows:

- $S_r = 1$ *for* $s \leq s_{aev}$, *and*
- $S_r \rightarrow 0$ *for* $s > s_{aev}$.

Such a type of water retention curve corresponds to a soil that has a uniformly sized grain distribution, which in turn suggests a possible uniform pore size distribution.

This example describes the methodology for estimating the progression of the saturation front produced by the infiltration that results from imposing a layer of water of negligible thickness on the surface of a soil that has a simplified water retention curve. The properties of the soil for a numerical evaluation are porosity $n = 0.4$, saturated hydraulic conductivity $k_w = 10^{-5}$ m/s, and suction corresponding to the air entry value $s_{aev} = 30$ kPa.

As mentioned above, a soil with a uniform pore size distribution remains saturated up to a suction pressure corresponding to the capillary pressure of the largest pore in the soil. This suction pressure is denoted as the air entry value, s_{aev}. Subsequently, due to its uniform pore size distribution, when the suction pressure increases by a small value, water loss occurs simultaneously in almost all pores of the soil. Therefore, the soil dries abruptly, leading to a water retention curve similar to that outlined in Figure 2.4. A similar behavior appears on wetting.

By superimposing a water retention curve similar to that shown in Figure 2.4 on a soil layer undergoing water infiltration, the profile of the degree of saturation becomes as shown in Figure 2.5. In fact, a saturated layer appears on top of the entire column of soil, and then, below this layer, the degree of saturation drops abruptly.

The first step for calculating the infiltration velocity is the evaluation of the gradient of potential between points A and B in Figure 2.5. Points A and B are located on the surface of the soil and the saturated front, respectively. The two components of water potential, expressed in units of length, are the position z and the pressure u_w/γ_w. Therefore, the potentials in points A and B are as follows:

- Potential in point A,

$$\psi_A = z_A + u_{wA}/\gamma_w, \text{ assuming } z_A = 0, \text{ then } \psi_A = 0, \text{ and}$$

- Potential in point B,

$$\psi_B = z_B + u_{wB}/\gamma_w, \text{ assuming } z_B = -z_{sat}, \text{ then } \psi_B = -z_{sat} - s_{aev}/\gamma_w.$$

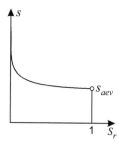

Figure 2.4 Simplified water retention curve of a soil that has a uniform pore size distribution.

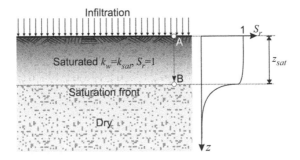

Figure 2.5 Progression of the saturation front in a soil with uniform pore size distribution.

Therefore, the gradient of potential i from $A \rightarrow B$ is

$$i = \frac{\psi_A - \psi_B}{z_{\text{sat}}} = -1 - \frac{s_{\text{aev}}}{\gamma_w z_{\text{sat}}}.$$

Furthermore, considering Darcy's law for water flowing in saturated soils, the infiltration flow is $q_w = -k_w i$ then

$$q_w = -k_w i = k_w \left(1 + \frac{s_{\text{aev}}}{\gamma_w z_{\text{sat}}} \right).$$

Moreover, the conservation condition for the mass of water implies that the infiltration volume through the surface must be equal to the volume required to fill the pores of the soil in the saturation front. Therefore, for a porosity n, this continuity equation is

$$q_w dt = n dz_{\text{sat}},$$

where t is the time.

Consequently, the continuity equation together with Darcy's equation is

$$k_w \left(1 + \frac{s_{\text{aev}}}{\gamma_w z_{\text{sat}}} \right) dt = n dz_{\text{sat}}.$$

Subsequently, reordering the variables leads to

$$\frac{dz_{\text{sat}}}{1 + \frac{s_{\text{aev}}}{\gamma_w z_{\text{sat}}}} = \frac{k_w}{n} dt.$$

The integration of both sides of this equation from $(0, z_{\text{sat}})$ and $(0, t)$ is

$$\int_0^{z_{\text{sat}}} \frac{dz_{\text{sat}}}{1 + \frac{s_{\text{aev}}}{\gamma_w z_{\text{sat}}}} = \frac{k_w}{n} \int_0^t dt.$$

Resulting in the following equation:

$$z_{\text{sat}} - \frac{s_{\text{aev}}}{\gamma_w} \ln \left(z_{\text{sat}} + \frac{s_{\text{aev}}}{\gamma_w} \right) + \frac{s_{\text{aev}}}{\gamma_w} \ln \frac{s_{\text{aev}}}{\gamma_w} = \frac{k_w}{n} t.$$

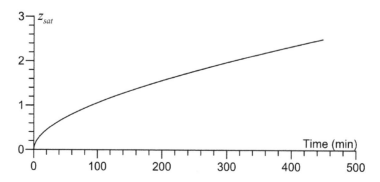

Figure 2.6 Progression over time of the infiltration front in a soil whose characteristics are $n = 0.4$, $k_w = 10^{-5} m/s$, and $s_{aev} = 30 kPa$.

Finally, the expression that gives the time required by the saturation front to reach a particular depth z_{sat} is

$$t = \frac{n}{k_w} \left(z_{sat} - \frac{s_{aev}}{\gamma_w} \ln \frac{z_{sat} + s_{aev}/\gamma_w}{s_{aev}/\gamma_w} \right).$$

When using the numerical values given for this the example which are $n = 0.4$, $k_w = 10^{-5}$ m/s, and $s_{aev} = 30$ kPa, the progression of the saturation front is given by

$$t = \frac{0.4}{10^{-5}} \left(z_{sat} - \frac{30}{1,000 \cdot 9.8} \ln \frac{z_{sat} + 30/(1,000 \cdot 9.8)}{30/(1,000 \cdot 9.8)} \right).$$

Figure 2.6 shows the results obtained with the above equation, and this figure illustrates how the advance velocity of the saturation front decreases as it goes deeper into the soil.

2.5 EXAMPLE 10: NUMERICAL CALCULATION OF WATER FLOW IN UNSATURATED MATERIALS, APPLICATION TO ROAD STRUCTURES

The purpose of this example is to simulate numerically the infiltration of water into a pavement structure, assuming that the water rests over the pavement for twelve hours. For this purpose, the example uses the explicit scheme of the FDM.
    The road structure is embedded in the subgrade without having a drainage system, the width of this road structure is 7.0 m, while the thicknesses of the layers are

- *the first layer of bituminous material 8.75 cm thick,*
- *a granular base of 20 cm thick,*
- *a granular sub-base of 30 cm thick, and*
- *the subgrade ∞ thick.*

Furthermore, the bituminous layer has a set of regular longitudinal cracks. These cracks can be schematized as longitudinal discontinuities in the asphalt layer, their width is 2 mm, the separation between them is 1 m, and they are filled with fine silty sand. Also, the longitudinal and transverse slopes of the pavement structure are close to zero and, therefore, when it rains, a thin layer of water rests on its surface and infiltrates the pavement structure.

Regarding the properties of the materials:

- *Table 2.6 describes the properties of the materials involved in the problem, their water retention curve is estimated using the empirical correlation suggested in the MEPDG, and their unsaturated hydraulic conductivity is assessed using the relationship $k_w = k_{sat} S_r^p$.*
- *Table 2.7 shows the experimental results of the water retention curve of the asphalt material given in Refs. [49,50]. In this example, these data are adjusted using the model proposed by Fredlund and Xing [30].*

Moreover, it is assumed as an initial condition that the entire structure is in equilibrium with the water table, which is at a depth of 7.5 m below the ground surface.

First of all, this example explains in a general way the use of the finite difference method, whose acronym is FDM, to solve the partial differential equation (PDE) that describes the flow of water in unsaturated materials.

Subsequently, the example uses the explicit FDM to analyze the flow of water.

Table 2.6 Properties of the materials of the road structure

Material	Water Retention Curve				Conductivity k_w	
	P_{200}	PI (%)	d_{60} (mm)	θ_{sat}	k_{sat} (m/s)	p
Asphalt layer				0.07	$2 \cdot 10^{-8}$	3
Cracks (fine sand)			0.1	0.50	$1 \cdot 10^{-4}$	3
Granular base			1.0	0.20	$5 \cdot 10^{-3}$	4
Granular sub-base	0.1	6		0.20	$1 \cdot 10^{-5}$	4
Subgrade	0.2	20		0.40	$5 \cdot 10^{-7}$	3

Table 2.7 Experimental data for the water retention curve of an asphalt material

Suction (MPa)	Sr
342	0.18
148	0.36
102	0.53
55	0.58
27	0.62
5	0.82

Source: Given in Refs. [49,50].

This problem is developed in two parts:

- **Part A:** describes, in general terms, the use of the explicit FDM to transform the PDE describing the flow of water into a system of linear equations.
- **Part B:** applies the explicit FDM, explained in Part A, for computing numerically the infiltration of water in the road structure of the example.

2.5.1   Part A. Numerical solution of the nonlinear partial differential equation describing the flow of water in unsaturated soils using the explicit finite difference method

As mentioned earlier, the flow of water in unsaturated materials that does not undergo volume changes during infiltration (*i.e.,* without expansion or collapse) is described by the following conservation equation:

$$\frac{\partial \theta_w}{\partial t} = -\nabla \cdot q_w, \tag{2.20}$$

where θ_w is the volumetric water content, t is the time, and q_w is the flux of water given by the Darcy equation, $q_w = -k_w \nabla \psi$, ψ being the potential of water. Therefore, the continuity equation for the flow of water becomes

$$\frac{\partial \theta_w}{\partial t} = \nabla \cdot (k_w \nabla \psi). \tag{2.21}$$

However, since the volumetric water content and the unsaturated hydraulic conductivity change in a nonlinear way depending on the change of suction, Equation 2.21 is a nonlinear differential equation whose solution is only achievable by using numerical methods.

Several numerical methods, for instance, the finite element method (FDM) or others are well adapted for solving Equation 2.21. Among them, the FDM is a straightforward technique for solving diffusion problems, mainly when the problem can be discretized using rectangular grids.

The FDM allows transforming the partial differential Equation 2.21 into a system of linear equations that can be solved more easily. This transformation is possible by following the steps:

1   Implement the space discretization (*i.e.,* the right side of Equation 2.21).
2   Implement the time discretization (*i.e.,* the left side of Equation 2.21).
3   Consider the boundary conditions.
4   Consider the initial values of the variables.
5   Implement the solution using the explicit FDM.

Although the problem developed in part B of this example does not include a drainage system, the description of the method presented here is made in a general way. Then, it uses Figure 2.7, which illustrates the case of a road that has a drainage system and undergoes water infiltration through its surface.

The FDM allows the problem to be described in full detail. Nevertheless, due to the symmetry of the problem, only the left block of Figure 2.7 is considered here for a simplified description of the method.

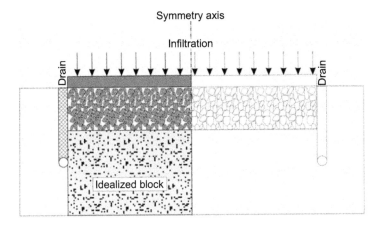

Figure 2.7 Schematic layout of a road with a drainage system.

2.5.1.1 Discretization in space

The right side of Equation 2.20, which is $-\nabla \cdot q_w$, represents the divergence operator of the flux of water q_w. This operator performs the balance of the water flux at a point within the analysis domain and then distributes it to the area or volume around that point.

Likewise, the meaning of the space discretization is transforming the continuous domain into a grid that has a finite number of nodes. Therefore, the space discretization consists of transforming the divergence operator, which applies to a continuum, into a set of linear equations that apply to the discrete domain made by the set of nodes of the grid.

To sum up, the discretization in space is easily achievable by creating a grid into the domain of analysis, which is a rectangular grid in this example, and calculating the balance of the water flux in each point of this domain.

Figure 2.8a exemplifies how a grid allows the space discretization, and then, for a node (i, j), transforming the right side of Equation 2.21 into a discrete form requires the following steps:

First, it is possible to calculate the flow of water in each face of the element of the grid (i, j) as

- The flow through the face AB of Figure 2.8b is $Q_{AB} = \Delta z q_w^{i,j-1/2}$.
- The flow through the face CD of Figure 2.8b is $Q_{CD} = \Delta z q_w^{i,j+1/2}$.
- The flow through the face AD of Figure 2.8b is $Q_{AD} = \Delta x q_w^{i-1/2,j}$.
- The flow through the face BC of Figure 2.8b is $Q_{BC} = \Delta x q_w^{i+1/2,j}$.

In the same way, it is possible to use Darcy's equation to find out the fluxes q_w as

$$q_w^{i,j-1/2} = k_w^{i,j-1/2} \frac{\psi_{i,j-1} - \psi_{i,j}}{\Delta x}, \tag{2.22}$$

$$q_w^{i,j+1/2} = k_w^{i,j+1/2} \frac{\psi_{i,j} - \psi_{i,j+1}}{\Delta x}, \tag{2.23}$$

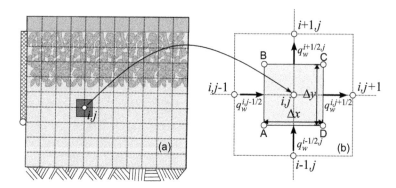

Figure 2.8 Discretization in space of Equation 2.21, (a) grid for the whole model, (b) close-up view of a node in the grid and its neighboring nodes.

$$q_w^{i-1/2,j} = k_w^{i-1/2,j} \frac{\psi_{i-1,j} - \psi_{i,j}}{\Delta z}, \tag{2.24}$$

$$q_w^{i+1/2,j} = k_w^{i+1/2,j} \frac{\psi_{i,j} - \psi_{i+1,j}}{\Delta z}, \tag{2.25}$$

where $k_w^{i,j-1/2}$ represents the mean hydraulic conductivity between points (i,j) and $(i, j-1)$. This mean conductivity can be evaluated using the solution for water conductivity when two materials are in series. However, the following equation is more stable in a numerical solution:

$$k_w^{i,j-1/2} = \sqrt{k_w^{i,j-1} \cdot k_w^{i,j}}.$$

Subsequently, since the divergence represents the net flux per unit of area or volume, then the divergence of q_w in a discrete form is

$$\nabla \cdot q_w \approx \frac{(Q_{AB} - Q_{CD}) + (Q_{AD} - Q_{BC})}{\Delta x \Delta z}. \tag{2.26}$$

Finally, from Equations 2.22 to 2.26, the right side of Equation 2.21 becomes

$$\nabla \cdot k_w \nabla \psi \approx \frac{k_w^{i,j-1/2}}{\Delta x^2} \psi_{i,j-1} + \frac{k_w^{i,j+1/2}}{\Delta x^2} \psi_{i,j+1} + \frac{k_w^{i-1/2,j}}{\Delta z^2} \psi_{i-1,j} + \frac{k_w^{i+1/2,j}}{\Delta z^2} \psi_{i+1,j}$$

$$- \psi_{i,j} \left(\frac{k_w^{i,j-1/2}}{\Delta x^2} + \frac{k_w^{i,j+1/2}}{\Delta x^2} + \frac{k_w^{i-1/2,j}}{\Delta z^2} + \frac{k_w^{i+1/2,j}}{\Delta z^2} \right). \tag{2.27}$$

2.5.1.2 Discretization in time

The left side of Equation 2.21 represents the evolution of the volume of water over time. Since the volumetric water content, θ_w, is given by the product of the porosity by the degree of saturation (*i.e.,* $\theta_w = nS_r$), the change in volumetric water content over time is

$$\frac{\partial \theta_w}{\partial t} = \frac{\partial nS_r}{\partial t}.$$

However, when considering a soil whose volume is constant, it implies that the porosity is also constant with time, then

$$\frac{\partial \theta_w}{\partial t} = n\frac{\partial S_r}{\partial t},$$

becoming

$$\frac{\partial \theta_w}{\partial t} = n\frac{\partial S_r}{\partial u_w}\frac{\partial u_w}{\partial t}.$$

Now, introducing the specific water capacity, C_θ, defined as $C_\theta = n\frac{\partial S_r}{\partial u_w}$, leads to

$$\frac{\partial \theta_w}{\partial t} = C_\theta \frac{\partial u_w}{\partial t}.$$

In the same way as the discretization in space, the discretization in time consists in transforming the derivative of the continuum over time into a discrete derivative which is

$$C_\theta \frac{\partial u_w}{\partial t} \approx C_\theta \frac{\psi_{i,j}^{t+\Delta t} - \psi_{i,j}^{t}}{\Delta t}, \tag{2.28}$$

where Δt is the time lapse of discretization. Also, keep in mind that, in a non-compressible material, z is constant in time, so $\frac{\partial u_w}{\partial t} = \frac{\partial \psi}{\partial t}$.

2.5.1.3   Implementation of the explicit Finite Difference Method

To sum up, Equation 2.28 must be equated with minus Equation 2.27 (because $\frac{\partial \theta_w}{\partial t} = -\nabla \cdot q_w$). Moreover, this equalization requires deciding whether the right side of Equation 2.21, which in discrete form is represented by Equation 2.27, must be evaluated at time t or time $t + \Delta t$. Both schemes are possible:

• The explicit solution evaluates the balance of water using potentials at time t for computing the potential of the node (i, j) at time $t + \Delta t$. The numerical implementation of this scheme is straightforward, but sometimes it requires very small Δt to avoid instabilities of the solution.
• In contrast, the implicit solution evaluates the balance of water using potentials at time $t + \Delta t$ for computing the potential of the node (i, j) at time $t + \Delta t$. The numerical implementation of this scheme requires solving a large system of simultaneous equations, but the solution is more stable than the explicit solution.

Both solutions permit to simulate the flow of water in unsaturated soils numerically. However, this example describes in detail the explicit solution, while the next example explains the implicit solution applied for the flow of heat in road structures.

To implement the explicit solution, it is first possible to introduce the variables $\bar{s}$ defined as

$$\bar{s}_{i,j-1} = -\frac{k_w^{i,j-1/2}\Delta t}{C_\theta \Delta x^2}, \qquad \bar{s}_{i,j+1} = -\frac{k_w^{i,j+1/2}\Delta t}{C_\theta \Delta x^2},$$

$$\bar{s}_{i-1,j} = -\frac{k_w^{i-1/2,j}\Delta t}{C_\theta \Delta z^2}, \qquad \bar{s}_{i+1,j} = -\frac{k_w^{i+1/2,j}\Delta t}{C_\theta \Delta z^2}, \text{ and}$$

$$\bar{s}_{i,j} = \bar{s}_{i,j-1} + \bar{s}_{i,j+1} + \bar{s}_{i-1,j} + \bar{s}_{i+1,j}$$

Then, the recursive equation which permits computing the evolution of water potential using the explicit solution is

$$\psi_{i,j}^{t+\Delta t} = \bar{s}_{i,j-1}\psi_{i,j-1}^{t} + \bar{s}_{i,j+1}\psi_{i,j+1}^{t} + \bar{s}_{i-1,j}\psi_{i-1,j}^{t} + \bar{s}_{i+1,j}\psi_{i+1,j}^{t} - \psi_{i,j}^{t}(\bar{s}_{i,j} - 1). \tag{2.29}$$

As C_θ is in the denominator of the expressions giving $\bar{s}$, then, when C_θ is small or tends to zero, Equation 2.29 produces significant increases in water potential, which later produce instabilities of the solution. Therefore, the stability of the explicit solution requires choosing time-lapses lower than

$$\Delta t < \frac{1}{2}\frac{C_\theta \min(\Delta x^2, \Delta z^2)}{\max k_w}. \tag{2.30}$$

Equation 2.29 applies for the unsaturated state because, when the soil is saturated and does not undergo volumetric changes, the change in volumetric water content over time is zero and therefore the balance of water entering into an elementary volume given by the divergence operator is $\nabla \cdot q_w = 0$. Therefore, Equation 2.27 must be equal to zero, leading to the following equation that applies to the saturated state:

$$\psi_{i,j} = \frac{\frac{k_w^{i,j-1/2}}{\Delta x^2}\psi_{i,j-1} + \frac{k_w^{i,j+1/2}}{\Delta x^2}\psi_{i,j+1} + \frac{k_w^{i-1/2,j}}{\Delta z^2}\psi_{i-1,j} + \frac{k_w^{i+1/2,j}}{\Delta z^2}\psi_{i+1,j}}{\frac{k_w^{i,j-1/2}}{\Delta x^2} + \frac{k_w^{i,j+1/2}}{\Delta x^2} + \frac{k_w^{i-1/2,j}}{\Delta z^2} + \frac{k_w^{i+1/2,j}}{\Delta z^2}}.$$

2.5.1.4 Boundary conditions

The continuity equation, which is developed in Equation 2.29, applies only to the internal nodes of the body. In fact, the potential of the nodes on the contour of the body depends on the boundary conditions.

Three types of boundary conditions are useful for analyzing the problem of water flow in unsaturated materials:

- The "Dirichlet" boundary condition corresponds to a condition of imposed potential.
- The "Neumann" boundary condition corresponds to an imposed flux through the contour of the body.
- The "drainage" boundary condition applies the "Dirichlet" or "Neumann" condition depending on the value of the pore pressure in the body contour.

Since for the Dirichlet condition, the potential is imposed on the contour, its calculation at the time step $t + \Delta t$ is not necessary. So, for nodes where this condition applies, the equation is

$$\psi_{i,j}^{t+\Delta t} = \psi_{\text{Imposed}}.$$

On the other hand, the Neumann boundary condition imposes the flux of water on a portion of the body contour. In other words, the potential gradient $\frac{\partial \psi}{\partial n}$ is known at the points where the flux is imposed ($\vec{n}$ being the direction of the flux vector in those points).

This condition requires the use of a set of imaginary nodes, as shown in Figure 2.9. For example, when the Neumann condition is imposed on the top of the body, the equation linking the potential of the boundary nodes and the gradient is

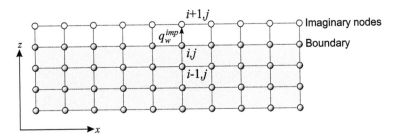

Figure 2.9 Imaginary nodes to analyze the Neumann boundary condition when imposed on the top of the body.

$$\psi^t_{i-1,j} = \psi^t_{i+1,j} + 2\Delta z \left(\frac{\partial \psi}{\partial z} \right)^t.$$

In the same way, considering Darcy's law, the gradient of potential is

$$\left(\frac{\partial \psi}{\partial z} \right) = -\frac{q_w^{\text{Imposed}}}{k_w^{1+1/2,j}}.$$

Therefore, the potential of the imaginary nodes is

$$\psi^t_{i+1,j} = \psi^t_{i-1,j} - 2\Delta z \frac{q_w^{\text{Imposed}}}{k_w^{1-1/2,j}}. \tag{2.31}$$

Although Equation 2.31 corresponds to a Neumann boundary at the upper boundary of the grid, the same analysis is applicable for the other faces.

Another type of boundary condition, which is useful to study the flow of water, is the drainage condition. This particular case requires combining the conditions of imposed potential and imposed flux, a combination which depends on the following considerations:

1   When the contour of the model is in contact with a drain system or directly in contact with the atmosphere, the pore pressure never undertakes positive values, so $u_w \leq 0$.
2   When the soil on this contour (*i.e.*, the drain system or the atmosphere) is unsaturated, the flux of water in the liquid phase is zero, so $q_w = 0$. Nevertheless, some water flux can occur in vapor phase, but it is not considered in this example.

These two conditions are schematized in Figure 2.10.

Because this particular boundary condition alternates between the imposed potential or imposed flow conditions, to decide which condition should be applied, the following iterative procedure needs to be applied to all nodes of the drainage boundary:

• First, the potential is computed assuming a condition of zero water flow on all nodes where the drainage condition is imposed.
• From the results of this computation, the pore water pressure is calculated and

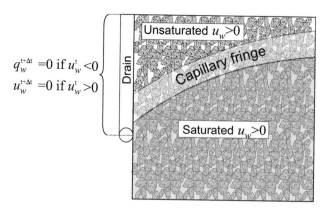

Figure 2.10 Combined boundary conditions for the drain face.

- If the pore pressure is negative, then the assumption of imposing zero flux on the node is correct.
- If the calculations lead to positive pore water pressures, then the boundary condition of the node must shift to a condition of imposed potential (*i.e.*, $u_w = 0$).

2.5.1.5   Initial conditions

Computing the potential for the first time step $\psi_{i,j}^{\Delta t}$ requires the knowledge of the initial value of potential $\psi_{i,j}^0$ in all nodes of the body.

Any method for imposing the initial conditions is possible. However, the two usual ways are

- Set the water content in all nodes; it permits to calculate the pore water pressure using the water retention curve and then to calculate the potential as $\psi = z + u_w/\gamma_w$.
- Set the potential directly, for example, assuming a hydrostatic equilibrium with the water table.

2.5.2   Part B. Numerical solution using the data of the example

After describing the explicit FDM, this part solves the water flow within the road structure defined at the beginning of the example.

Solving the problem numerically using the explicit scheme of the FDM requires the following steps:

1   Assess the water retention curve and the water capacity using the Fredlund and Xing model and the correlations proposed in the MEPDG.
2   Discretize the problem in space.
3   Discretize the problem in time and choose the proper time step.
4   Consider the initial and boundary conditions.
5   Obtain recurrent equations to find out the evolution over time of the water potential throughout the model.

2.5.2.1 *Water retention curves*

Similar to Example 7, the correlations proposed for the MEPDG, described in Section 2.1.1, allow obtaining the set of parameters required to assess the water retention curve (WRC) using the Fredlund and Xing model. These parameters are shown in Table 2.8.

Therefore, using Equation 2.1, the relationships giving the water retention curves of the materials (excepting the bituminous material) are

- for the fine sand filling the cracks,

$$S_r = \left[1 - \frac{\ln(1 + s/17.51)}{\ln(1 + 10^6/17.51)}\right] \frac{1}{\left\{\ln\left[\exp(1) + (s/4.86)^{7.50}\right]\right\}^{0.37}}, \qquad (2.32)$$

- for the granular base,

$$S_r = \left[1 - \frac{\ln(1 + s/0.73)}{\ln(1 + 10^6/0.73)}\right] \frac{1}{\left\{\ln\left[\exp(1) + (s/0.86)^{7.5}\right]\right\}^{0.77}}, \qquad (2.33)$$

- for the granular sub-base,

$$S_r = \left[1 - \frac{\ln(1 + s/436.60)}{\ln(1 + 10^6/436.60)}\right] \frac{1}{\left\{\ln\left[\exp(1) + (s/13.40)^{1.54}\right]\right\}^{0.54}}, \text{ and} \qquad (2.34)$$

- for the subgrade,

$$S_r = \left[1 - \frac{\ln(1 + s/956.76)}{\ln(1 + 10^6/956.76)}\right] \frac{1}{\left\{\ln\left[\exp(1) + (s/27.38)^{1.31}\right]\right\}^{0.60}}. \qquad (2.35)$$

Regarding the water retention curve of bituminous materials, the results are scarce. Still, the data in the example, taken from Refs. [49,50], allow a water retention curve to be used for the bituminous layer.

Adjusting the Fredlund and Xing model for the WRC is an arduous process because it requires adjusting four independent parameters. However, as the pairs suction-saturation given in this example for bituminous material resemble the WRC of a high-plasticity soil, an alternative method consists in trying to fit the experimental data using

Table 2.8 Parameters of the water retention curve of the materials involved in the example, shaded column gives the parameters of the bituminous material.

Parameters	Material				
	Cracks	Bituminous	Base	Sub-base	Subgrade
PI				6	20
P_{200}				0.1	0.2
$P_{200}PI$		80		0.6	4
d_{60}	0.10		1.00		
a	4.86	8969.79	0.86	13.40	27.38
n	7.50	0.65	7.50	1.54	1.31
m	0.37	0.89	0.77	0.54	0.60
s_{res} (kPa)	17.51	$1.29 \cdot 10^6$	0.73	439.60	956.76

the correlations given for fine soils. In other words, it consists in searching the value of $P_{200}PI$ that leads to the lower error in the adjustment.

The mean square error (MSE), that is, the average squared difference between estimated and experimental values, permits to evaluate the error of the fitting. The MSE is defined as

$$MSE = \frac{1}{n} \sum_{i=1}^{n} \left(S_r^{exp} - S_r^{Model} \right)^2,$$

where n is the number of measurements, while S_r^{exp} and S_r^{Model} are the measured and predicted degrees of saturation. Figure 2.11 shows that a value of $P_{200}PI = 80$ leads to the lowest error, and then, the model parameters given in the shaded column of the Table 2.8, which are obtained using the correlations for fine soils, lead to the following equation for the water retention curve of the bituminous material:

$$S_r = \left[1 - \frac{\ln(1 + s/1.26 \cdot 10^6)}{\ln(1 + 10^6/1.26 \cdot 10^6)} \right] \frac{1}{\left\{ \ln \left[\exp(1) + (s/8969.79)^{0.65} \right] \right\}^{0.89}}. \tag{2.36}$$

This equation allows a very good agreement between the experimental and predicted results, as it is shown in Table 2.9.

Finally, Figure 2.12 shows the water retention curves of all the materials involved in the example.

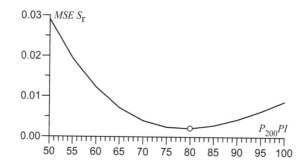

Figure 2.11   Mean square error of the model for different values of $P_{200}PI$.

Table 2.9   Degrees of saturation of the bituminous material measured experimentally and obtained using $P_{200}PI = 80$.

s (kPa)	S_r^{Model}	S_r^{Exp}
342,000	0.25	0.18
148,000	0.40	0.36
102,000	0.46	0.53
55,000	0.55	0.58
27,000	0.65	0.62
5,000	0.83	0.82

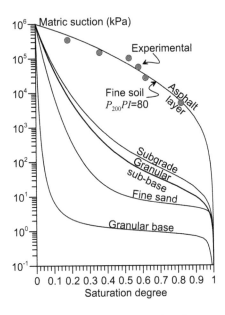

Figure 2.12 Water retention curves of the materials involved in the example.

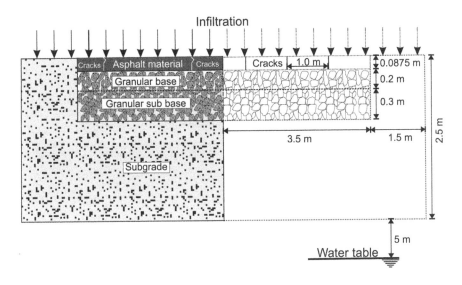

Figure 2.13 Layout of the pavement structure of the example.

2.5.2.2 Discretization in space

As it was already described in part A of this example, the purpose of the space discretization is to transform the continuum model into a discrete grid that enables the use of the FDM. Such a grid is easily obtained based on the layout shown in Figure 2.13.

The first consideration that is useful to create a simplified grid is to identify the plane of symmetry, which divides the road into two half problems with identical behavior.

Therefore, it is only necessary to simulate the flow of water in one half of the model, and adopting a boundary condition of zero flow through the symmetry plane. The discretized space is schematized in Figure 2.14; in this discrete model, the distance between nodes is 0.1 m in the horizontal direction and 0.025 m in the vertical direction (*i.e.,* $\Delta x = 0.1$ m, $\Delta z = 0.0.025$ m). As a result, the model has 101 and 51 nodes toward the horizontal and vertical directions, respectively.

Nevertheless, since the cracks have a 2 mm gap, rendering these cracks in full detail could require a very high number of nodes. On the other hand, since in this example, the distance between the nodes is 0.1 m, to represent the cracks, it is necessary to create a hybrid material that has 0.098 m of bituminous material and 0.002 m of fine sand filling the cracks. The average hydraulic conductivity of such hybrid material, $k_{w_{av}}$, can be calculated using the relationship that averages the hydraulic conductivity when two materials are in parallel to the water flow, that is,

$$k_{w_{av}} = \frac{1}{0.1}(0.002k_{w_{\text{sand}}} + 0.098k_{w_{\text{bitu}}})$$

$$= \frac{1}{0.1}(1 \cdot 10^{-4} \cdot 0.002 + 2 \cdot 10^{-8} \cdot 0.098) = 2.02 \cdot 10^{-6} \text{ m/s.}$$

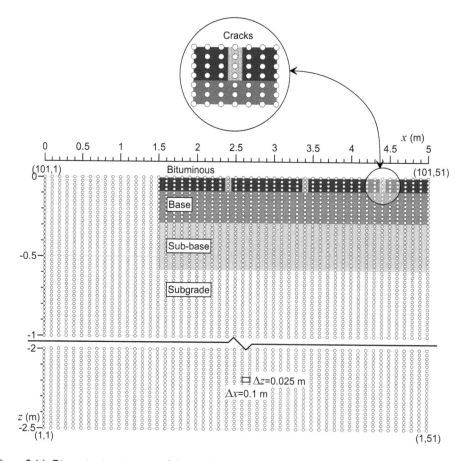

Figure 2.14 Discretization in space of the road structure.

It is also possible to obtain a water retention curve for this hybrid material by adding the volume of water of each material for all suction pressures and then dividing this sum by the volume of voids in the hybrid material. However, in this example, this refinement is unnecessary because both materials reach a saturation degree of 100% in the early stages of infiltration. Therefore, this hybrid material behaves like a saturated material during most of the simulation.

To sum up, the model has $51 \times 101 = 5,151$ nodes whose locations are

- Bottom face,   nodes $(1, 1, ..., 51)$.
- Top face,   nodes $(101, 1, ..., 51)$.
- Left face,   nodes $(1, ..., 101, 1)$.
- Right face,   nodes $(1, ..., 101, 51)$.

Regarding materials, their locations are

- Bituminous material,   nodes $(98, ..., 101, 25)$, $(98..101, 35)$, $(98..101, 45)$.
- Cracks,   nodes $(98, ..., 101, 16, ..., 24)$, $(98, ..., 101, 26, ..., 34)$,
-                             $(98, ..., 101, 36, ..., 44)$, $(98, ..., 101, 46, ..., 51)$.
- Granular base,   nodes $(90, ..., 97, 16, ..., 51)$.
- Granular sub-base,   nodes $(78, ..., 96, 16, ..., 51)$.
- Subgrade,   nodes $(1, ..., 101, 1, ..., 15)$, $(1, ..., 77, 16, ..., 51)$.

2.5.2.3 Discretization in time

As it was remarked earlier, the explicit solution could lead to instabilities. Then, it is necessary to choose a time step in agreement with the inequality given in Equation 2.30. Since the space discretization described in the previous section uses a smaller length in the vertical direction, the requirement for the time step is

$$\Delta t < \frac{1}{2} \frac{C_\theta(S_r)\Delta z^2}{k_w(S_r)}. \tag{2.37}$$

However, when water flows in an unsaturated material, both C_θ and k_w change depending on the saturation degree. Therefore, it is necessary to verify the stability condition for all materials and all the range of saturation degrees.

First, the verification requires to calculate the unsaturated hydraulic conductivity of all materials, which is calculated using the relationship $k_w(S_r) = k_{\text{sat}}S_r^p$ using the data given in Table 2.6 for all materials, excepting the cracks for which the saturated hydraulic conductivity was calculated previously for the system in parallel. Then, the expressions giving the unsaturated hydraulic conductivity are

- for the bituminous material,

$$k_w(S_r) = 2 \cdot 10^{-8} S_r^3 \text{ m/s}, \tag{2.38}$$

- for the fine sand filling the cracks,

$$k_w(S_r) = 2.02 \cdot 10^{-6} S_r^3 \text{ m/s}, \tag{2.39}$$

- for the granular base,

$$k_w(S_r) = 5 \cdot 10^{-3} S_r^4 \text{ m/s},$$ (2.40)

- for the granular sub-base,

$$k_w(S_r) = 1 \cdot 10^{-3} S_r^4 \text{ m/s},$$ (2.41)

- for the subgrade,

$$k_w(S_r) = 5 \cdot 10^{-7} S_r^3 \text{ m/s}.$$ (2.42)

Figure 2.15 shows the evolution of the unsaturated hydraulic conductivity of all the materials involved in the example.

Subsequently, the unsaturated water capacity, denoted as $C_\theta = \theta'(s)$, is the derivative of the volumetric water content regarding suction which, for the water retention curve proposed in Ref. [30], is obtained combining Equations 2.1, 2.11, and 2.12 and leads to

$$C_\theta = \frac{-1}{(s_{\text{res}} + s) \ln \left(1 + \frac{10^6}{s_{\text{res}}}\right)} \frac{\theta_{\text{sat}}}{\{\ln [e + (s/a)^n]\}^m}$$

$$- \left[1 - \frac{\ln(1 + s/s_{\text{res}})}{\ln(1 + 10^6/s_{\text{res}})}\right] \frac{\theta_{\text{sat}}}{\{\ln [e + (s/a)^n]\}^{m+1}} \frac{mn \left(\frac{s}{a}\right)^{n-1}}{a [e + (s/a)^n]}.$$

Then, using the parameters given in Table 2.6, the expressions for C_θ for the materials of the example are

- bituminous material

$$C_\theta = \frac{-1}{(1.29 \cdot 10^6 + s) \ln \left(1 + \frac{10^6}{1.29 \cdot 10^6}\right)} \frac{0.07}{\{\ln [e + (s/8969.7)^{0.65}]\}^{0.89}}$$

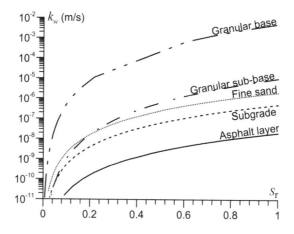

Figure 2.15   Unsaturated hydraulic conductivity of the materials involved in the example.

$$-\left[1-\frac{\ln(1+s/1.29\cdot10^6)}{\ln(1+10^6/1.29\cdot10^6)}\right]\frac{0.07}{\left\{\ln\left[e+(s/8969.79)^{0.65}\right]\right\}^{1.89}}$$

$$\cdot\frac{0.5785\left(\frac{s}{8969.7}\right)^{-0.35}}{8969.7\left[e+(s/8969.7)^{0.65}\right]},\tag{2.43}$$

- cracks,

$$C_\theta=\frac{-1}{(17.51+s)\ln\left(1+\frac{10^6}{17.51}\right)}\frac{0.5}{\left\{\ln\left[e+(s/4.86)^{7.5}\right]\right\}^{0.37}}$$

$$-\left[1-\frac{\ln(1+s/17.51)}{\ln(1+10^6/17.51)}\right]\frac{0.5}{\left\{\ln\left[e+(s/4.86)^{7.5}\right]\right\}^{1.37}}$$

$$\cdot\frac{2.775\left(\frac{s}{4.86}\right)^{6.5}}{4.86\left[e+(s/4.86)^{7.5}\right]},\tag{2.44}$$

- granular base

$$C_\theta=\frac{-1}{(0.73+s)\ln\left(1+\frac{10^6}{0.73}\right)}\frac{0.2}{\left\{\ln\left[e+(s/0.86)^{7.5}\right]\right\}^{0.77}}$$

$$-\left[1-\frac{\ln(1+s/0.73)}{\ln(1+10^6/0.73)}\right]\frac{0.2}{\left\{\ln\left[e+(s/0.86)^{7.5}\right]\right\}^{1.77}}$$

$$\cdot\frac{5.775\left(\frac{s}{0.86}\right)^{6.5}}{0.86\left[e+(s/0.86)^{7.5}\right]},\tag{2.45}$$

- granular sub-base

$$C_\theta=\frac{-1}{(439.6+s)\ln\left(1+\frac{10^6}{439.6}\right)}\frac{0.2}{\left\{\ln\left[e+(s/13.4)^{1.54}\right]\right\}^{0.54}}$$

$$-\left[1-\frac{\ln(1+s/439.6)}{\ln(1+10^6/439.6)}\right]\frac{0.2}{\left\{\ln\left[e+(s/13.4)^{1.54}\right]\right\}^{1.54}}$$

$$\cdot\frac{0.8316\left(\frac{s}{13.4}\right)^{0.54}}{13.4\left[e+(s/13.4)^{1.54}\right]},\tag{2.46}$$

- subgrade

$$C_\theta=\frac{-1}{(956.76+s)\ln\left(1+\frac{10^6}{956.76}\right)}\frac{0.4}{\left\{\ln\left[e+(s/27.38)^{1.31}\right]\right\}^{0.6}}$$

$$-\left[1-\frac{\ln(1+s/956.76)}{\ln(1+10^6/956.76)}\right]\frac{0.4}{\left\{\ln\left[e+(s/27.38)^{1.31}\right]\right\}^{1.6}}$$

$$\cdot\frac{0.786\left(\frac{s}{27.38}\right)^{0.31}}{27.38\left[e+(s/27.38)^{1.31}\right]}.\tag{2.47}$$

Figure 2.16 shows the evolution of C_θ for all materials involved in the example.

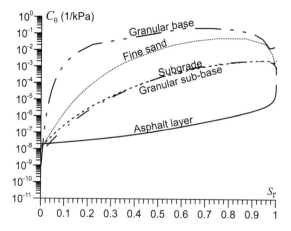

Figure 2.16 Unsaturated water capacity of the materials involved in the example.

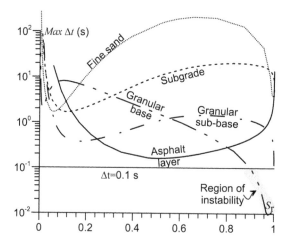

Figure 2.17 Maximum time step allowed for each material when using the fully explicit method.

Finally, combining k_w and C_θ in Equation 2.37 leads to the values of Δt indicated in Figure 2.17.

Note that the granular base requires to use a very small time step, mainly when this material approaches saturation. Nevertheless, instead of using a very short time step, which is an option that certainly has a very high computational cost, this example uses a time step of $\Delta t = 0.1$ s. Then, to cope with the eventual instability, the computational procedure includes a verification of the saturation degree of the granular base switching this material to a saturated state when it approaches the instability.

2.5.2.4 Boundary and initial conditions

Each face of the model has different boundary conditions. First, a thin layer of water on 1 mm is imposed to simulate the infiltration through the top face; therefore, the pore pressure on this face is $u_w = 0.001\gamma_w$. Moreover, on the bottom face, the pore pressure is controlled by the position of the water table, which is 5 m below this face; therefore, the imposed pore pressure is $u_w = -5.0\gamma_w$. Finally, the left and right boundaries have conditions of zero water flow. Thus, in summary, the boundary conditions are

- Top face,   imposed pore pressure $u_w = 0.001\gamma_w$.
- Bottom face,   imposed pore pressure $u_w = -5.0\gamma_w$.
- Right face,   zero flow.
- Left face,   zero flow.

These conditions are schematized in Figure 2.18.

Regarding the initial conditions, it is reasonable to assume that the whole model is in equilibrium with the water table at the beginning of the simulation. For instance, when considering the bottom face, which is at -2.5 m deep, the potential on this face is

$$\psi(t = 0) = -2.5 - \frac{5\gamma_w}{\gamma_w} = -7.5 \text{ m.}$$

This water potential is the same throughout the model because of the assumption of initial equilibrium with the water table.

2.5.2.5 Simulation

After all, the procedure to perform the simulation is as follows:

1   Initialize the water potential at $t = 0$

$$\psi_{(i=1..101, j=1..51)}^{t=0} = -7.5 \text{ m.}$$

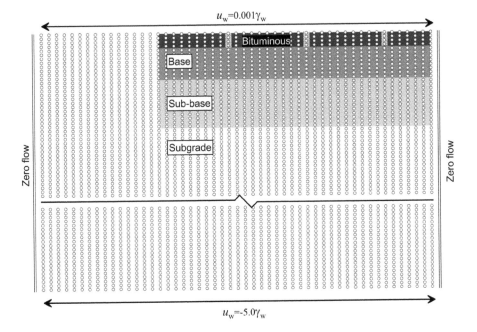

Figure 2.18 Boundary conditions around the model.

2 Compute the nonlinear variables S_r, k_w, and C_θ for all the nodes using the following equations:

- Degree of saturation, Equations 2.32–2.36.
- Unsaturated hydraulic conductivity, Equations 2.38– 2.42.
- Unsaturated water capacity, Equations 2.43–2.47.

3 Verify the saturation state of each node setting the saturation degree to 1 when it is higher than 0.95. This verification intends to avoid instabilities mainly in the granular base.

$$\text{if} \quad S_r^{i,j} > 0.95 \quad \longrightarrow \quad S_r^{i,j} = 1.0, \text{ and } C_\theta^{i,j} = 1.0$$

4 Compute the ancillary variables $\bar{s}$

$$\bar{s}_{i,j-1} = -\frac{10}{C_\theta^{i,j}}\sqrt{k_w^{i,j} k_w^{i,j-1}}, \qquad \bar{s}_{i,j+1} = -\frac{10}{C_\theta^{i,j}}\sqrt{k_w^{i,j} k_w^{i,j+1}},$$

$$\bar{s}_{i-1,j} = -\frac{160}{C_\theta^{i,j}}\sqrt{k_w^{i,j} k_w^{i-1,j}}, \qquad \bar{s}_{i+1,j} = -\frac{160}{C_\theta^{i,j}}\sqrt{k_w^{i,j} k_w^{i+1,j}},$$

$$\bar{s}_{i,j} = \bar{s}_{i,j-1} + \bar{s}_{i,j+1} + \bar{s}_{i-1,j} + \bar{s}_{i+1,j}$$

5 Compute the potential in $t + \Delta t$ in all internal nodes $(i, j) = (2, ..., 100, 2, ..., 50)$ as

$$\bar{s}_{i,j} = \bar{s}_{i,j-1} + \bar{s}_{i,j+1} + \bar{s}_{i-1,j} + \bar{s}_{i+1,j}$$

$$\text{if} \quad S_r^{i,j} < 1.0 \quad \text{then}$$

$$\psi_{i,j}^{t+\Delta t} = \bar{s}_{i,j-1}\psi_{i,j-1}^t + \bar{s}_{i,j+1}\psi_{i,j+1}^t + \bar{s}_{i-1,j}\psi_{i-1,j}^t + \bar{s}_{i+1,j}\psi_{i+1,j}^t - \psi_{i,j}^t(\bar{s}_{i,j} - 1)$$

$$\text{if} \quad S_r^{i,j} = 1.0 \quad \text{then}$$

$$\psi_{i,j}^{t+\Delta t} = \frac{\bar{s}_{i,j-1}\psi_{i,j-1}^t + \bar{s}_{i,j+1}\psi_{i,j+1}^t + \bar{s}_{i-1,j}\psi_{i-1,j}^t + \bar{s}_{i+1,j}\psi_{i+1,j}^t}{\bar{s}_{i,j}}$$

6 Compute the potential in $t + \Delta t$ of the boundary nodes

Left nodes $(i, j) = (1, ..., 101, 1)$

$$\bar{s}_{i,j} = 2\bar{s}_{i,j+1} + \bar{s}_{i-1,j} + \bar{s}_{i+1,j}$$

$$\text{if} \quad S_r^{i,j} < 1.0 \quad \text{then}$$

$$\psi_{i,j}^{t+\Delta t} = 2\bar{s}_{i,j+1}\psi_{i,j+1}^t + \bar{s}_{i-1,j}\psi_{i-1,j}^t + \bar{s}_{i+1,j}\psi_{i+1,j}^t - \psi_{i,j}^t(\bar{s}_{i,j} - 1)$$

$$\text{if} \quad S_r^{i,j} = 1.0 \quad \text{then}$$

$$\psi_{i,j}^{t+\Delta t} = \frac{2\bar{s}_{i,j+1}\psi_{i,j+1}^t + \bar{s}_{i-1,j}\psi_{i-1,j}^t + \bar{s}_{i+1,j}\psi_{i+1,j}^t}{\bar{s}_{i,j}}$$

Right nodes $(i, j) = (1, ..., 101, 51)$

$$\bar{s}_{i,j} = 2\bar{s}_{i,j-1} + \bar{s}_{i-1,j} + \bar{s}_{i+1,j}$$

$$\text{if} \quad S_r^{i,j} < 1.0 \quad \text{then}$$

$$\psi_{i,j}^{t+\Delta t} = 2\bar{s}_{i,j-1}\psi_{i,j-1}^t + \bar{s}_{i-1,j}\psi_{i-1,j}^t + \bar{s}_{i+1,j}\psi_{i+1,j}^t - \psi_{i,j}^t(\bar{s}_{i,j} - 1)$$

if $S_r^{i,j} = 1.0$ then

$$\psi_{i,j}^{t+\Delta t} = \frac{2\bar{s}_{i,j-1}\psi_{i,j-1}^t + \bar{s}_{i-1,j}\psi_{i-1,j}^t + \bar{s}_{i+1,j}\psi_{i+1,j}^t}{\bar{s}_{i,j}}$$

Top nodes $(i,j) = (101, 1, ..., 51)$ $\psi_{i,j}^{t+\Delta t} = 0.001$ m

Bottom nodes $(i,j) = (1, 1, ..., 51)$ $\psi_{i,j}^{t+\Delta t} = -7.5$ m

7   Return to step 2 for calculating the value of the nonlinear variables in the next step of the computation.

The MATLAB script provided with this book uses the procedure described above. Some of the results obtained with this script are presented in Figures 2.19–2.21.

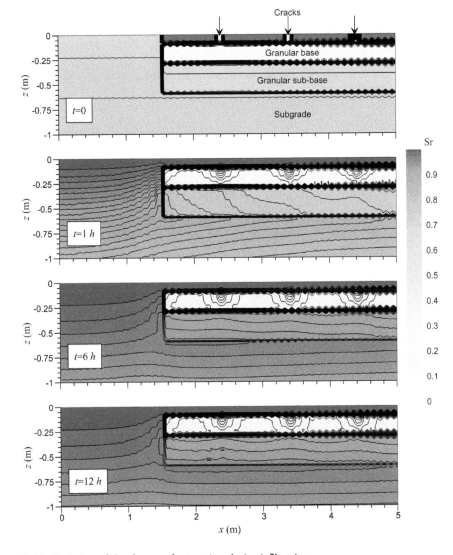

Figure 2.19 Evolution of the degree of saturation during infiltration.

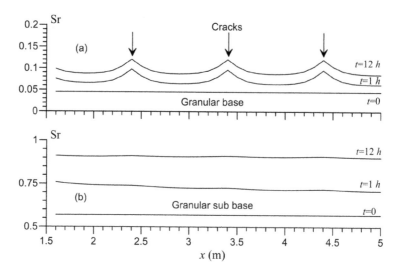

Figure 2.20 Horizontal profiles of the degree of saturation in the granular base and sub-base for different elapsed times.

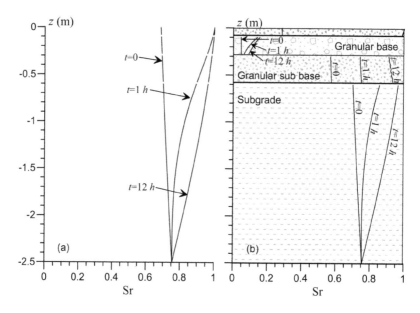

Figure 2.21 Vertical profiles of the degree of saturation in the middle axis of the berm and below a crack.

First, Figure 2.19 shows maps of the degree of saturation for different elapsed times of the simulation. It is interesting to note in this figure how the degree of saturation of the granular base increases below the cracks. Also, the degree of saturation of the granular sub-base increases due to the combined effects of lateral flow of water coming from the berm and some vertical flow of water below the cracks that reach the sub-base after crossing the granular base.

Coupled with the saturation map, the profiles of the degree of saturation in the horizontal direction allow evidencing the significant effect of the cracks in the granular base, as shown in Figure 2.20a, and also the horizontal flow of water in the granular sub-base in Figure 2.20b. This horizontal flow is identifiable because, for $t = 1$ h, the saturation degree in the sub-base shows higher values toward the berm side of the model.

Finally, Figure 2.21 shows the vertical profiles of the degree of saturation in the middle of the berm in Figure 2.21a and below a crack in Figure 2.21b. These figures also give an idea of the increase in the saturation degree due to rain.

2.6 EXAMPLE II: NUMERICAL SOLUTION OF THE HEAT FLOW IN ROAD STRUCTURES

This example describes the use of the implicit scheme of the FDM for calculating the heat propagation within a pavement structure. In this example, the evolution of temperature and the heat flow on the top boundary is imposed, and it is only a first approximation of a complete problem presented in Example 19, in which the heat transfer depends on the environmental variables.

The road's structure is the same as in Example 10 without considering the cracks. Table 2.10 gives the properties of the base, sub-base, and subgrade layers (i.e., dry density ρ_d, water content w, density of solid grains ρ_s, and quartz content q_c), properties which are necessary to assess the thermal properties of those layers. On the other hand, the thermal conductivity of the asphalt layer is $k_H = 1.5$ W/m °C, and its volumetric heat capacity is $c_{H_v} = 2 \cdot 10^6$ J/($m^3 K$).

The temperature within the road structure changes as a result of imposing at the surface of the road and the berm two different types of boundary conditions: imposed heat flow or imposed temperature. These boundary conditions evolve following sinusoidal curves having a period of 24 h whose amplitudes and mean values are

- *Road, the imposed heat flow, which goes to the road structure, is characterized by a mean value of 50 W/m^2 and an amplitude of 300 W/m^2.*

- *Berm, the imposed temperature, is characterized by a mean temperature of 15° C and an amplitude of 10° C.*

Moreover, the temperature imposed at a depth of 2.5 m is 15° C, which is also the initial condition throughout the road's structure, including the subgrade.

The purpose of this example is to simulate the temperature evolution in the pavement structure numerically.

Since the initial conditions influence the result obtained for times near the start of the simulation, in this example, the temperature is calculated for seven days of simulation, and only the results of the last day are analyzed.

Table 2.10 Properties of the materials in the road structure

Material	ρ_d (kg/m^3)	w (%)	ρ_s (kg/m^3)	q_c (%)
Granular base (sandstone)	2,200	6	2,600	25
Granular sub-base (sandstone)	1,900	12	2,600	25
Subgrade (clayey soil)	1,600	24	2,700	0

Similar to the previous example, this problem is developed in two parts:

- **Part A:** describes the use of the implicit FDM for transforming the PDE that describes the heat flow into a set of linear equations.
- **Part B:** describes the numerical solution of the heat flow in the road structure of the example.

2.6.1 Part A. Numerical solution of the diffusion equation using the implicit finite difference method

Similar to water, heat flow is also described by the following diffusion equation:

$$c_{H_v}\frac{\partial T}{\partial t} = -\nabla \cdot q_H,\tag{2.48}$$

where c_{H_v} is the volumetric heat capacity, T is the temperature, t is the time, and q_H is the flux of heat given by the Fourier equation, $q_H = -k_H\nabla T$, k_H being the thermal conductivity. Therefore, the continuity equation for the heat flow becomes

$$c_{H_v}\frac{\partial T}{\partial t} = \nabla \cdot (k_H\nabla T).\tag{2.49}$$

As in the case of water flow, the FDM allows transforming the partial differential Equation 2.49 into a system of linear equations by applying the following steps:

1. Implement the space discretization (*i.e.,* right side of Equation 2.48).
2. Implement the time discretization (*i.e.,* left side of Equation 2.48).
3. Implement the solution using the implicit FDM.
4. Consider the boundary conditions.
5. Consider the initial values of the variables.

2.6.1.1 Discretization in space

In the same way that was analyzed in the previous example, the divergence of q_H in discrete form is

$$\nabla \cdot k_H\nabla T \approx \frac{k_H^{i,j-1/2}}{\Delta x^2}T_{i,j-1} + \frac{k_H^{i,j+1/2}}{\Delta x^2}T_{i,j+1} + \frac{k_H^{i-1/2,j}}{\Delta z^2}T_{i-1,j} + \frac{k_H^{i+1/2,j}}{\Delta z^2}T_{i+1,j}$$
$$- T_{i,j}\left(\frac{k_H^{i,j-1/2}}{\Delta x^2} + \frac{k_H^{i,j+1/2}}{\Delta x^2} + \frac{k_H^{i-1/2,j}}{\Delta z^2} + \frac{k_H^{i+1/2,j}}{\Delta z^2}\right),\tag{2.50}$$

where $k_H^{i,j-1/2}$ represents the thermal conductivity between points i,j and $i,j-1$. This mean conductivity can also be evaluated using the following equation:

$$k_H^{i,j-1/2} = \sqrt{k_H^{i,j-1}\cdot k_H^{i,j}}.$$

However, this method of evaluating average conductivity is approximate. A more rigorous method, which is convenient when there are layers with high contrast in their thermal conductivities, is to include a continuity equation at the interfaces between layers. This method is explained in detail in Example 19.

In a condensed form, Equation 2.50 can be written as

$$\nabla \cdot k_H \nabla T \approx \Sigma k_{Hxy} T^{t+\Delta t} - T_{i,j}^{t+\Delta t} \Sigma k_{Hxy}, \tag{2.51}$$

where

$$\Sigma k_{Hxy} T^{t+\Delta t} = \frac{k_H^{i,j-1/2}}{\Delta x^2} T_{i,j-1}^{t+\Delta t} + \frac{k_H^{i,j+1/2}}{\Delta x^2} T_{i,j+1}^{t+\Delta t} + \frac{k_H^{i-1/2,j}}{\Delta z^2} T_{i-1,j}^{t+\Delta t} + \frac{k_H^{i+1/2,j}}{\Delta z^2} T_{i+1,j}^{t+\Delta t},$$

and

$$\Sigma k_{Hxy} = \frac{k_H^{i,j-1/2}}{\Delta x^2} + \frac{k_H^{i,j+1/2}}{\Delta x^2} + \frac{k_H^{i-1/2,j}}{\Delta z^2} + \frac{k_H^{i+1/2,j}}{\Delta z^2}.$$

2.6.1.2 Discretization in time

The left side of Equation 2.49, which represents the evolution of temperature over time, in discrete form becomes

$$c_{H_v} \frac{\partial T}{\partial t} \approx c_{H_v} \frac{T_{i,j}^{t+\Delta t} - T_{i,j}^t}{\Delta t},$$

where Δt is the time-lapse of the discretization.

2.6.1.3 Implementation of the FDM using the implicit solution

As described earlier, the fully implicit scheme is unconditionally stable. Nevertheless, the numerical solution is less intuitive than the explicit scheme, and also it requires constructing a large system of linear equations.

The following equation presents the development of the continuity equation in a discrete form for a fully implicit scheme:

$$c_{H_v} \frac{T_{i,j}^{t+\Delta t} - T_{i,j}^t}{\Delta t} = \left(\Sigma k_{Hxy} T^{t+\Delta t} - T_{i,j}^{t+\Delta t} \Sigma k_{Hxy} \right),$$

becoming

$$T_{i,j}^t = -\frac{\Delta t}{C_{H_v}} \left(\Sigma k_{Hxy} T^{t+\Delta t} - T_{i,j}^{t+\Delta t} \Sigma k_{Hxy} \right) + T_{i,j}^{t+\Delta t}. \tag{2.52}$$

The construction of the system of linear equations is easier for a rectangular grid, such as in Figure 2.22. Nodes are consecutively numbered in a grid having n_x columns in the x-direction and n_z rows in the z-direction.

For a node (i, j), denoted as the node n, the enumeration of the neighboring nodes are

- Node i, j: node n
- Node $i - 1, j$:   node $n - n_x$
- Node $i, j - 1$:   node $n - 1$
- Node $i, j + 1$:   node $n + 1$
- Node $i + 1, j$:   node $n + n_x$

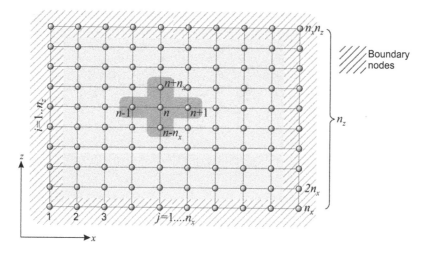

Figure 2.22 Rectangular grid for constructing a matrix equation of the fully implicit scheme.

It is also possible to use the variable $\bar{s}$ defined as

$$\bar{s}_{n,n-1} = -\frac{k_H^{i,j-1/2}\Delta t}{c_{H_v}\Delta x^2}, \qquad \bar{s}_{n,n+1} = -\frac{k_H^{i,j+1/2}\Delta t}{c_{H_v}\Delta x^2},$$

$$\bar{s}_{n,n-n_x} = -\frac{k_H^{i-1/2,j}\Delta t}{c_{H_v}\Delta z^2}, \qquad \bar{s}_{n,n+n_x} = -\frac{k_H^{i+1/2,j}\Delta t}{c_{H_v}\Delta z^2}, \text{ and}$$

$$\bar{s}_n = \bar{s}_{n,n-1} + \bar{s}_{n,n+1} + \bar{s}_{n,n-n_x} + \bar{s}_{n,n+n_x}.$$

Therefore, for the rectangular arrangement of Figure 2.22, Equation 2.52 becomes

$$T_n^t = \bar{s}_{n,n-1}T_{n-1}^{t+\Delta t} + \bar{s}_{n,n+1}T_{n+1}^{t+\Delta t} + \bar{s}_{n,n-n_x}T_{n-n_x}^{t+\Delta t} + \bar{s}_{n,n+n_x}T_{n+n_x}^{t+\Delta t} - T_n^{t+\Delta t}(\bar{s}_n - 1). \quad (2.53)$$

Equation 2.53 leads to a system of linear equations represented in matrix form as

$$
\begin{bmatrix}
\cdots & \cdots & \cdots & \cdots & \cdots & \cdots & \cdots & \cdots & \cdots \\
\cdots & \cdots & \cdots & \cdots & \cdots & \cdots & \cdots & \cdots & \cdots \\
\cdots & \bar{s}_{n,n-n_x} & \cdots & \bar{s}_{n,n-1} & 1-\bar{s}_n & \bar{s}_{n,n+1} & \cdots & \bar{s}_{n,n+n_x} & \cdots \\
\cdots & \cdots & \cdots & \cdots & \cdots & \cdots & \cdots & \cdots & \cdots \\
\cdots & \cdots & \cdots & \cdots & \cdots & \cdots & \cdots & \cdots & \cdots
\end{bmatrix}
\begin{bmatrix}
\vdots \\ T_{n-n_x} \\ \vdots \\ T_{n-1} \\ T_n \\ T_{n+1} \\ \vdots \\ T_{n+n_x} \\ \vdots
\end{bmatrix}^{t+\Delta t}
=
\begin{bmatrix}
\vdots \\ T_{n-n_x} \\ \vdots \\ T_{n-1} \\ T_n \\ T_{n+1} \\ \vdots \\ T_{n+n_x} \\ \vdots
\end{bmatrix}^{t}
$$

This equation has the form

$$[S][T]^{t+\Delta t} = [T]^t.$$

Then, the temperature T at time $t + \Delta t$ is computed as

$$[T]^{t+\Delta t} = [S]^{-1}[T]^t.$$

The matrix $[S]$ is a five-band matrix. Nevertheless, even though the fully implicit scheme is unconditionally stable, two bands are located far from the diagonal (*i.e.*, at points $n - n_x$ and $n + n_x$), which makes the use of efficient numerical treatment difficult.

The alternating direction implicit, known as the ADI method, combines the stability of the implicit scheme with the simplicity of the explicit. In fact, the ADI method is unconditionally stable, and the system of linear equations is tridiagonal and diagonally dominant.

The idea of the ADI method is to alternate directions and to solve two one-dimensional problems at each time step.

In the first step $(t + \Delta t)$, the x-direction is in explicit form, and then

$$C_{H_v} \underbrace{\frac{T_{i,j}^{t+\Delta t} - T_{i,j}^t}{\Delta t}}_{\text{Computing } T^{t+\Delta t}} = \underbrace{\frac{k_H^{i,j-1/2} T_{i,j-1}^t - (k_H^{i,j-1/2} + k_H^{i,j+1/2}) T_{i,j}^t + k_H^{i,j+1/2} T_{i,j+1}^t}{\Delta x^2}}_{T \text{ at time } t}$$

$$+ \underbrace{\frac{k_H^{i-1/2,j} T_{i-1,j+}^{t+\Delta t} - (k_H^{i-1/2,j} + k_H^{i+1/2,j}) T_{i,j}^{t+\Delta t} + k_H^{i+1/2,j} T_{i+1,j}^{t+\Delta t}}{\Delta z^2}}_{T \text{ at time } t+\Delta t}.$$

$$(2.54)$$

In the second step $(t + 2\Delta t)$, the y direction is in explicit, and then

$$C_{H_v} \underbrace{\frac{T_{i,j}^{t+2\Delta t} - T_{i,j}^{t+\Delta t}}{\Delta t}}_{\text{Computing } T^{t+2\Delta t}} = \underbrace{\frac{k_H^{i,j-1/2} T_{i,j-1}^{t+2\Delta t} - (k_H^{i,j-1/2} + k_H^{i,j+1/2}) T_{i,j}^{t+2\Delta t} + k_H^{i,j+1/2} T_{i,j+1}^{t+2\Delta t}}{\Delta x^2}}_{T \text{ at time } t+2\Delta t}$$

$$+ \underbrace{\frac{k_H^{i-1/2,j} T_{i-1,j+}^{t+\Delta t} - (k_H^{i-1/2,j} + k_H^{i+1/2,j}) T_{i,j}^{t+\Delta t} + k_H^{i+1/2,j} T_{i+1,j}^{t+\Delta t}}{\Delta z^2}}_{T \text{ at time } t+\Delta t}.$$

$$(2.55)$$

2.6.1.4 *Boundary conditions*

Because the temperature of the nodes on the contour of the body depends on the boundary conditions, the continuity equation, given in Equations 2.53 or 2.54 and 2.55 for the implicit or ADI schemes of solution, applies only to the internal nodes.

Three types of boundary conditions are useful for analyzing the problem of heat flow in road structures:

- The "Dirichlet" condition which corresponds to a condition of imposed temperature.
- The "Neumann" condition which corresponds to an imposed heat flux through the contour of the body.

- A boundary condition for atmospheric interaction, which allows the effect of environmental variables to be included in the calculation of heat flow. Without a doubt, it is the best option for calculating the temperature in a road structure. However, this condition is not explained in this example because it is developed in detail in Example 19.

For the Dirichlet condition, the temperature is imposed on the contour. Therefore, its calculation at the time step $t + \Delta t$ is unnecessary and then

$$T_n^{t+\Delta t} = T_{\text{Imposed}}.$$

The Neumann boundary condition consists of imposing the heat flux, which is similar to imposing the gradient of temperature as $\frac{\partial T}{\partial \bar{n}}$, where $\bar{n}$ is the direction of the flux vector.

This condition requires the use of a set of imaginary nodes (as shown in Figure 2.23). Therefore, the temperature of the imaginary nodes for the rectangular grid of Figure 2.22 is

$$T_{n+n_x}^{t+\Delta t} = T_{n-n_x}^{t+\Delta t} + 2\Delta z \frac{q_H^{\text{Imposed}}}{k_H^{1+1/2,j}}.$$

Subsequently, introducing the Neumann condition into the implicit continuity equation, which is Equation 2.53, leads to

$$T_n^t = \bar{s}_{n,n-1} T_{n-1}^{t+\Delta t} + \bar{s}_{n,n+1} T_{n+1}^{t+\Delta t} + \bar{s}_{n,n-n_x} T_{n-n_x}^{t+\Delta t} + \bar{s}_{n,n+n_x} \left(T_{n-n_x}^{t+\Delta t} + 2\Delta z \frac{q_H^{\text{Imposed}}}{k_H^{i+1/2,j}} \right)$$

$$- T_n^{t+\Delta t}(\bar{s}_n - 1). \tag{2.56}$$

Finally, rearranging Equation 2.56, the implicit equation for computing the temperature at the top face, when imposing the heat flux, is

$$T_n^t - \frac{2\Delta t q_H^{\text{Imposed}}}{C_{H_v}\Delta z} = \bar{s}_{n,n-1} T_{n-1}^{t+\Delta t} + \bar{s}_{n,n+1} T_{n+1}^{t+\Delta t}$$

$$+ (\bar{s}_{n,n-n_x} + \bar{s}_{n,n+n_x}) T_{n-n_x}^{t+\Delta t} - T_n^{t+\Delta t}(\bar{s}_n - 1). \tag{2.57}$$

Although Equation 2.57 corresponds to a Neumann boundary on the top boundary of the grid, the same analysis is applicable for the other faces.

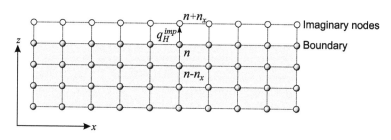

Figure 2.23 Imaginary nodes for analyzing the Neumann boundary condition imposed at the top of the body.

2.6.1.5 Initial conditions

Starting the computation requires knowing the temperature for the first time step T_n^0 in all nodes of the body. Any value of temperature can be allocated at time $t = 0$. However, it is necessary to run the model during a certain time, called "heating time", to reach the equilibrium conditions. These equilibrium conditions are identified when the temperature evolution in a day reaches a periodic pattern. Of course, as far the initial temperature is from the equilibrium conditions as larger the "heating time" must be.

2.6.2 Part B. Numerical solution using the data of the example

The simulation of the problem requires the following steps:

1. Assess the thermal conductivity and heat capacity of the different materials involved in the road's structure.
2. Discretize the problem in space.
3. Discretize the problem in time.
4. Consider the initial and boundary conditions.
5. Perform the simulation using the explicit FDM.

2.6.2.1 Thermal conductivity and heat capacity

The Côté and Konrad method [19], described in Figure 2.1, permits to assess the thermal conductivity of the granular materials and the subgrade by using Equations 2.15–2.17 and the empirical parameters given in Table 2.1. These equations require estimating the porosity n of the materials and their degree of saturation S_r, which are obtained using the classical phase diagram shown in Figure 2.24.

Then, from the phase diagram, the porosity and the saturation degree become

$$n = 1 - \frac{\rho_d}{\rho_s}, \quad \text{and} \quad nS_r = \frac{w\rho_d}{\rho_w}.$$

According to these phase relationships and the properties of the materials given in Table 2.10, the porosity and the saturation degree of each layer are

- Granular base, $n = 0.154, \ S_r = 0.858$.
- Granular sub-base, $n = 0.269, \ S_r = 0.847$.
- Subgrade, $n = 0.407, \ S_r = 0.942$.

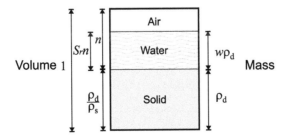

Figure 2.24 Phase diagram to calculate porosity and degree of saturation.

Also, according to Table 2.1, the empirical parameters χ_H, η, and κ_H for the granular layers and the subgrade are

- Granular layers, $\chi_H = 1.70$ W/(mK), $\eta = 1.8$, and $\kappa_H = 4.6$.
- Subgrade, $\chi_H = 0.75$ W/(mK), $\eta = 1.2$, and $\kappa_H = 1.9$.

Finally, Table 2.11 describes the procedure for calculating the thermal conductivity of the granular layers and the subgrade. Note that the thermal conductivity of water is 0.59 W/(mK).

Assessing the heat capacity of the different layers is possible using Equation 2.18. This equation requires the heat capacity of the solid grains which, according to Table 2.2, can be adopted as 0.92 kJ/(kg °C) for the granular materials (*i.e.*, base and sub-base) and 0.9 kJ/(kg °C) for the clayey subgrade. Moreover, the heat capacity of water is 4.2 kJ/(kg°C). Then, considering the water content of each material, Equation 2.18 leads to the following heat capacities c_H:

- Granular base, $c_H = 1.106$ kJ/(kg°C) .
- Granular sub-base, $c_H = 1.271$ kJ/(kg°C).
- Subgrade, $c_H = 1.539$ kJ/(kg°C) .

However, to guarantee the dimensional compatibility, Equation 2.48 requires the use of the volumetric heat capacity c_{H_v}, which is given in J/(m^{3}K), and it can be computed using the total density of each material as $c_{H_v} = \rho_d(1 + w)c_H$. Therefore, the volumetric heat capacity of each layer becomes

- Granular base, $c_{H_v} = 2.578 \cdot 10^6$ J/(m^{3}K).
- Granular sub-base, $c_{H_v} = 2.706 \cdot 10^6$ J/(m^{3}K).
- Subgrade, $c_{H_v} = 3.053 \cdot 10^6$ J/(m^{3}K)).

2.6.2.2 Discretization in space

Similarly to the previous example, this problem is also symmetrical. Therefore, it is only necessary to simulate the heat flow in one half of the model, imposing a boundary condition of zero flow through the symmetry plane. The discretized space is schematized in Figure 2.25, and in this discrete model, the distance between nodes is 0.1 m in the horizontal direction and 0.05 m in the vertical direction (*i.e.*, $\Delta x = 0.1$ m, $\Delta z = 0.05$ m).

Table 2.11 Thermal conductivities of the granular layers and the subgrade

Material	Solid grains k_{Hs} $2^{1-q_c\,7.7q_c}$ or $3^{1-q_c\,7.7q_c}$	Water k_{Hw}	Saturated k_{Hsat} $k_{Hs}^{1-n}k_{Hw}^{n}$	Dry k_{Hdry} $\chi_H 10^{-\eta n}$	K_e $\frac{\kappa_H S_r}{1+(\kappa_H-1)S_r}$	Unsaturated k_H $(k_{Hsat} - k_{Hdry})K_e$ $+k_{Hdry}$
Base	2.802	0.59	2.205	0.899	0.965	2.159
Sub-base	2.802	0.59	1.842	0.557	0.962	1.793
Subgrade	3.000	0.59	1.547	0.243	0.969	1.506

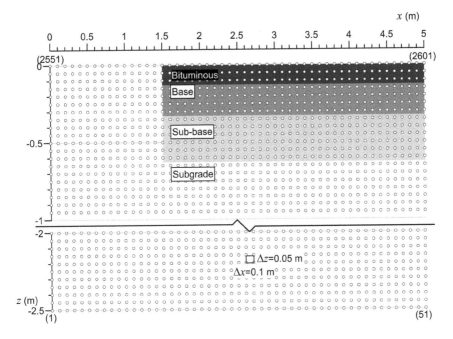

Figure 2.25 Discretization in space of the road structure.

As a result, the model has 51×51 nodes in the horizontal and vertical directions. The locations of those nodes are

- Bottom face,   nodes $1, ..., 51$.
- Top face,   nodes $2, 551, ..., 2, 601$.
- Left face,   nodes $51(i - 1) + 1 \ \ 2 \leq i \leq 51$.
- Right face,   nodes $51i \ \ 1 \leq i \leq 51$.

Regarding materials, their location is

- Bituminous material,   nodes $51(i - 1) + 16, ..., .51i \ 49 \leq i \leq 51$.
- Granular base,   nodes $51(i - 1) + 16, ..., 51i \ 45 \leq i \leq 48$.
- Granular sub base,   nodes $51(i - 1) + 16, ..., 51i \ 39 \leq i \leq 44$.
- Subgrade,   nodes $51(i - 1), ..., 51i \ 1 \leq i \leq 38$,
-                         and $51(i - 1) + 1, ..., 51(i - 1) + 15 \ 39 \leq i \leq 51$

2.6.2.3 Discretization in time

Since the implicit scheme of the FDM is unconditionally stable, it is less restrictive regarding the time step. Nevertheless, choosing a time steep too large moves the result away from the real solution, because a large time step implies a large linearization of the temperature evolution. Then, this example uses a time step of $\Delta t = 1, 800$ s (*i.e.,* half an hour).

2.6.2.4 Boundary and initial conditions

Each face of the model has different boundary conditions as follows:

- Top face, berm   imposed temperature $T(t) = T_{\mathrm{mean}} - T_{\mathrm{amp}}\cos(2\pi t/86,400)$.
- Top face, road   imposed heat flow $q_H(t) = q_{H_{\mathrm{mean}}} - q_{H_{\mathrm{amp}}}\cos(2\pi t/86,400)$.
- Bottom face,   imposed temperature $T = 15.0°\mathrm{C}$.
- Right face,   zero heat flow.
- Left face,   zero heat flow.

These conditions are schematized in Figure 2.26.

Regarding the initial conditions, a temperature of 15°C is imposed throughout the model, which is similar to the imposed temperature on its bottom face.

$$T(t = 0) = 15°\mathrm{C}.$$

Moreover, the simulation adopts a "heating time" of six days. Therefore, the results of the simulation are presented for the 7th day.

2.6.2.5 Simulation

After all, the procedure to perform the simulation is as follows:

1   Initialize the temperature at $t = 0$

$$T_{1..2601}^{t=0} = 15°\mathrm{C}.$$

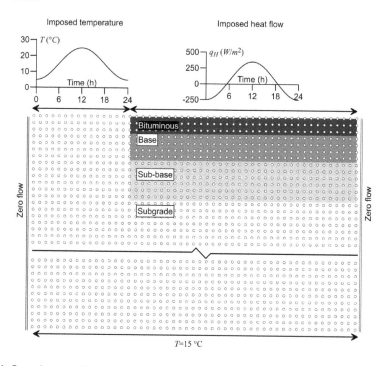

Figure 2.26 Boundary conditions around the model.

2 Allocate for all nodes the thermal conductivity, k_H, and the volumetric heat capacity, c_{H_v}.

3 Calculate the ancillary variables $\bar{s}$ for all internal nodes n, and there are five variables for each node

$$\bar{s}_{n,1} = -\frac{1.8 \cdot 10^5}{c_{H_v}^n}\sqrt{k_H^n k_H^{n+n_x}}, \qquad \bar{s}_{n,2} = -\frac{7.2 \cdot 10^5}{c_{H_v}^n}\sqrt{k_w^n k_w^{n+1}},$$

$$\bar{s}_{n,3} = -\frac{1.8 \cdot 10^5}{c_{H_v}^n}\sqrt{k_w^n k_w^{n-n_x}}, \qquad \bar{s}_{n,4} = -\frac{7.2 \cdot 10^5}{c_{H_v}^n}\sqrt{k_w^n k_w^{n-1}},$$

$$\bar{s}_{n,5} = \bar{s}_{n,1} + \bar{s}_{n,2} + \bar{s}_{n,3} + \bar{s}_{n,4}$$

Variables $\bar{s}$ must be adapted to take into account the lack of neighbors of the nodes on the boundaries.

4 Construction of the matrix $[S]$ and the vector $[T]^t$ as

Bottom boundary :

$n = 1, ..., n_x \rightarrow S[n, n] = 1$, and $T^t[n] = 15$,

Top boundary, berm :

$n = 2,551, ..., 2,565 \rightarrow S[n, n] = 1$, and $T^t[n] = T_{\mathrm{mean}} - T_{\mathrm{amp}}\cos(2\pi t/86, 400)$,

Top boundary, road :

$n = 2,566, ..., 2,600 \rightarrow q_H(t) = q_{H\mathrm{mean}} - q_{H\mathrm{amp}}\cos(2\pi t/86, 400)$ and

$T^t[n] = T^t[n] + 2\Delta t q_H(t)/(c(k)\Delta z)$,

$\qquad S[n, n - n_x] = 2\bar{s}_{n,3}$,

$\qquad S[n, n - 1] = \bar{s}_{n,4}$,

$\qquad S[n, n + 1] = \bar{s}_{n,2}$,

$\qquad S[n, n] = -\bar{s}_{n,2} - 2\bar{s}_{n,3} - \bar{s}_{n,4} + 1$,

$n = 2,601$

$\qquad S[n, n - 1] = 2\bar{s}_{n,4}$,

$\qquad S[n, n - n_x] = 2\bar{s}_{n,3}$,

$\qquad S[n, n] = -2\bar{s}_{n,3} - 2\bar{s}_{n,4} + 1$,

Internal nodes: $i = 2, ..., n_z - 1$, $j = 2, ..., n_x - 1$ $n = n_x(i - 1) + j$,

$\qquad S[n, n + n_x] = \bar{s}_{n,1}$,

$\qquad S[n, n + 1] = \bar{s}_{n,2}$,

$\qquad S[n, n - n_x] = \bar{s}_{n,3}$,

$\qquad S[n, n - 1] = \bar{s}_{n,4}$,

$\qquad S[n, n] = -\bar{s}_{n,5} + 1$,

Left boundary: $i = 2, ..., n_z - 1$, $n = n_x(i - 1)$,

$\qquad S[n, n + n_x] = \bar{s}_{n,1}$,

$\qquad S[n, n + 1] = 2\bar{s}_{n,2}$,

$\qquad S[n, n - n_x] = \bar{s}_{n,3}$,

$\qquad S[n, n] = -\bar{s}_{n,1} - 2\bar{s}_{n,2} - \bar{s}_{n,3} + 1$,

Right boundary: $i = 2, ..., n_z - 1, \ n = n_x i,$

$$S[n, n + n_x] = \bar{s}_{n,1},$$

$$S[n, n - n_x] = \bar{s}_{n,3},$$

$$S[n, n - 1] = 2\bar{s}_{n,4},$$

$$S[n, n] = -\bar{s}_{n,1} - \bar{s}_{n,3} - 2\bar{s}_{n,4} + 1.$$

5   Compute the temperature in $t + \Delta t$ as

$$[T]^{t+\Delta t} = [S]^{-1}[T]^t.$$

6   Set $[T]^t = [T]^{t+\Delta t}$ and return to step 4 for calculating the temperature in the next time step.

Similar to the previous example, the MATLAB script, which is provided with this book, allows obtaining the results shown in Figures 2.27 and 2.28.

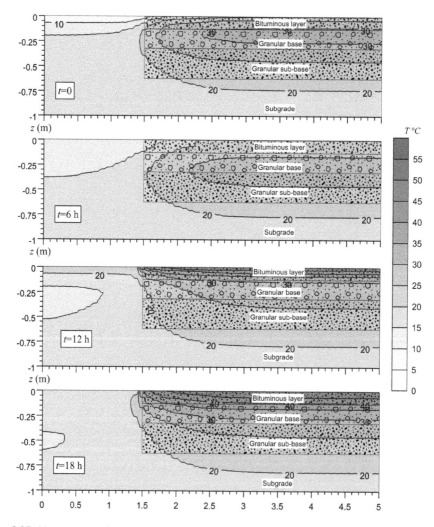

Figure 2.27 Heat maps in the road structure at different elapsed times of the simulation.

Figure 2.27, which shows the temperature maps resulting from the simulation, illustrates how the isotherms move downward and upward depending on the imposed temperature or heat flow. For these boundary conditions, the maximum temperature at the surface of the road occurs at 18 h, while the minimum temperature occurs at 6 h. Figure 2.28, which shows the vertical profiles of temperature below the berm (*i.e.*, at $x = 0.8$ m in the horizontal axis) and below the road structure (*i.e.*, at $x = 3.5$ m in the horizontal axis), also permits to appreciate the extreme values of temperature.

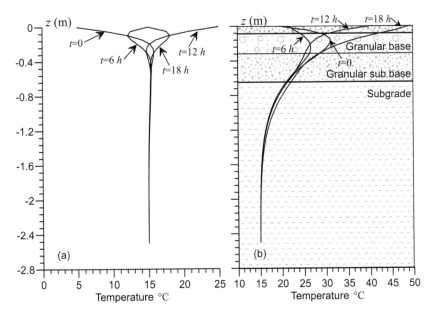

Figure 2.28 Vertical profiles of the temperature in the middle of the berm $x = 0.8$ m and below the road structure $x = 3.5$ m.

Chapter 3

Compaction

3.1 RELEVANT EQUATIONS

This chapter explains the modeling of soil compaction through two examples. The first one consists in using the Barcelona Basic Model (BBM) to analyze compaction from a mechanistic point of view. Then, the second example uses a methodology to link the soil's density with its grain size distribution.

3.1.1 Summary of the equations describing the BBM

The Barcelona Basic Model, also known as the BBM, is an elastoplastic model developed for unsaturated soils. The model uses the net stress, σ_{net}, as the stress variable, which is defined as

$$\sigma_{net} = \sigma - u_a \quad \text{net stress, and}$$

$$s = u_a - u_w \quad \text{suction,}$$

where σ is the total stress, u_a is the pore air pressure, and u_w is the pore water pressure.

Regarding volume, the BBM uses the specific volume v, defined as $v = 1 + e$, where e is the void ratio.

In summary, the set of state variables used in the BBM are the specific volume v, the matric suction s, the mean net stress p, and the deviatoric stress q.

As shown in Figure 3.1, the BBM requires three planes to describe the soil's behavior:

- A three-dimensional plane (p, q, s), depicted in Figure 3.1a, is used to describe the yield surfaces.
- A second plane, shown in Figure 3.1b, relates the specific volume and the mean net stress (v, p).
- The third plane, shown in Figure 3.1c, relates the specific volume and the matric suction (v, s).

The yield surface defined on the three-dimensional plane (p, q, s) has two limits. The first surface is defined by ellipses, with a similar shape to that defined in the modified Cam-Clay model, but growing with suction because of the capillary bonding between soil's particles. The second limit in the three-dimensional plane is a flat surface that defines the maximum value of suction that the soil has suffered, and this surface is denoted as the *SI* (*i.e.*, suction increase surface).

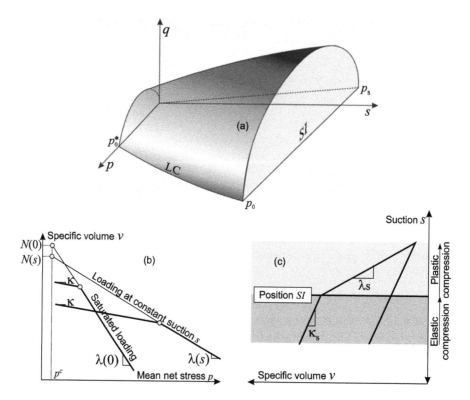

Figure 3.1 Representation of the BBM in the planes (a)(p, q, s), (b)(v, p), and (c)(v, s).

For the elliptical surface, the BBM proposes that the tensile strength of the soil, p_s, increases linearly with suction, $p_s = k_c s$, so that the equation for these ellipses is

$$q^2 - M^2(p + p_s)(p_0 - p) = 0, \tag{3.1}$$

where M is the slope of the critical state line, which assumed as constant regardless of suction, and p_0 denotes the overconsolidation mean stress that depends on suction. The set of p_0 values defines a curve in the (p, s) plane, and this curve is known as the Loading Collapse curve, LC, whose equation is

$$\left(\frac{p_0}{p^c}\right) = \left(\frac{p_0^*}{p^c}\right)^{\frac{\lambda(0)-\kappa}{\lambda(s)-\kappa}}, \tag{3.2}$$

where κ defines the compressibility of the soil in the elastic domain, also assumed as constant regardless of the suction, $\lambda(0)$ defines the compressibility of the soil in the saturated state in the elastoplastic domain, p_0^* is the overconsolidation mean stress in the saturated state, and p^c is a reference mean stress.

Regarding the change in specific volume, the BBM predicts that it can occur either changing the mean stress p or the matric suction s, and these changes are defined in Figures 3.1b and c. First, the BBM uses a relationship to calculate volumetric changes when the mean stress p increases. This relationship is analogous to the one used for saturated soils in the Cam-Clay model but using coefficients N and λ depending on suction. Therefore, as the mean stress p increases, the reduction of the specific volume v along the virgin compression line becomes

$$v = N(s) - \lambda(s) \ln\left(\frac{p}{p^c}\right), \tag{3.3}$$

where $N(s)$ is the specific volume for the mean reference stress p^c, and $\lambda(s)$ is the slope, in logarithmic scale, of the virgin compression line in the unsaturated state.

To describe the increase of soil stiffness depending on the suction, *i.e.*, the reduction of $\lambda(s)$, the BBM uses the following expression:

$$\lambda(s) = \lambda(0)\left[(1-r)e^{-\beta s} + r\right], \tag{3.4}$$

where β is a shape parameter, and r is a constant that relates the maximum reduction of $\lambda(s)$ when the suction grows to infinity,

$$r = \frac{\lambda(s \rightarrow \infty)}{\lambda(0)}.$$

On unloading and reloading, the material remains in the elastic domain, and the specific volume follows a straight line (in logarithmic scale) with slope κ, so that the change of the specific volume in the elastic domain for both the saturated and unsaturated state is

$$dv^e = -\kappa \ln\left(\frac{p+dp}{p}\right). \tag{3.5}$$

The BBM suggests another mechanism that produces irreversible strains. This mechanism is activated when the suction increases beyond the maximum suction experienced by the soil (see Figure 3.1c). The logarithmic relationships that relate volumetric changes due to suction in the elastic and elastoplastic domains are as follows:

$$dv^e = \kappa_s \ln\left(\frac{s+p_{atm}}{p_{atm}}\right) \qquad \text{elastic domain,} \tag{3.6}$$

$$dv^{ep} = \lambda_s \ln\left(\frac{s+p_{atm}}{p_{atm}}\right) \qquad \text{elastoplastic domain,} \tag{3.7}$$

where κ_s and λ_s define the volumetric changes, reversible and elastoplastic, that result from the change in suction, respectively.

Finally, the plane (v, s) allows obtaining the following relationship between the specific volumes $N(s)$ and $N(0)$ as

$$N(0) = N(s) + \kappa_s \ln\frac{s+p_{atm}}{p_{atm}}. \tag{3.8}$$

To sum up, the set parameters required for the BBM are

- Parameters κ, $\lambda(0)$, and $N(0)$ define the compressibility of the soil in the saturated state.
- Parameters r, β, and p^c define how the stiffness of the soil increases as suction grows.
- Parameters κ_s and λ_s define the expansion or compression resulting from the change in suction, in the elastic or the elastoplastic domains.
- Parameter M is the slope of the critical state line.
- Parameter k_c defines the increase of tensile strength resulting from increased suction.

3.1.2   Effect of cyclic loading

During compaction, density grows as the number of loading cycles increases. The following equation, proposed in Ref. [33], describes the effect of the number of loading cycles on compaction:

$$\ln\left(\frac{\sigma_{v_{N_C}}}{\sigma_{v_{N_C=1}}}\right) = K_N \ln(N_C), \tag{3.9}$$

where $\sigma_{v_{N_C}}$ is the vertical stress required to reach a specific density by applying N_C loading cycles, $\sigma_{v_{N_C=1}}$ is the vertical stress required to reach the same density after one loading cycle, and K_N is a coefficient that depends on the degree of saturation.

From Equations 3.9 and 3.3, the change in specific volume depending on the number of loading cycles can be obtained by the following equation:

$$v = N(s) - \lambda(s) \ln\left(\frac{p}{p^c} N_C^{K_N}\right). \tag{3.10}$$

3.1.3   Evolution of the water retention curve during compaction

The water retention curve of a soil evolves during compaction because the sizes of the pores change. Several models have been proposed to describe the evolution of the water retention curve for soils undergoing compaction. Examples of this book use Equation 3.11, which was proposed in Ref. [34]:

$$S_r = \left\{\frac{1}{1 + [\phi_W s e^{\psi_W}]^{n_W}}\right\}^{m_W}, \tag{3.11}$$

where e is the void ratio, and ϕ_W, ψ_W, n_W, and m_W are the parameters of the water retention model.

Moreover, during compaction, the soil undergoes loading-unloading-reloading cycles; for these processes, the evolution of the matric suction is given in Equation 3.12, which was proposed in Ref. [55]:

$$S_r = S_{r0} - k_{s_{WR}}(s - s_0), \tag{3.12}$$

where s_0 and S_{r0} are the suction and degree of saturation at the beginning of unloading, and k_s is the slope of the unloading-reloading curve in the space relating matric suction to the degree of saturation.

3.1.4   A linear packing model for establishing the relationship between grain size distribution and density

Reaching high density in materials made of particles is only possible if the material has a proper grain size distribution.

Models dealing with binary mixtures of clay and granular particles have been proposed by several authors [60,62,69]. On the other hand, models working with materials having a broad grain size distribution, known as polydisperse mixtures, have been proposed for concrete technology to find out mixtures of granular materials minimizing void spaces.

Table 3.1   Definitions used in the linear packing model

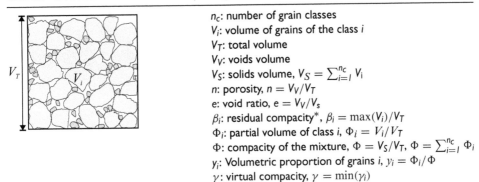

n_c: number of grain classes
V_i: volume of grains of the class i
V_T: total volume
V_V: voids volume
V_S: solids volume, $V_S = \sum_{i=1}^{nc} V_i$
n: porosity, $n = V_V/V_T$
e: void ratio, $e = V_V/V_s$
β_i: residual compacity*, $\beta_i = \max(V_i)/V_T$
Φ_i: partial volume of class i, $\Phi_i = V_i/V_T$
Φ: compacity of the mixture, $\Phi = V_S/V_T$, $\Phi = \sum_{i=1}^{nc} \Phi_i$
y_i: Volumetric proportion of grains i, $y_i = \Phi_i/\Phi$
γ: virtual compacity, $\gamma = \min(\gamma_i)$

(*) The maximum volume of fraction i, $\max(V_i)$, is obtained taking each fraction separately and compacting it to its maximum density as described in Section 3.1.7.3.

The linear packing model denoted as the LPM, which is presented in this section, is based on the model for polydisperse mixtures proposed by De Larrard [21,22,54]. Caicedo et al. [12] proved the good performance of the LPM for granular road materials.

The derivation of the equations that allow calculating the density of a granular mixture, based on its grain size distribution, requires the definitions of the variables given in Table 3.1.

The compacity of a mixture of particles, Φ, is defined as the volume of grains within the mixture by unit volume, and it is directly related to the volume of voids in a granular material given by the porosity n, or the void ratio e as follows:

$$n = 1 - \Phi, \quad e = \frac{1}{\Phi} - 1.$$

The residual compacity β_i is defined as the maximum compacity, obtained experimentally for each granular fraction. Moreover, the virtual compacity, γ, is the maximum compacity theoretically reachable in the granular mixture. Therefore, the LPM permits obtaining the actual compacity of a granular mixture, Φ, knowing the volumetric proportion, y_i, of particles of size d_i, and the residual compacities of each grain size β_i.

3.1.4.1   *Virtual compacity of binary mixtures*

The most straightforward case of granular mixtures occurs when the mixture has only two classes of grain sizes, also known as binary mixtures. This case is unrealistic for granular materials. However, its understanding is useful before describing the more general case of polydisperse mixtures.

For binary mixtures with grain sizes d_1 and d_2, three cases could be considered depending upon the interaction between the two classes of grains [21,22,54]:

- **No interaction:** in this case, one of the grain classes is substantially bigger than the other, $d_1 \gg d_2$ or $d_2 \gg d_1$. As a result, the local structure of each type of grains is unaltered by the presence of the other. Two cases are possible, and Figure 3.2a shows the case of a mixture supported by the arrangement of big grains, the smaller

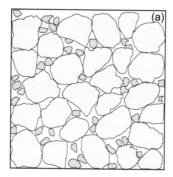 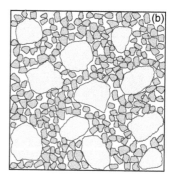

Figure 3.2 Grains in a binary mixture without interaction: (a) bigger grains are dominant, and (b) smaller grains are dominant.

grains occupying the voids. In contrast, in Figure 3.2b, the mixture is supported by the smaller grains and the big grains are immersed in the space of small particles.

- **Total interaction:** occurs when the size of the particles in the binary mixture is identical, but the residual compacity is different: $d_1 = d_2$; $\beta_1 \neq \beta_2$, as in Figure 3.4a.
- **Partial interaction:** occurs in the case of binary mixtures with $d_1 > d_2$. In this case, two physical effects appear: decompaction effect created by the small grains filling voids, as in Figure 3.5a; and boundary effect appearing in the contact between small and big grains, as in Figure 3.5b.

3.1.5   Virtual compacity of binary mixtures without interaction

The virtual compacity γ for the case of without interaction can be obtained knowing the volumetric proportions, y_1 and y_2, and the corresponding residual compacities, β_1 and β_2.

As shown in Figure 3.2, there are two possible cases of mixtures without interaction. First, when the bigger grains are dominant, the virtual compacity is γ_1, and the big grains fill up the volume without interaction with the small grains. On the other hand, for the second case, the smaller grains fill the space existing between the bigger grains at their maximum individual compacity β_2.

According to the definition, the virtual compacity, γ, is

$$\gamma = \Phi_1 + \Phi_2. \tag{3.13}$$

Also, the volumetric proportions of each class of grains are

$$y_1 = \frac{\Phi_1}{\Phi_1 + \Phi_2}, \tag{3.14}$$

$$y_2 = \frac{\Phi_2}{\Phi_1 + \Phi_2}, \tag{3.15}$$

$$y_1 + y_2 = 1. \tag{3.16}$$

When the bigger grains are dominant, these grains fill up the volume without interacting with the small grains. Then, the compacity of grains 1 within the mixture, Φ_1, is the same as the residual compacity β_1 (*i.e.*, $\Phi_1 = \beta_1$). Considering Equations 3.13–3.16, the virtual compacity γ_1 becomes

$$\gamma_1 = \Phi_1 + \Phi_2 = \beta_1 + \Phi_2,$$
$$\gamma_1 = \beta_1 + (\Phi_1 + \Phi_2)y_2,$$
$$\gamma_1 = \beta_1 + \gamma_1 y_2.$$

And then,

$$\gamma_1 = \frac{\beta_1}{1 - y_2}.$$

On the other hand, when the small grains are dominant, these grains fill the space between the bigger grains, that is, $1 - \Phi_1$, at their maximum residual compacity β_2, and then,

$$\Phi_2 = \beta_2(1 - \Phi_1),$$
$$\gamma_2 = \Phi_1 + \Phi_2 = \Phi_1 + \beta_2(1 - \Phi_1),$$
$$\gamma_2 = (\Phi_1 + \Phi_2)\gamma_1 + \beta_2 - \beta_2(\Phi_1 + \Phi_1)\gamma_1,$$
$$\gamma_2 = \gamma_2 \gamma_1(1 - \beta_2) + \beta_2.$$

As a result, the virtual compacity, γ_2, becomes

$$\gamma_2 = \frac{\beta_2}{1 - (1 - \beta_2)\gamma_1}.$$

For the binary mixture, only the minimum between γ_1 and γ_2 is possible. In fact, if γ is higher than γ_1, grains 2 penetrate into grains 1 and vice versa, as described in Figure 3.3. This is the impenetrability condition which is defined as

$$\gamma = \min(\gamma_1, \gamma_2).$$

3.1.6 Virtual compacity of binary mixtures with total interaction

A mixture with total interaction is defined as a mixture having particles of the same size, $d_1 = d_2$, but having different residual compacity $\beta_1 \neq \beta_2$, as shown in Figure 3.4.

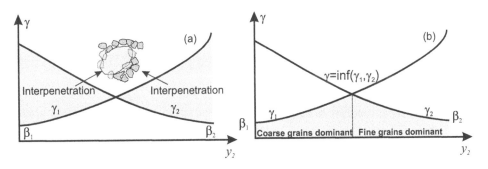

Figure 3.3 Schematic drawing of the impenetrability condition, the shaded areas in (a) represent the interpenetration between grains and in (b) the area of no interpenetration.

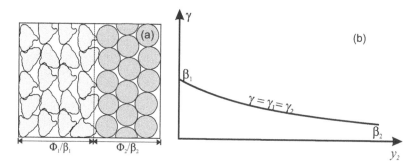

Figure 3.4 (a) Schematic drawing of a binary mixture with total interaction, (b) evolution of the virtual compacity as the proportions of grains 2 increases.

For total interaction, the virtual compacity of the mixture is calculated assuming that the compacity γ remains invariable after a total segregation, as shown in Figure 3.4a, [21,22]. This condition becomes

$$\frac{\Phi_1}{\beta_1} + \frac{\Phi_2}{\beta_2} = 1. \tag{3.17}$$

Equation 3.17 allows calculating each partial volume in terms of the other one as

$$\Phi_1 = \beta_1 \left(1 - \frac{\Phi_2}{\beta_2}\right), \quad \text{and}$$

$$\Phi_2 = \beta_2 \left(1 - \frac{\Phi_1}{\beta_1}\right).$$

Since the virtual compacity γ is $\gamma = \Phi_1 + \Phi_2$,

$$\gamma_1 = \Phi_1 + \beta_2 \left(1 - \frac{\Phi_1}{\beta_1}\right), \quad \text{and}$$

$$\gamma_2 = \beta_1 \left(1 - \frac{\Phi_2}{\beta_2}\right) + \Phi_2.$$

and then becoming

$$\gamma_1 = \beta_2 + \Phi_1 \left(1 - \frac{\beta_2}{\beta_1}\right), \quad \text{and}$$

$$\gamma_2 = \beta_1 + \Phi_2 \left(1 - \frac{\beta_1}{\beta_2}\right).$$

Also, as the partial volumes Φ are $\Phi_1 = \gamma y_1$ and $\Phi_2 = \gamma y_2$,

$$\gamma_1 = \beta_2 + \gamma y_1 \left(1 - \frac{\beta_2}{\beta_1}\right), \quad \text{and}$$

$$\gamma_2 = \beta_1 + \Phi_2 = \gamma y_2 \left(1 - \frac{\beta_1}{\beta_2}\right).$$

Finally,

$$\gamma_1 = \frac{\beta_2}{1 - \left(1 - \frac{\beta_2}{\beta_1}\right)y_1}, \quad \text{and}$$

$$\gamma_2 = \frac{\beta_1}{1 - \left(1 - \frac{\beta_1}{\beta_2}\right)y_2}.$$

Since $y_1 + y_2 = 1$, it is possible to prove that in the case of total interaction the virtual compacities are identical, $\gamma_1 = \gamma_2$. Furthermore, in this case, the curve of virtual compacity depending on the proportion of grains 2 decreases monotonically, as shown in Figure 3.4b.

3.1.7   Virtual compacity of binary mixtures with partial interaction

In the case of partial interaction, two effects may appear:

- **Decompaction effect:** occurs when the voids between the big grains don't have enough volume to admit small grains. In that case, the only possibility to arrange the small grains without alteration of their original shape is separating the bigger grains, reducing the compacity of the whole mixture, as shown in Figure 3.5a.
- **Wall effect:** occurs at the contact between small and big grains, as indicated in Figure 3.5b.

As in the other cases, the virtual compacity γ is calculated adding the partial volume of each class:

$$\gamma_1 = \Phi_1 + \Phi_2.$$

However, in this case, the partial volume Φ_1 is affected by the decompaction effect, and it is possible to assume that the decompaction effect is a linear function depending on the partial volume of small grains, as

$$\Phi_1 = \beta_1(1 - f_{1,2}\Phi_2).$$

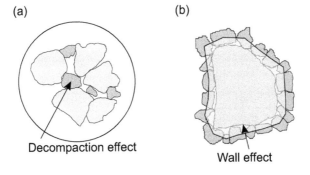

Figure 3.5 Effects of: (a) decompaction and (b) wall.

Then, the virtual compacity γ_1 becomes

$$\gamma_1 = \Phi_1 + \Phi_2 = \beta_1(1 - f_{1,2}\Phi_2) + \Phi_2, \quad \text{or}$$
$$\gamma_1 = \beta_1 + \Phi_2(1 - f_{1,2}\beta_1).$$

Also, considering the volumetric relationship, $\Phi_2 = \gamma y_2$,

$$\gamma_1 = \beta_1 + \gamma_1(1 - f_{1,2}\beta_1)y_2,$$

and then becoming

$$\gamma_1 = \frac{\beta_1}{1 - \gamma_1(1 - f_{1,2}\beta_1)y_2}. \tag{3.18}$$

Equation 3.18 must be valid for the extreme cases of no interaction and total interaction, and this condition is achieved if the function $f_{1,2}$ is $f_{1,2} = a_{1,2}/\beta_2$:

$$\gamma_1 = \frac{\beta_1}{1 - \left(1 - a_{1,2}\frac{\beta_1}{\beta_1}\right)y_2}. \tag{3.19}$$

Moreover, Equation 3.19 is a general equation that must be valid for the cases of total and no interaction. This restriction leads to $a_{1,2} = 0$ for no interaction and $a_{1,2} = 1$ for total interaction.

The wall effect occurs at the contact between big and small grains when the small grains are dominant. Under these circumstances, the compacity equation becomes

$$\gamma_2 = \Phi_1 + \Phi_2 = \Phi_1 + \beta_2(1 - \Phi_1). \tag{3.20}$$

Nevertheless, the small grains can't reach the maximum compacity β_2 because of the wall effect. This effect is more significant as the volume of big grains increases. Assuming that this effect is proportional to the relationship $\Phi_1/(1 - \Phi_1)$, Equation 3.20 becomes

$$\gamma_2 = \Phi_1 + \Phi_2 = \beta_2\left(1 - g_{2,1}\frac{\Phi_1}{1 - \Phi_1}\right)(1 - \Phi_1).$$

However, since $\Phi_1 = \gamma y_1$,

$$\gamma_2 = \beta_2 + \gamma_2\left[1 - \beta_2(1 + g_{1,2})\right]y_1, \quad \text{so}$$
$$\gamma_2 = \frac{\beta_2}{1 - \left[1 - \beta_2 - \beta_2 g_{2,1}\right]y_1}. \tag{3.21}$$

Also, in this case, the wall effect function, $g_{2,1}$, must be valid for the two extreme cases of no interaction and total interaction. A convenient equation guaranteeing this condition is $g_{2,1} = b_{2,1}(1/\beta_1 - 1)$. Therefore, Equation 3.21 becomes

$$\gamma_2 = \frac{\beta_2}{1 - \left[1 - \beta_2 - \beta_2 b_{2,1}\right)(1 - 1/\beta_1)\right]y_1}.$$

Moreover, to guarantee the continuity of the solution in the two extreme cases, $b_{2,1}$ must be $b_{2,1} = 1$ for total interaction and $b_{2,1} = 0$ for no interaction.

Figure 3.6 shows the evolution of the virtual compacity for all categories of interaction.

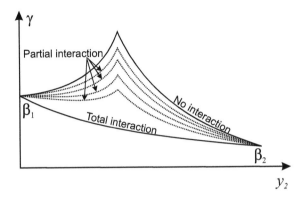

Figure 3.6 Virtual compacities for the different cases of interaction.

Decompaction and wall coefficients $a_{1,2} = 1$ and $b_{2,1} = 1$ were calibrated experimentally by De Larrard [21,22] for different combinations of shape and size of particles. Expressions proposed in Refs. [21,22] are

$$a_{i,j} = \sqrt{1 - \left(1 - \frac{d_j}{d_i}\right)^{1.02}}, \text{ and} \tag{3.22}$$

$$b_{j,i} = 1 - \left(1 - \frac{d_i}{d_j}\right)^{1.5}. \tag{3.23}$$

3.1.7.1 Virtual compacity of polydisperse mixtures

The case of a polydisperse mixture with n_c granular classes with $d_1 \gg d_2, ..., \gg d_{nc}$ was analyzed in Refs. [21,22].

Computing the virtual compacity of the mixture requires the assumption that each class i could be dominant in the granular mixture. Assuming that the class i is dominant means that grains i occupy the available volume at its maximum compacity β_i, and then

$$\Phi_i = \beta_i \left(1 - \sum_{j=1}^{i-1} \Phi_j\right).$$

Since the virtual compacity results from the addition of all partial volumes of the different classes,

$$\gamma = \sum_{j=1}^{n_c} \Phi_j, \quad \text{and then}$$

$$\gamma_i = \sum_{j=1, j\neq i}^{n_c} \Phi_j + \beta_i \left(1 - \sum_{j=1}^{i-1} \Phi_j\right).$$

Becoming

$$\gamma_i = \beta_i + (1 - \beta_i) \sum_{j=1}^{i-1} \Phi_j + \sum_{j=i+1}^{n_c} \Phi_j.$$

Moreover, since $\Phi_i = \gamma y_i$,

$$\gamma_i = \beta_i + \gamma_i \left[(1 - \beta_i) \sum_{j=1}^{i-1} y_j + \sum_{j=i+1}^{n_c} y_j \right].$$

Finally, when assuming the class i as dominant, the virtual compacity becomes

$$\gamma_i = \frac{\beta_i}{1 - (1 - \beta_i) \sum_{j=1}^{i-1} y_j + \sum_{j=i+1}^{n_c} y_j}.$$

Similar to the binary mixtures, in the case of a polydisperse mixture with partial interaction, grains with size $d > d_i$ undergo the decompaction effect due to the smaller grains whose size is $d < d_i$. Also, grains whose size is $d < d_i$ undergo the wall effect. Analyzing the case of a ternary mixture, De Larrard [21,22] generalized the virtual compacity considering grains i as dominant as

$$\gamma_i = \frac{\beta_i}{1 - \sum_{j=1}^{i-1} \left[1 - \beta_i + b_{i,j}\beta_i \left(1 - 1/\beta_j \right) \right] y_j - \sum_{j=i+1}^{n_c} \left[1 - a_{i,j}\beta_i/\beta_j \right] y_j}. \tag{3.24}$$

To find out which is the dominant class of the mixture, denoted as d_D, it is possible to calculate the virtual compacity for each class, assuming that this particular class corresponds to the dominant grains. Under these circumstances, grains $d > d_D$ undergo a decompaction effect due to the grains with size $d < d_D$, and the mixture undergoes a boundary effect due to the grains with size $d < d_D$. From a geotechnical point of view, all grains with $d < d_D$ are the matrix of the mixture, and grains with $d > d_D$ are dispersed grains within the mixture. From this point of view, the dominant class D is the class that supports the granular arrangement.

Since the impenetrability restriction is still applicable, the dominant size of grains corresponds to the class that leads to the minimum value of the virtual compacity, and so

$$\gamma = \min_{1 \leq i \leq n_c} \gamma_i.$$

3.1.7.2 *Actual compacity of granular mixtures*

The virtual compacity γ is a theoretical maximum value of compacity. However, this value is unreachable experimentally. In fact, the actual compacity of the mixture, Φ, is more or less close to the virtual compacity depending on the compaction method ($\Phi < \gamma$). A packing coefficient, K, was proposed in Refs. [21,22,46] to relate the virtual compacity, γ, with the actual compacity Φ.

For a real mixture, the actual compacity of the mixture is obtained by summation of the partial volumes of each grain size, Φ_i, as

$$\Phi = \sum_{i=1}^{n_c} \Phi_i.$$

Computing the actual compacity of the mixture, Φ, requires to introduce a new variable Φ_i^*. This new variable is the maximum compacity of class i taking into account the presence of the other classes. Therefore, a virtual mixture is a mixture where partial volumes of each class are $\Phi_0, ..., \Phi_{i-1}, \Phi_i^*, \Phi_{i+1}, ..., \Phi_{n_c}$.

The packing coefficient K is based on the viscosity models proposed in Ref. [46]. This coefficient has the additivity property regarding the packing coefficient of each class of the mixture, K_i, and then

$$K = \sum_{i=1}^{n_c} K_i, \tag{3.25}$$

where K_i relates the packing state of grains i within the mixture.

The filling coefficient, F_c, describes the ratio between the partial volume of grains i in the actual mixture and the volume that grains i can occupy when these grains become dominant, and then $F_c = \Phi_i/\Phi_i^*$, as depicted in Figure 3.7.

The value of K_i can be related with the filling coefficient through a function that must accomplish the additivity property presented in Equation 3.25. The following function that depends on the ratio $F_c/(1 - F_c)$ was proposed in Refs. [21,22]:

$$K = \sum_{i=1}^{n_c} \frac{\Phi_i/\Phi_i^*}{1 - \Phi_i/\Phi_i^*} = \sum_{i=1}^{n_c} \frac{\Phi_i}{\Phi_i^* - \Phi_i}.$$

Therefore, Φ_i^* becomes

$$\Phi_i^* = \beta_i \left[1 - \sum_{j=1}^{i-1} \left(1 - b_{i,j} \left(1 - 1/\beta_j \right) \right) \Phi_j - \sum_{j=i+1}^{n_c} \frac{a_{i,j}}{\beta_i} \Phi_j \right].$$

Also, since $y_i = \Phi_i/\Phi$,

$$K_i = \frac{y_i \Phi}{\beta_i \left[1 - \sum_{j=1}^{i-1} \left(1 - b_{i,j} \left(1 - 1/\beta_j \right) \right) y_j \Phi - \sum_{j=i+1}^{n_c} \frac{a_{i,j}}{\beta_j} y_j \Phi \right] - y_i \Phi},$$

leading to

$$K_i = \frac{y_i/\beta_i}{\frac{1}{\Phi} - \sum_{j=1}^{i-1} \left(1 - b_{i,j} \left(1 - 1/\beta_j \right) \right) y_j - \sum_{j=i+1}^{n_c} \frac{a_{i,j}}{\beta_j} y_j - \frac{y_i}{\beta_i}}.$$

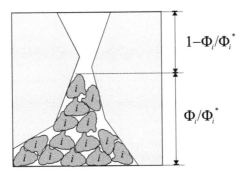

Figure 3.7 Schematic drawing of the filling coefficient $F_c = \Phi_i/\Phi_i^*$.

Furthermore, considering that $\sum y_i = 1$, it is possible to demonstrate that

$$\frac{1}{\gamma_i} = \sum_{j=1}^{i-1} \left(1 - b_{i,j} \left(1 - 1/\beta_j\right)\right) y_j - \sum_{j=i+1}^{n_c} \frac{a_{i,j}}{\beta_j} y_j - \frac{y_i}{\beta_i}, \quad \text{and therefore}$$

$$K_i = \frac{y_i/\beta_i}{1/\Phi - 1/\gamma_i}.$$

To sum up, the actual compacity of the mixture, Φ, depends directly on the packing coefficient of the mixture, K, and, therefore, on its compaction state. The following section describes the methodology to use the LPM and analyzes its performance.

3.1.7.3 Assessment of compacted densities using the linear packing model

The experimental measures presented in Refs. [21,22] that were carried out compacting granular mixtures with different methods, and then back-calculating K based on the experimental compacity lead to the packing coefficients presented in Table 3.2.

Moreover, the residual compacity of each grain size, β_i, is a fundamental variable for the LPM. Based on the packing coefficients presented in Table 3.2, two methods have been proposed to measure the residual compacity of each fraction:

- A vibration method that consists of placing a mass of 7 kg of a sample of each fraction in a cylindrical mold in three successive layers and applying a vertical vibration of 0.2 mm at 50 Hz for 20 s. Afterward, a steel plate applying 10 kPa is placed over the third layer, and the vibration is applied for 90 s.
- The second method is similar to the first one, but the compaction energy is applied by shocks: 20 shocks from a height of 10 mm for each one of the three layers, and then 40 shocks from the same height placing the steel plate over the third layer.

For both methods, the compacity is calculated by measuring the volume of the compacted sample.

Using the LPM, it is possible to study the effect of variables such as the shape of the particles, the grain size distribution, the level of compaction, and the density reached after compaction.

The effect of the shape of the particles is captured in the LPM by the residual compacity β_i, which could be different for each grain size. In fact, Figure 3.8 presents a comparison between the results of the maximum and minimum void ratios presented in Ref. [70], and the values calculated using the LPM for a different shape of particles

Table 3.2 Values of packing coefficients K obtained for different compaction methods

Method	K
Pouring	4.1
Sticking with a rod	4.0
Vibration, free surface	4.75
Vibration + compression 10 kPa	9.0
Chock table	9.0

Source: From Refs. [21,22].

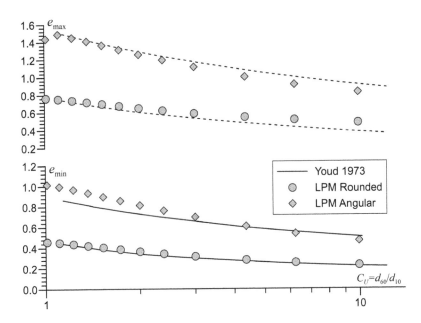

Figure 3.8 Comparison between the maximum and minimum void ratios given in Ref. [70] and calculated using the LPM.

and different coefficient of uniformity of the mixture. Results of the LPM were obtained using packing coefficients of $K = 4.1$ for the loose state and $K = 9$ for the dense state; residual compacities, β_i for all particles within the mixture were $\beta_i = 0.74$ for rounded particles, which correspond to a cubical arrangement of spheres, and $\beta_i = 0.55$ for angular particles.

The good agreement between the LPM and the results presented in Ref. [70] shows the capability of the LPM for calculating the loose and dense state of granular materials that are key values to calculate the relative density.

Regarding compacted materials, an extensive experimental work presented in Ref. [48] shows that the dry density of the modified Proctor test corresponds to a packing coefficient of $K = 12$.

3.2 EXAMPLE 12: SIMULATION OF FIELD COMPACTION USING THE BBM

The purpose of this example is to use the BBM for simulating field compaction. Field data is based on the results presented in Ref. [43] obtained using a tire compactor. The example requires the analysis of several cases: (i) two stress distribution on the soil–tire contact (uniform distribution over a circular area and nonuniform over a super-ellipse), (ii) two models for the stress distribution below the tire (Boussinesq and Fröhlich), (iii) two stress paths (oedometric and triaxial), and (iv) the effect of different water contents.

The soil is a medium-plasticity clay having the following characteristics:

- *Liquid Limit $w_L = 70\%$, plasticity index $PI = 40\%$, density of solid particles $\frac{\gamma_s}{\gamma_w} = 2.70$.*

- *Proctor compaction characteristics for the standard test are $\rho_{d_{max}} = 1.42 \, Mg/m^3$, $w_{opt} = 28\%$.*

- *Table 3.3 provides the parameters of the BBM and the water retention curve of this soil.*

- *For the oedometric path, the coefficient of horizontal stress at rest for the optimum water content is $K_0 = 0.45$.*

- *The parameter K_N, to include Equation 3.10, which permits to assess the effect of the number of passes of the compactor on the increase in density, depends on the water content. This example adopts the following values based on the data presented in Ref. [33]: $K_N = 0.15$ for $w = 28\%$, $K_N = 0.09$ for $w = 22\%$, and $K_N = 0.01$ for $w = 36\%$.*

- *Finally, the occlusion saturation degree is $Sr_{oc} = 97.5\%$.*

Table 3.3 shows the results of field compaction presented in Ref. [43] corresponding to 32 passes of the heavy tire compactor depicted in Figure 3.9 in which each tire carries 100 kN.

For the uniform stress distribution over a circular area, the contact pressure is $p_i = 770 \, kN$, while Table 3.5 provides the characteristics of the nonuniform stress distribution over a super-elliptical area described in Section 1.1.4.

Regarding the stress distribution within the soil parameters are Poisson ratio, $\nu = 0.3$, for the Boussinesq stress distribution, and concentration factor, $\xi = 5$, for the Fröhlich stress distribution.

Calculating the increase in density produced by the tire compactor requires the following steps:

1   Calculate the stress distribution on the soil's surface produced by one tire.
2   Calculate the stress distribution below a single tire using the Boussinesq and the Fröhlich stress distributions.
3   Calculate the effect of the adjacent tires of the compactor.

Table 3.3 Parameters of the BBM and the water retention curve for the soil of the example

Parameter	Meaning	Value	Parameter	Meaning	Value
$N(0)$	Compressibility	4.4	r	Compressibility	0.35
$\lambda(0)$	in saturated	0.32	β	in unsaturated	$0.5 \, MPa^{-1}$
κ	state	0.032	p^c	state	1 kPa
κ_s	Compressibility	0.35	ϕ_{wr}		$0.013 \, kPa^{-1}$
λ_s	due to suction		ψ_{wr}	Water	1.9
M	Critical state line	1.0	n_{wr}	retention	1.243
k_c	Tensile strength	0.8	m_{wr}	curve	0.414
			k_{swr}		$0.00025 \, kPa^{-1}$

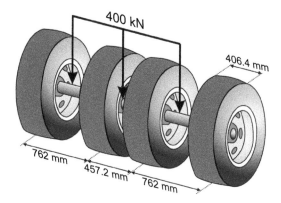

Figure 3.9 Layout of the heavy tire compactor used in Ref. [43].

4   Use the BBM to calculate the density profile for either the oedometric or triaxial stress paths and the optimum water content of the Proctor standard test.

5   Assess the increase of density due to the repetition of loading cycles (*i.e.,* 32 passes of the compactor), and compare the results with the field results given in Table 3.4.

6   Repeat the process for two different water contents: one above and the other below the optimum water content (*i.e.,* $w = 22\%$ and $w = 36\%$).

3.2.1   Stress distribution produced by one tire on the surface of the soil

The first point of the problem consists in computing the stress distribution on the contact tire–soil. The uniform stress distribution on a circular area requires the calculation of the radius a of the circular area which, considering the contact pressure of $p_i = 770$ kN is

$$a = \sqrt{\frac{F}{\pi p_i}} = \sqrt{\frac{100}{\pi \cdot 770}} = 0.203 \text{ m.}$$

On the other hand, the second case of this example regarding the stress distribution over the soil considers that the tire acts over a super-elliptical loaded area given by

Table 3.4 Field results obtained in Ref. [43] using a heavy tire compactor

Depth (m)	0.05	0.10	0.15	0.20	0.25	0.30	0.35	0.40
$\rho_d/\rho_{d\max}\%$	105.0	105.0	105.0	103.0	101.0	98.5	96.1	94.0

Table 3.5   Characteristics of the tire to obtain the nonuniform stress distribution over a super-elliptical area

Parameter	Value
Exponent of the super-elliptical area	$n = 4$
Width of the tire footprint	$w_T = 0.406$ m
Aspect ratio of the footprint area	$w_T/L_T = 0.8$
Coefficients for the nonuniform contact stress distribution	$\delta_k = 3, \alpha_k = 2$

Equation 1.27. For this case, the unit step function $\langle\rangle^0$ (which returns 0 for $\langle x\rangle^0 < 0$ and 1 for $\langle x\rangle^0 \geq 0$), applied to Equation 1.28, permits to define a function $\Omega(x, y)$ which returns 1 for a point (x, y) inside the loaded area or zero for points outside the loaded area.

$$\Omega(x, y) = \left(1 - \left(|x/a|^n + |y/b|^n\right)\right)^0. \tag{3.26}$$

For the data of the example, Equation 3.26 becomes

$$\Omega(x, y) = \left(1 - |x/0.204|^4 - |y/0.254|^4\right)^0.$$

Then, Equation 1.29 permits to compute the stress distribution on the loaded area as

$$\sigma_z(x, y) = C_{AK}\Omega(x, y)\left(0.5 - \frac{y}{w_T(x)}\right)e^{-3(0.5-y/w_T(x))}\left[1 - \left(\frac{x}{l_T(y)/2}\right)^2\right],$$

where $w_T(x)$ and $l_T(y)$ are the width and the length of the contact area, δ_K and α_K are shape parameters, and C_{AK} is a proportionality factor accounting for the total load.

Equation 1.27, which represents the equation of the super-ellipse, permits to compute the width and the length of the contact area $w_T(x)$ and $l_T(y)$ as

$$l_T(y) = 0.204\left[1 - \left(\frac{y}{0.254}\right)^4\right]^{1/4}, \text{ and}$$

$$w_T(x) = 0.254\left[1 - \left(\frac{x}{0.204}\right)^4\right]^{1/4}.$$

A discrete solution is a straightforward method to compute the stress distribution. However, the smaller the discretization intervals, the more accurate this type of solution. In this example, the length and the width of the footprint area are divided into 24 intervals to simplify the presentation, leading to the following lengths $\Delta x = 0.406/24 = 0.0169$ m and $\Delta y = 0.5075/24 = 0.0211$ m. Then, the coordinates (x, y) of any point of the surface are $x = i\Delta x$ and $y = j\Delta y$ for $-12 \leq i, j \leq 12$.

The value of the constant C_{AK} results in integrating, in a discrete form, the stress distribution function over the loaded area and equating this integral to the tire load of 100 kN, as follows:

$$C_{AK} = \frac{100/(\Delta x \Delta y)}{\sum\limits_{i=-12}^{12}\sum\limits_{j=-12}^{12}\left\{\Omega(i\Delta x, j\Delta y)\left(0.5 - \frac{j\Delta y}{w_T(i\Delta x)}\right)e^{-3(0.5-j\Delta y/w_T(i\Delta x))}\left[1 - \left(\frac{i\Delta x}{l_T(j\Delta y)/2}\right)^2\right]\right\}}.$$

The previous equation leads to a value of $C_{AK} = 7470.1$ kPa, which permits to compute the stress distribution presented in Table 3.6, corresponding to the positive quadrant of the loaded area since the results for the other quadrants are symmetrical with respect to the positive one.

Discretizing the uniform loaded area is possible using small areas. However, obtaining the actual area of a circle using discrete square areas is only possible if the number of areas increases to infinity. Therefore, obtaining the same total load of a circularly loaded area as a sum of discrete elementary areas requires either increasing the contact stress or increasing the size of the circle.

Table 3.6 Results of the nonuniform stress distribution on the super-elliptical loaded area, in kPa. Shaded cells are points outside the loaded area.

		0	1	2	3	4	5	*j* 6	7	8	9	10	11	12
	12	553.4	0.0	0.0	0.0	0.0	0.0	0.0	0.0	0.0	0.0	0.0	0.0	0.0
	11	667.6	657.2	625.9	573.8	500.7	406.8	292.1	156.5	0.0	0.0	0.0	0.0	0.0
	10	755.1	747.6	724.9	687.2	634.3	566.4	483.3	385.1	271.9	143.5	0.0	0.0	0.0
	9	819.9	811.7	787.1	746.1	688.7	615.0	524.8	418.2	295.2	155.8	0.0	0.0	0.0
	8	865.6	858.4	836.9	801.2	751.1	686.7	608.0	515.0	407.7	286.1	150.2	0.0	0.0
	7	895.1	887.7	865.5	828.5	776.7	710.1	628.8	532.6	421.6	295.9	155.3	0.0	0.0
i	6	911.1	903.6	881.0	843.4	790.7	722.9	640.1	542.2	429.2	301.2	158.1	0.0	0.0
	5	916.0	908.5	885.8	847.9	794.9	726.8	643.5	545.1	431.5	302.8	159.0	0.0	0.0
	4	911.7	904.2	881.6	843.9	791.2	723.4	640.5	542.5	429.5	301.4	158.2	0.0	0.0
	3	900.0	892.5	870.2	833.0	781.0	714.0	632.2	535.5	424.0	297.5	156.2	0.0	0.0
	2	882.2	874.9	853.0	816.6	765.5	699.9	619.7	524.9	415.6	291.6	153.1	0.0	0.0
	1	859.6	852.5	831.2	795.7	746.0	682.0	603.9	511.5	405.0	284.2	149.2	0.0	0.0
	0	833.4	827.6	810.3	781.3	740.8	688.7	625.1	549.8	463.0	364.6	254.7	133.1	0.0

This example uses the second option. Then, inscribing a discrete area having 24×24 square sections into a circle requires increasing the radius of the circle from $a = 0.203$ m to $a = 0.206$ m. This small increment in radius guaranties a load of 100.07 kN, which is close to 100 kN applied by each tire. Therefore, each square section of the discretized area has $\Delta x = \Delta y = 0.412/24$ m.

Figure 3.10 compares the results of the uniform and nonuniform stress distribution along a transverse section located at $y = 0$. The results show a maximum value of vertical stress for the nonuniform distribution of 916 kPa, which is 19% higher than the uniform contact pressure of 770 kPa. On the other hand, the maximum value of contact pressure for the nonuniform distribution is not at the center of the loaded area; in fact, the maximum value of 916 kPa is located on the point $x = 5\Delta x$ and $y = 0$.

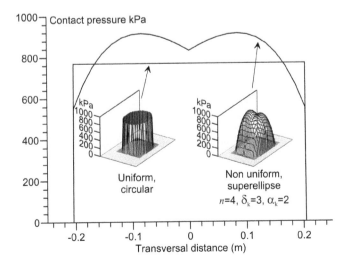

Figure 3.10 Stress distribution at the interface tire–soil, comparison along a transverse section.

Table 3.7 Results of the uniform stress distribution on a circularly loaded area, in kPa. Shaded cells are points outside the loaded area.

		0	1	2	3	4	5	6	7	8	9	10	11	12
								j						
	12	770.0	0.0	0.0	0.0	0.0	0.0	0.0	0.0	0.0	0.0	0.0	0.0	0.0
	11	770.0	770.0	770.0	770.0	770.0	0.0	0.0	0.0	0.0	0.0	0.0	0.0	0.0
	10	770.0	770.0	770.0	770.0	770.0	770.0	770.0	0.0	0.0	0.0	0.0	0.0	0.0
	9	770.0	770.0	770.0	770.0	770.0	770.0	770.0	770.0	0.0	0.0	0.0	0.0	0.0
	8	770.0	770.0	770.0	770.0	770.0	770.0	770.0	770.0	770.0	0.0	0.0	0.0	0.0
	7	770.0	770.0	770.0	770.0	770.0	770.0	770.0	770.0	770.0	770.0	0.0	0.0	0.0
i	6	770.0	770.0	770.0	770.0	770.0	770.0	770.0	770.0	770.0	770.0	770.0	0.0	0.0
	5	770.0	770.0	770.0	770.0	770.0	770.0	770.0	770.0	770.0	770.0	770.0	0.0	0.0
	4	770.0	770.0	770.0	770.0	770.0	770.0	770.0	770.0	770.0	770.0	770.0	770.0	0.0
	3	770.0	770.0	770.0	770.0	770.0	770.0	770.0	770.0	770.0	770.0	770.0	770.0	0.0
	2	770.0	770.0	770.0	770.0	770.0	770.0	770.0	770.0	770.0	770.0	770.0	770.0	0.0
	1	770.0	770.0	770.0	770.0	770.0	770.0	770.0	770.0	770.0	770.0	770.0	770.0	0.0
	0	770.0	770.0	770.0	770.0	770.0	770.0	770.0	770.0	770.0	770.0	770.0	770.0	770.0

The next point of this example analyzes the effect of the uniform and nonuniform contact pressures applied at the soil surface on the stress distribution within it (Table 3.7).

3.2.2 Stress distribution within the soil mass

The example uses two approaches for computing the stress distribution within the soil mass: the Boussinesq approach with a Poisson's ratio $\nu = 0.3$ and the Fröhlich approach with a stress concentration factor $\xi = 5$. Both approaches are applied for the uniform and nonuniform contact stress distribution calculated in the previous point.

According to the Boussinesq approach, Equation 1.3 allows computing the vertical stress in a point (x, y, z) as

$$\sigma_z(x, y, z) = \sum_{i=-12}^{12} \sum_{j=-12}^{12} \frac{3\sigma_0 \Delta x \Delta y}{2\pi} \frac{z^3}{\left[(x - i\Delta x)^2 + (y - j\Delta y)^2 + z^2\right]^{5/2}}.$$

On the other hand, the Fröhlich approach requires calculating the radial stress $\sigma_R(x, y, z)_{i,j}$ produced by a load located at the surface on a point $(i\Delta x, j\Delta y)$ using Equation 1.19:

$$\sigma_R(x, y, z)_{i,j} = \frac{\xi \sigma_0 \Delta x \Delta y}{2\pi \left[(x - i\Delta x)^2 + (y - j\Delta y)^2 + z^2\right]^2}$$

$$\cdot \cos\left(\frac{z}{\sqrt{(x - i\Delta x)^2 + (y - j\Delta y)^2 + z^2}}\right)^{\xi - 2}.$$

Afterward, using Equation 1.23, the vertical stress becomes

$$\sigma_z(x, y, z) = \sum_{i=-12}^{12} \sum_{j=-12}^{12} \sigma_R(x, y, z)_{i,j} \frac{z^2}{(x - i\Delta x)^2 + (y - j\Delta y)^2 + z^2}.$$

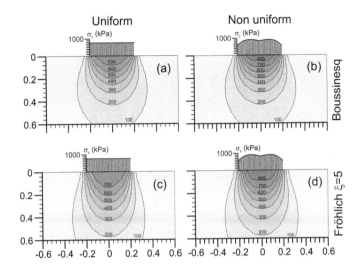

Figure 3.11 Stress distribution within the soil layer (a,b) Boussinesq (c,d) Fröhlich.

Figure 3.11 illustrates the results of the distribution of vertical stresses along a transverse section of the tire, $y = 0$, for the Boussinesq and Fröhlich approaches and the stress distributions uniform and nonuniform. Two points are evident in this figure:

- The nonuniform stress distribution leads to higher stresses mainly at the shallow layers of the soil $z < 0.1$ m; however, when comparing the effect of the uniform or the nonuniform contact stress at deeper layers, the vertical stress is almost the same.
- The Fröhlich stress distribution produces higher penetration of the vertical stress within the soil.

3.2.3 Stress distribution produced by the whole compactor

Equations 1.1–1.6 and 1.21–1.26 provide the complete stress tensor for the Boussinesq and Fröhlich approaches, respectively. Since, as shown in Figure 3.9, the whole compactor has four tires, the stress below central tires results from the superposition of stresses produced by each one of them (Figure 3.12).

The superposition method for stresses assumes an elastic response of the soil, which is only an approximation of the problem. However, accepting that the assumption of elasticity is valid, the stress along directions x, y, z below the tire B is

$$\sigma_{x,y,z} = \sigma_{x,y,z}(B - A) + \sigma_{x,y,z}(B - B) + \sigma_{x,y,z}(B - C) + \sigma_{x,y,z}(B - D).$$

Then, considering the distances of each tire to tire B,

$$\sigma_{x,y,z} = \sigma_{x,y,z}(x = 0.762, y = 0) + \sigma_{x,y,z}(x = 0, y = 0)$$
$$+ \sigma_{x,y,z}(x = 0.4572, y = 0) + \sigma_{x,y,z}(x = 1.2192, y = 0).$$

Figure 3.13 shows the results of the stress distribution below the center of a central tire. This figure compares the results of (i) uniform contact stress with a Boussinesq stress

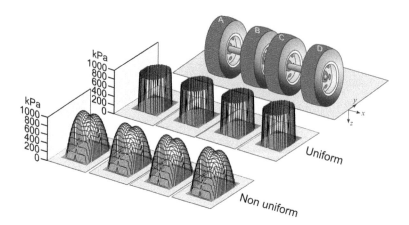

Figure 3.12 Arrangement of the contact stresses produced by the heavy tire compactor.

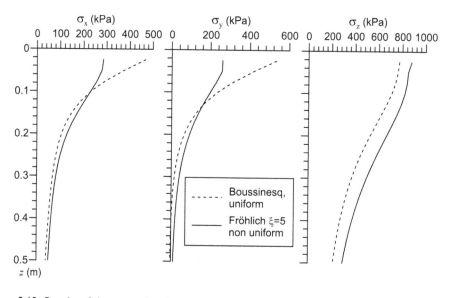

Figure 3.13 Results of the stress distribution below the center of a central tire.

distribution and (ii) nonuniform contact stress with Fröhlich stress distribution showing that this last option leads to higher vertical stresses within the whole layer. However, the Boussinesq approach produces higher horizontal stresses at the upper part of the soil, decreasing the deviator stress in this zone.

The same analysis is valid for computing shear stresses. However, the example neglects shear stresses because the point below a central tire is close to the axis of symmetry. Therefore, stresses along the x, y, z axis could be assumed as the principal stresses σ_1, σ_2, σ_3. Subsequently, Equations 1.104 and 1.105 allow computing the mean and deviator stresses as follows:

$$p = \frac{\sigma_x + \sigma_y + \sigma_z}{3},$$

$$q = \sqrt{\frac{1}{2}\left[(\sigma_x - \sigma_y)^2 + (\sigma_y - \sigma_z)^2 + (\sigma_x - \sigma_z)^2\right]}.$$

Figure 3.14 shows the results of mean and deviator stresses for both approaches: Boussinesq and Fröhlich, as well as for the uniform and nonuniform contact stresses. It is important to remark that the maximum contact stress for the nonuniform stress distribution is located at the point $(x = 5\Delta x, y = 0)$, and therefore, for shallow depths, $z < 0.1$ m, the maximum deviator stress appears below the point of maximum contact stress, which is located at $x = 0.0846$, as indicated in Figure 3.14.

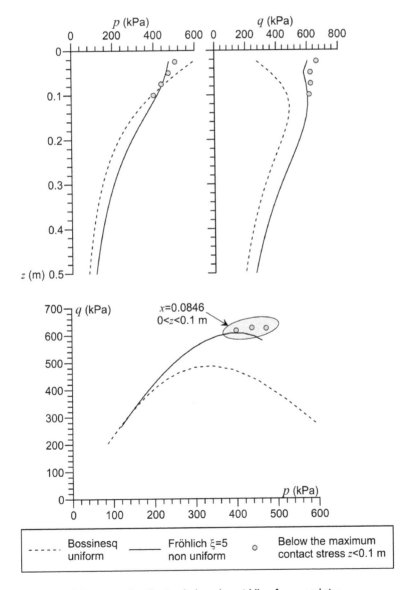

Figure 3.14 Results of the stress distribution below the middle of a central tire.

3.2.4 Compaction profiles calculated using the BBM

The next stage of the example consists in calculating the density profile of the compacted layer using the BBM. The first step in the process of compaction is placing the soil in a loose state and at a specific water content, it leads to an initial saturation degree and an initial suction pressure. Afterward, the compaction process reduces the air voids, increasing the saturation degree and modifying the suction pressure.

Equation 3.11 permits to compute the evolution of suction s as the compaction process progresses and the void ratio e decreases as follows:

$$s = \frac{1}{\phi_{wr}} \left(S_r^{-1/m_{wr}} - 1 \right)^{1/n_{wr}} e^{-\psi_{wr}}.$$

However, for a constant water content w_0, the saturation degree is $S_r = \frac{w_0}{e} \frac{\gamma_s}{\gamma_w}$, and moreover, the void ratio e, in terms of the specific volume v, is $e = v - 1$, and then the previous equation becomes

$$s = \frac{1}{\phi_{wr}} \left[\left(\frac{w_0}{v-1} \frac{\gamma_s}{\gamma_w} \right)^{-1/m_{wr}} - 1 \right]^{1/n_{wr}} (v - 1)^{-\psi_{wr}}. \tag{3.27}$$

On the other hand, for a stress state represented by the invariants (p, q), the equation of the elliptic yield curve of the BBM is given in Equation 3.1 as

$$q^2 - M^2(p + p_s)(p_0 - p) = 0.$$

This equation permits to calculate the overconsolidation stress p_0 for the unsaturated state, which is the point where the ellipse crosses the positive side of the p axis, as shown in Figure 3.15. Therefore, p_0 is

$$p_0 = p + \frac{q^2}{M^2(p + p_s)} = p \left[1 + \frac{\alpha^2}{M^2 \left(1 + \frac{p_s}{p} \right)} \right].$$

Designating with α the relationship between the deviator and the mean stress $\alpha = q/p$, the overconsolidation stress p_0 becomes

$$p_0 = p \left[1 + \frac{\alpha^2}{M^2 \left(1 + \frac{p_s}{p} \right)} \right]. \tag{3.28}$$

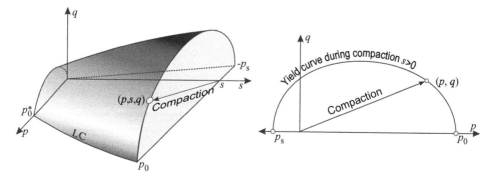

Figure 3.15 Interpretation of the compaction process within the BBM framework.

The value of α depends on the compression path assumed for the compaction process. For the oedometric stress path, the horizontal stress is proportional to the vertical stress with the at-rest coefficient as the proportionality factor, $\sigma_h = K_0\sigma_v$, while for the triaxial path, the value of α depends on the stress path followed in the field. These considerations lead to the following values for α:

Oedometric path

$$\alpha_{oedo} = \frac{q}{p} = \frac{3(1 - K_0)}{1 + 2K_0}, \quad and$$

$$p = \frac{\sigma_v(1 + 2K_0)}{3},$$

Triaxial path

$$\alpha_{triax} = \frac{q}{p}.$$

After calculating the value of the overconsolidation stress, p_0, Equations 3.3, 3.4, and 3.8 allow computing the evolution of the specific volume as follows:

$$v = N(0) - \kappa_s \ln \frac{s + p_{atm}}{p_{atm}} - \lambda(0) \left[(1 - r)e^{-\beta s} + r \right] \ln \frac{p_0}{p^c}. \tag{3.29}$$

To sum up, the evolution of suction and specific volume, or void ratio, of a soil during compaction at a constant water content results from the solution of the system of non linear equations given in Table 3.8.

Considering the values of the parameters given in Table 3.3, the equations that allow calculating the specific volume for a given vertical stress are

Saturation degree

$$S_r = \frac{2.7w_0}{v - 1}, \tag{3.30}$$

Suction in kPa

$$s = 76.92(v - 1)^{-1.9} \left[S_r^{-2.415} - 1 \right]^{0.805}, \tag{3.31}$$

Oedometric path

$$\alpha = 0.8684, \quad and \quad p = \frac{1.9}{3}\sigma_v, \tag{3.32}$$

Overconsolidation stress

Table 3.8 System of nonlinear equations for computing suction and specific volume during compaction

Equation	Expression
3.27	$s = \frac{1}{\phi_{wr}} \left[\left(\frac{w_0}{v-1} \frac{\gamma_s}{\gamma_w} \right)^{-1/m_{wr}} - 1 \right]^{1/n_{wr}} (v - 1)^{-\psi_{wr}}$
3.28	$p_0 = p \left[1 + \frac{\alpha^2}{M^2 \left(1 + \frac{p_s}{p}\right)} \right]$
3.29	$v = N(0) - \kappa_s \ln \frac{s + p_{atm}}{p_{atm}} - \lambda(0) \left[(1 - r)e^{-\beta s} + r \right] \ln \frac{p_0}{p^c}$

$$p_0 = p \left(1 + \frac{\alpha^2}{1 + \frac{0.8s}{p}} \right),$$

(3.33)

Specific volume

$$v = 4.4 - 0.35 \ln \frac{s + 101.3}{101.3} - 0.32 \left(0.65 e^{-0.0005s} + 0.35 \right) \ln \left(\frac{p_0}{p^c} \right).$$

(3.34)

The following iterative procedure allows computing the void ratio at any depth of the compacted layer (Table 3.9):

1   Find the value of the vertical stress or p, q for a specific depth calculated using the Fröhlich approach.
2   Assume a value of specific volume v.
3   Compute the suction, s, using Equation 3.31.
4   Compute p_0 using Equation 3.33.
5   Compute v using Equation 3.34.
6   Verify the convergence by comparing the assumed and computed specific volumes.
7   Return to step 2 until convergence.
8   Choose another depth in the compacted layer.

3.2.5   Effect of the loading cycles

Equation 3.10 permits to assess the effect of the loading cycles on the increase in density. For the data of the problem and for 32 passes of the compactor, Equation 3.34 becomes

$$v = 4.4 - 0.35 \ln \frac{s + 101.3}{101.3} - 0.32 \left(0.65 e^{-0.0005s} + 0.35 \right) \ln \left(\frac{p_0}{p^c} 32^{K_N} \right)$$

(3.35)

However, Equation 3.35 is only valid providing that the saturation degree is lower than the value that produces occlusion of air within the voids of the soil, and this value is denoted as $S_{r_{oc}}$. Otherwise, the maximum dry density is controlled by $S_{r_{oc}}$. The phase diagram presented in Figure 3.16 permits to obtain the maximum density when the air is occluded. According to Figure 3.16, for a saturation degree $S_r = S_{r_{oc}}$, the volume of water is $V_w = S_{r_{oc}} e$. On the other hand, the mass analysis leads to a volume of water of $V_w = w_0 \frac{\rho_s}{\rho_w}$.

Therefore, equating the volumes of water obtained from the phase diagram of Figure 3.16, the void ratio at the occlusion point becomes

$$S_{r_{oc}} e_{oc} = w_0 \frac{\rho_s}{\rho_w} \quad \text{then} \quad e_{oc} = \frac{w_0}{S_{r_{oc}}} \frac{\rho_s}{\rho_w}.$$

As a result, the maximum dry density at the occlusion point, $\rho_{d_{oc}}$, is

$$\rho_{d_{oc}} = \frac{\rho_s}{1 + \frac{w_0}{S_{r_{oc}}} \frac{\rho_s}{\rho_w}}.$$

(3.36)

Finally, the dry density is the minimum value between the dry density obtained with the model using Equation 3.35 and the dry density at the occlusion point given by Equation 3.36.

Table 3.9 Results of the simulation of the compaction process obtained using the BBM

z (m)	Oedometric path							Triaxial path						
	σ_v (kPa)	p (kPa)	v	S_r	s (kPa)	p_0 (kPa)	$\frac{\gamma_d}{\gamma_{dmax}}$ (%)	p (kPa)	q (kPa)	v	S_r	s (kPa)	p_0 (kPa)	$\frac{\gamma_d}{\gamma_{dmax}}$ (%)
0.025	946.8	599.6	2.039	0.727	80.4	1008.0	93.2	504.9	662.9	1.975	0.775	70.7	1287.6	96.3
0.050	888.5	562.7	2.057	0.715	82.7	942.4	92.4	472.4	624.1	1.992	0.762	73.5	1205.6	95.5
0.075	854.5	541.2	2.069	0.708	84.1	904.2	91.9	437.4	625.6	1.987	0.766	72.8	1227.0	95.7
0.100	820.2	519.5	2.080	0.700	85.5	865.6	91.4	399.5	616.1	1.987	0.766	72.7	1228.9	95.7
0.125	789.9	500.3	2.091	0.693	86.7	831.6	90.9	386.3	606.2	1.991	0.763	73.3	1212.2	95.5
0.150	751.8	476.1	2.106	0.684	88.3	788.8	90.3	354.6	596.6	1.990	0.764	73.1	1216.0	95.6
0.175	708.7	448.8	2.123	0.673	90.1	740.4	89.5	323.8	578.2	1.994	0.761	73.8	1197.0	95.4
0.200	663.3	420.1	2.143	0.661	92.1	689.6	88.7	295.0	553.4	2.003	0.754	75.1	1157.5	94.9
0.225	618.0	391.4	2.165	0.649	94.1	638.9	87.8	268.8	524.9	2.015	0.745	77.0	1102.4	94.3
0.250	574.2	363.7	2.187	0.637	96.0	590.1	86.9	245.4	494.5	2.032	0.733	79.3	1037.3	93.6
0.275	533.0	337.6	2.210	0.625	97.8	544.2	86.0	224.6	464.1	2.050	0.720	81.8	967.2	92.7
0.300	494.9	313.4	2.233	0.613	99.6	501.9	85.1	206.2	434.6	2.071	0.706	84.4	896.2	91.8
0.325	460.1	291.4	2.256	0.602	101.2	463.3	84.3	190.1	406.7	2.093	0.692	86.9	827.1	90.9
0.350	428.5	271.4	2.279	0.591	102.7	428.5	83.4	175.9	380.7	2.116	0.678	89.3	761.8	89.9
0.375	399.9	253.3	2.301	0.581	104.1	397.0	82.6	163.4	356.8	2.139	0.664	91.6	701.2	88.9
0.400	374.2	237.0	2.322	0.572	105.4	368.8	81.9	152.4	334.9	2.162	0.651	93.8	645.7	88.0
0.425	351.0	222.3	2.343	0.563	106.5	343.5	81.2	142.5	315.0	2.185	0.638	95.8	595.3	87.0
0.450	330.2	209.1	2.363	0.555	107.6	320.8	80.5	133.8	296.9	2.207	0.626	97.6	549.9	86.1
0.475	311.3	197.2	2.382	0.547	108.6	300.4	79.8	125.9	280.5	2.229	0.615	99.3	509.0	85.3
0.500	294.3	186.4	2.401	0.540	109.5	282.0	79.2	118.9	265.6	2.251	0.604	100.8	472.3	84.5

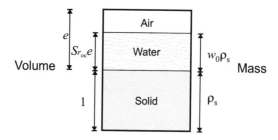

Figure 3.16 Phase diagram for the occlusion saturation degree.

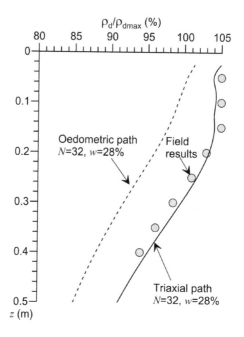

Figure 3.17 Comparison between the density profiles computed for the oedometric and triaxial stress paths and the field results.

Table 3.10 and Figure 3.17 show the results of the model when assuming the oedometric or the triaxial paths. The results show that the triaxial path leads to a better agreement with the field results compared with the oedometric path. The better agreement of the triaxial path occurs because this path produces higher deviatoric stresses leading to higher values of p_0.

Table 3.10 Comparison between the model and the field results

Depth (m)	0.05	0.10	0.15	0.20	0.25	0.30	0.35	0.40
$\rho_d/\rho_{d\text{max}}$% Oedometric path	100.3	99.1	97.8	95.9	93.9	91.8	89.8	88.0
$\rho_d/\rho_{d\text{max}}$% Triaxial path	103.9	104.2	104.1	103.3	101.7	99.6	97.3	95.1
$\rho_d/\rho_{d\text{max}}$% Field	105.0	105.0	105.0	103.0	101.0	98.5	96.1	94.0

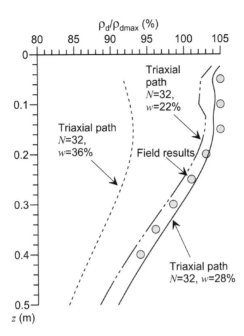

Figure 3.18 Density profiles computed for three different water contents.

3.2.6 Effect of the water content

Figure 3.18 presents the results obtained following the analysis presented above for other water contents ($w_0 = 36\%$, and $w_0 = 22\%$). The results show the capabilities of the BBM in assessing the effect of the water content on the profile of densities in a layer undergoing compaction.

3.3 EXAMPLE 13: USE OF THE LINEAR PACKING MODEL TO COMPUTE THE DENSITY OF A COMPACTED MATERIAL BASED ON ITS GRAIN SIZE DISTRIBUTION

The purpose of this example is to use the LPM to compute the density of a granular material compacted using the modified Proctor test. Tables 3.11 and 3.12 present the characteristics of the granular material studied in Ref. [18] regarding its grain size distribution and the results of the modified Proctor compaction test. The main questions of this example are as follows:

1 Compute the virtual compacity, considering the grain size distribution of the granular material.

2 Assume different values of the actual compacity and compute the corresponding packing coefficient for each one of them.

3   Compute the dry density of each actual compacity and obtain the dry density corresponding to a packing coefficient of $K = 12$.

4   Compare the dry density computed with the LPM with the experimental dry density obtained in the modified Proctor test.

The results presented in Ref. [23] show that the values of residual compacity could be $\beta_i = 0.59$ for grains whose size is $d_i > 6.3$ mm and $\beta_i = 0.63$ for grains with sizes $d_i \leq 6.3$ mm. Moreover, the densities of the particles are $\rho_s = 2.65$ Mg/m^3 for grains with $d_i > 6.3$ mm and $\rho_s = 2.70$ Mg/m^3 for grains with sizes $d_i \leq 6.3$ mm.

3.3.1   Virtual compacity

Equation 3.24 allows computing the virtual compacity of the granular mixture, and in a simplified form, this equation becomes

$$\gamma_i = \frac{\beta_i}{1 - S_{1,i} - S_{2,i}},$$

where

$$S_{1,i} = \sum_{j=1}^{i-1} \left[1 - \beta_i + b_{i,j}\beta_i \left(1 - 1/\beta_j \right) \right] y_j, \quad \text{and}$$

$$S_{2,i} = \sum_{j=i+1}^{n_c} \left[1 - a_{i,j}\beta_i/\beta_j \right] y_j.$$

Computing γ_i requires assuming successively grain size i as the dominant grain size and then choosing the minimum value of γ_i. For example, if assuming as dominant the class size denoted with $i = 6$, the computation of $S_{1,6}$ and $S_{2,6}$ is as follows:

$$S_{1,6} = \sum_{j=1}^{5} \left[1 - \beta_i + b_{i,j}\beta_i \left(1 - 1/\beta_j \right) \right] y_j = \begin{pmatrix} 1 - \beta_6 + b_{6,1}(1 - 1/\beta_1)\beta_6 \\ 1 - \beta_6 + b_{6,2}(1 - 1/\beta_2)\beta_6 \\ 1 - \beta_6 + b_{6,3}(1 - 1/\beta_3)\beta_6 \\ 1 - \beta_6 + b_{6,4}(1 - 1/\beta_4)\beta_6 \\ 1 - \beta_6 + b_{6,5}(1 - 1/\beta_5)\beta_6 \end{pmatrix} \begin{pmatrix} y_1 \\ y_2 \\ y_3 \\ y_4 \\ y_5 \end{pmatrix}$$

Table 3.11   Grain size distribution of the material

Sieve (mm)	25.4	19.0	12.7	9.51	6.30	4.76	2.38	1.19	0.595	0.211	0.074	<0.074
% Retained	0	5.0	19.4	9.6	10.4	9.0	13.9	9.4	6.8	6.2	3.3	7.0

Source: From Ref. [18].

Table 3.12   Results of the modified Proctor compaction test

ρ_d (Mg/m³)	2.17	2.20	2.20	2.19	2.17
w (%)		4.48	5.20	6.16	6.70	7.29

Source: From Ref. [18].

$$S_{2,6} = \sum_{j=7}^{11} [1 - a_{6,j}\beta_6/\beta_j] y_j = \begin{pmatrix} 1 - a_{6,7}\beta_6/\beta_7 \\ 1 - a_{6,8}\beta_6/\beta_8 \\ 1 - a_{6,9}\beta_6/\beta_9 \\ 1 - a_{6,10}\beta_6/\beta_{10} \\ 1 - a_{6,11}\beta_6/\beta_{11} \end{pmatrix} \begin{pmatrix} y_7 \\ y_8 \\ y_9 \\ y_{10} \\ y_{11} \end{pmatrix}$$

For each grain class, the medium size of the grains is chosen as the mean size of consecutive sieves: $d_i = (d_{\text{sieve}_i} + d_{\text{sieve}_{i-1}})/2$. Then, from Equations 3.22 and 3.23, coefficients $a_{i,j}$ and $b_{i,j}$ are

$$\begin{pmatrix} a_{6,7} \\ a_{6,8} \\ a_{6,9} \\ a_{6,10} \\ a_{6,11} \end{pmatrix} = \begin{pmatrix} 1 - (1 - d_7/d_6)^{1.02} \\ 1 - (1 - d_8/d_6)^{1.02} \\ 1 - (1 - d_9/d_6)^{1.02} \\ 1 - (1 - d_{10}/d_6)^{1.02} \\ 1 - (1 - d_{11}/d_6)^{1.02} \end{pmatrix}^{0.5} = \begin{pmatrix} 0.7120 \\ 0.5043 \\ 0.3391 \\ 0.2017 \\ 0.1028 \end{pmatrix}$$

$$\begin{pmatrix} b_{6,1} \\ b_{6,2} \\ b_{6,3} \\ b_{6,4} \\ b_{6,5} \end{pmatrix} = \begin{pmatrix} 1 - (1 - d_6/d_1)^{1.5} \\ 1 - (1 - d_6/d_2)^{1.5} \\ 1 - (1 - d_6/d_3)^{1.5} \\ 1 - (1 - d_6/d_4)^{1.5} \\ 1 - (1 - d_6/d_5)^{1.5} \end{pmatrix} = \begin{pmatrix} 0.2312 \\ 0.3180 \\ 0.4411 \\ 0.5939 \\ 0.7890 \end{pmatrix}$$

Therefore, S_1 and S_2 become

$$S_{1,6} = \begin{pmatrix} 1 - 0.63 + 0.2312(1 - 1/0.59)0.63 \\ 1 - 0.63 + 0.3180(1 - 1/0.59)0.63 \\ 1 - 0.63 + 0.4411(1 - 1/0.59)0.63 \\ 1 - 0.63 + 0.5939(1 - 1/0.59)0.63 \\ 1 - 0.63 + 0.7890(1 - 1/0.63)0.63 \end{pmatrix} \begin{pmatrix} 0.05 \\ 0.194 \\ 0.096 \\ 0.104 \\ 0.09 \end{pmatrix} = 0.09365$$

$$S_{2,6} = \begin{pmatrix} 1 - 0.7120 \cdot 0.63/0.63 \\ 1 - 0.5043 \cdot 0.63/0.63 \\ 1 - 0.3391 \cdot 0.63/0.63 \\ 1 - 0.2017 \cdot 0.63/0.63 \\ 1 - 0.1028 \cdot 0.63/0.63 \end{pmatrix} \begin{pmatrix} 0.094 \\ 0.068 \\ 0.062 \\ 0.033 \\ 0.07 \end{pmatrix} = 0.1909$$

Finally, γ_6 is

$$\gamma_6 = \frac{0.63}{1 - 0.09365 - 0.1909} = 0.8806$$

This procedure must be repeated assuming each grain class as dominant, and Table 3.13 presents the results of the virtual compacity γ after such assumption.

Finally, because of the impenetrability condition, the virtual compacity of the granular material is the minimum value of γ_i (i.e., min $\gamma_{1 \leq i \leq 11}$), leading to $\gamma_i = 0.8806$.

3.3.2 Actual compacity

The virtual compacity is unreachable by using either laboratory or field techniques. Only values of actual compacity $\Phi < \gamma$ are accessible. As described in Section 3.1.7.2, virtual

Table 3.13 Table for computing the virtual compacity of the granular mixture

Class i	Sieve (mm)	Retained (%)	Passing (%)	d_i (mm)	y_i	β_i	S_1	S_2	γ_i	
	25.400	0.0	100.0							
1				22.200	0.050	0.59	0.0000	0.5081	1.1993	
	19.000	5.0	95.0							
2				15.850	0.194	0.59	0.0031	0.4282	1.0375	
	12.700	19.4	75.6							
3				11.105	0.096	0.59	0.0203	0.3652	0.9600	
	9.510	9.6	66.0							
4				7.905	0.104	0.59	0.0449	0.3113	0.9164	
	6.300	10.4	55.6							
5				5.530	0.090	0.63	0.0512	0.2443	0.8942	
	4.760	9.0	46.6							
6				3.570	0.139	0.63	0.0937	0.1909	0.8806	Dominant size
	2.380	13.9	32.7							
7				1.785	0.094	0.63	0.1603	0.1353	0.8943	
	1.190	9.4	23.3							
8				0.893	0.068	0.63	0.2147	0.0953	0.9131	
	0.595	6.8	16.5							
9				0.403	0.062	0.63	0.2614	0.0618	0.9308	
	0.211	6.2	10.3							
10				0.143	0.033	0.63	0.3033	0.0340	0.9507	
	0.074	3.3	7.0							
11				0.037	0.070	0.63	0.3319	0.0000	0.9430	
	0.000	7.0								
	Σ	100						$\min(\gamma_i)$	0.8806	

and actual compacities are related through the packing coefficient K, which is calculated using the following equation:

$$K = \Sigma K_i = \sum_{i=1}^{11} \frac{y_i/\beta_i}{1/\Phi - 1/\gamma_i} = \begin{pmatrix} y_1/\beta_1 \\ y_2/\beta_2 \\ y_3/\beta_3 \\ y_4/\beta_4 \\ y_5/\beta_5 \\ y_6/\beta_6 \\ y_7/\beta_7 \\ y_8/\beta_8 \\ y_9/\beta_9 \\ y_{10}/\beta_{10} \\ y_{11}/\beta_{11} \end{pmatrix} \begin{pmatrix} 1/\Phi - 1/\gamma_1 \\ 1/\Phi - 1/\gamma_2 \\ 1/\Phi - 1/\gamma_3 \\ 1/\Phi - 1/\gamma_4 \\ 1/\Phi - 1/\gamma_5 \\ 1/\Phi - 1/\gamma_6 \\ 1/\Phi - 1/\gamma_7 \\ 1/\Phi - 1/\gamma_8 \\ 1/\Phi - 1/\gamma_9 \\ 1/\Phi - 1/\gamma_{10} \\ 1/\Phi - 1/\gamma_{11} \end{pmatrix}^{-1} .$$

For instance, for an actual compacity of $\Phi = 0.8$, the packing coefficient is

$$K(\Phi = 0.8) = \begin{pmatrix} 0.05/0.59 \\ 0.194/0.59 \\ 0.096/0.59 \\ 0.104/0.59 \\ 0.09/0.63 \\ 0.139/0.63 \\ 0.094/0.63 \\ 0.068/0.63 \\ 0.062/0.63 \\ 0.033/0.63 \\ 0.07/0.63 \end{pmatrix} \begin{pmatrix} 1/0.8 - 1/1.1993 \\ 1/0.8 - 1/1.0375 \\ 1/0.8 - 1/0.96 \\ 1/0.8 - 1/0.9164 \\ 1/0.8 - 1/0.8942 \\ 1/0.8 - 1/0.8806 \\ 1/0.8 - 1/0.8943 \\ 1/0.8 - 1/0.9131 \\ 1/0.8 - 1/0.9308 \\ 1/0.8 - 1/0.9507 \\ 1/0.8 - 1/0.943 \end{pmatrix}^{-1} = 9.50.$$

Table 3.14 presents the results of different packing coefficients K corresponding to different actual compacities Φ.

3.3.3 Dry density

After computing the actual compacity Φ, calculating the dry density is straightforward by multiplying the actual compacity by the mean density of solids $\rho_{s_{mean}}$, as follows:

$$\rho_d = \Phi \sum_{i=1}^{11} \rho_{s_i} y_i$$

Figure 3.19 presents the dry density of the granular material as a function of the packing coefficient, K.

Table 3.14 Table for computing the actual compacity and the packing coefficient of the granular mixture

Class i	γ_i		Φ					
			0.60	0.65	0.70	0.75	0.80	0.85
1	1.1993		0.102	0.120	0.142	0.170	0.204	0.247
2	1.0375		0.468	0.572	0.708	0.890	1.149	1.546
3	0.9600		0.260	0.328	0.421	0.558	0.781	1.207
4	0.9164		0.306	0.394	0.522	0.728	1.110	2.067
5	0.8942		0.261	0.340	0.460	0.664	1.085	2.456
6	0.8806	K_i	0.415	0.548	0.753	1.116	1.929	5.401
7	0.8943		0.272	0.355	0.481	0.693	1.132	2.559
8	0.9131		0.189	0.243	0.324	0.453	0.697	1.327
9	0.9308		0.166	0.212	0.278	0.380	0.560	0.963
10	0.9507		0.085	0.108	0.139	0.186	0.264	0.420
11	0.9430		0.183	0.232	0.302	0.407	0.586	0.957
		$K = \Sigma K_i$	3.308	4.102	5.230	6.995	10.297	20.003

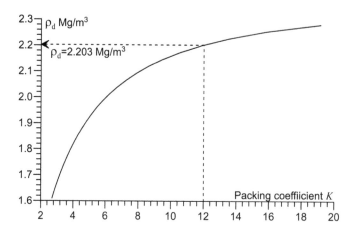

Figure 3.19 Dry density computed for different packing coefficients.

3.3.4   Results of the model and comparison with the Proctor test

As presented in Ref. [48], a packing coefficient of $K = 12$ leads to the dry density of the modified Proctor test. In fact, in the case of this example, the packing coefficient of $K = 12$ leads to a dry density of $\rho_d = 2.203$ Mg/m^3, which has a very good agreement with the maximum dry density measured experimentally (see Figure 3.20).

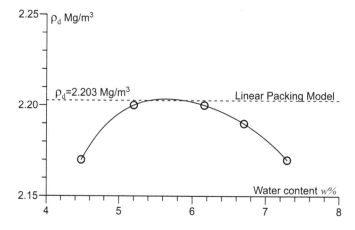

Figure 3.20   Comparison between the results of the modified Proctor test and the dry density computed using the LPM.

Chapter 4

Embankments

4.1 RELEVANT EQUATIONS

This chapter analyzes the behavior of embankments in two particular cases. First, the case of an embankment built on soft soil, and then the collapse of an embankment made of compacted materials as a result of soaking. This second case is analyzed using the Barcelona Basic Model (BBM) considering and without considering the effect of the microstructure of compacted soils.

4.1.1 Stress components due to triangular loads

The Boussinesq solution permits to obtain the stress components for a central symmetrically distributed finite triangular load as

$$\sigma_x = \frac{2q_0 \nu}{\pi a}\left[a(\varepsilon_1 + \varepsilon_2) + y(\varepsilon_1 - \varepsilon_2) - z\ln\frac{R_1 R_2}{R_0^2}\right], \tag{4.1}$$

$$\sigma_y = \frac{q_0}{\pi a}\left[a(\varepsilon_1 + \varepsilon_2) + y(\varepsilon_1 - \varepsilon_2) - 2z\ln\frac{R_1 R_2}{R_0^2}\right], \tag{4.2}$$

$$\sigma_z = \frac{q_0}{\pi a}[a(\varepsilon_1 + \varepsilon_2) + y(\varepsilon_1 - \varepsilon_2)], \tag{4.3}$$

$$\tau_{yz} = -\frac{q_0 z}{\pi a}(\varepsilon_1 - \varepsilon_2), \tag{4.4}$$

where ν is Poisson's ratio and a, ε_1, ε_2, y, and z are the geometrical variables shown in Figure 4.1.

4.1.2 Immediate settlements

Assuming that the horizontal strains are negligible compared with the vertical, the elastic stresses allow computing the immediate settlement as

$$\rho_i = \sum \frac{\sigma_z}{E_u}\Delta z, \tag{4.5}$$

where σ_z is the vertical stress, E_u is the undrained Young's modulus, and $\nu_u = 0.5$ is the Poisson's ratio for the saturated state.

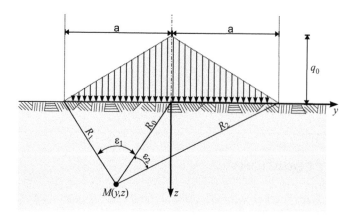

Figure 4.1 Geometric layout for a triangular load. (Adapted from Ref. [39].)

4.1.3 Primary consolidation

Settlement resulting from one-dimensional primary consolidation is

$$\rho_c = \sum_{i=1}^{n_L} \left[\frac{C_r}{1+e_0} \log \frac{\sigma_c'}{\sigma_{z0}'} + \frac{C_c}{1+e_0} \log \frac{\sigma_{z0}' + \Delta\sigma_{zi}'}{\sigma_c'} \right] H_i, \tag{4.6}$$

where ρ_c is the magnitude of settlement due to primary consolidation, n_L is the number of sublayers, σ_{z0}' is the initial effective stress, $\Delta\sigma_{zi}'$ is the increase in vertical stress produced by the load on sublayer i, σ_c' is the overconsolidation stress, H_i is the thickness of sublayer i, C_r is the recompression coefficient, C_c is the coefficient of compressibility, and e_0 is the initial void's ratio.

Moreover, the consolidation equation when water flows vertically, and the layer has constant total stress is

$$c_v \frac{\partial^2 u_{wc}}{\partial z^2} = \frac{\partial u_{wc}}{\partial t}, \tag{4.7}$$

where c_v is the coefficient of vertical consolidation.

For bidimensional conditions, the partial differential equation (PDE) to analyze consolidation is

$$c_h \frac{\partial^2 u_{wc}}{\partial x^2} + c_v \frac{\partial^2 u_{wc}}{\partial z^2} = \frac{\partial u_{wc}}{\partial t} - \frac{\partial \sigma_v}{\partial t}, \quad \text{where} \tag{4.8}$$

$$c_h = \frac{k_{wh}}{m_v \gamma_w}, \quad c_v = \frac{k_{wz}}{m_v \gamma_w}, \quad \text{and} \quad m_v = \frac{a_v}{1+e_0}.$$

The analytical solution of Equation 4.7 permits to obtain the following mean degree of consolidation of a layer undergoing consolidation:

$$U_v(t) = 1 - \sum_{m=0}^{\infty} \frac{2}{M^2} e^{-M^2 T_v}, \quad \text{where} \tag{4.9}$$

$$T_v = \frac{c_v t}{h^2}, \quad \text{and} \quad M = \frac{\pi}{2}(2m+1) \quad \text{with} \quad m = 0, 1, 2, ..., \infty,$$

and h is the drainage length of the layer (*i.e.*, $h = H/2$ for a layer with double drainage or $h = H$ for single drainage layer).

However, an approximate evaluation of the degree of consolidation is given by

$$U_v(t) = \left(\frac{T_v^3}{T_v^3 + 0.5} \right)^{1/6}. \tag{4.10}$$

Furthermore, in the case of stratified soils, an approximate evaluation of the average coefficient of consolidation is

$$\overline{c_v} = \frac{\left(\sum_i H_i \right)^2}{\left(\sum_i H_i / \sqrt{c_{vi}} \right)^2}. \tag{4.11}$$

4.1.4 Radial consolidation

Radial consolidation can be analyzed by solving Equation 4.8 in cylindrical coordinates. Two particular solutions are possible: free vertical settlement or equal vertical strain, which are described by the following PDEs:

$$\frac{\partial u_{wc}}{\partial t} = c_v \frac{\partial^2 u_{wc}}{\partial z^2} + c_h \left(\frac{\partial^2 u_{wc}}{\partial r^2} + \frac{1}{r} \frac{\partial u_{wc}}{\partial r} \right) \quad \text{free vertical settlement,} \tag{4.12}$$

$$\frac{\partial \overline{u}_{wc}}{\partial t} = c_v \frac{\partial^2 u_{wc}}{\partial z^2} + c_h \left(\frac{\partial^2 u_{wc}}{\partial r^2} + \frac{1}{r} \frac{\partial u_{wc}}{\partial r} \right) \quad \text{equal vertical strain,} \tag{4.13}$$

where $\overline{u}_{wc}$ is the average excess pore water pressure at any depth, and c_v and c_h are the coefficients of vertical and horizontal consolidation, respectively.

Equations 4.12 and 4.13 can be solved separately in the vertical and radial directions and then, the combined degree of consolidation (*i.e.*, in the vertical and radial directions U_{vr}) is computed using Carrillo's theorem as follows:

$$(1 - U_{vr}) = (1 - U_v)(1 - U_r). \tag{4.14}$$

4.1.5 Increase of shear strength for staged construction

The undrained shear strength produced by an increment of vertical stress $\Delta\sigma_v$ for any degree of consolidation $U(t)$, as shown in Figure 4.2, is

$$c_{uu}(t) = c_{cu} + \Delta c_u = \frac{\sin \phi_{cu}}{1 - \sin \phi_{cu}} \left(\frac{c_{cu}}{\tan \phi_{cu}} + U(t)\Delta\sigma_v \right). \tag{4.15}$$

4.1.6 Generalized bearing capacity

The safety factor for evaluating the bearing capacity when a soft soil rests over a rigid layer, which was proposed in Ref. [45], is

$$F_s = \frac{(\pi + 2)c_u}{h_F \gamma_F} \qquad \text{For } B/H \leq 1.49, \text{ or} \tag{4.16}$$

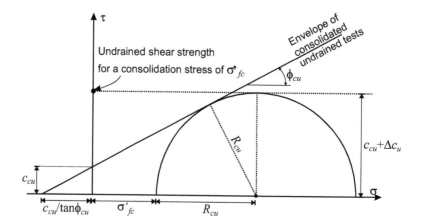

Figure 4.2 Evaluation of the shear strength after consolidation using the results of consolidated undrained tests.

$$F_s = \frac{(\pi + 2) + 0.475(B/H - 1.49)}{h_F \gamma_F} c_u \quad \text{For } B/H > 1.49, \tag{4.17}$$

where h_F and γ_F are the height and the unit weight of the material of the fill of the embankment, B is the width of the base of the foundation, and H is the thickness of the soft layer (Figure 4.3).

4.1.7   The BBM including the effect of soil's microstructure

The model proposed by Alonso et al. [4] extends the BBM to include the soil microstructure. The model assumes that the matric suction affects only the macropores and that the water, that occupies the micropores, has a small effect on stresses [2].

Defining the "effective" degree of saturation requires a new state variable, ξ_m, that represents the ratio between the volume of micropores and the total void's space as

$$\xi_m = \frac{e_m}{e}, \tag{4.18}$$

where e_m is the microstructural void ratio, and e is the total void ratio.

Moreover, the effective degree of saturation, $\overline{S}_r$, is defined as

$$\overline{S}_r = \frac{S_r - \xi_m}{1 - \xi_m} + \frac{1}{n_{sm}} \ln\left[1 + e^{-n_{sm}\frac{S_r - \xi_m}{1 - \xi_m}}\right], \tag{4.19}$$

where n_{sm} is a parameter that defines the degree of smoothing of the function representing the effective degree of saturation.

The effective degree of saturation permits to define the following constitutive state variables:

$$\overline{\sigma} = \sigma - u_a + \overline{S}_r s \quad \text{Constitutive stress,} \tag{4.20}$$

$$\overline{s} = \overline{S}_r s \qquad\qquad \text{Effective suction,} \tag{4.21}$$

where σ is the total stress and u_a is the air pressure.

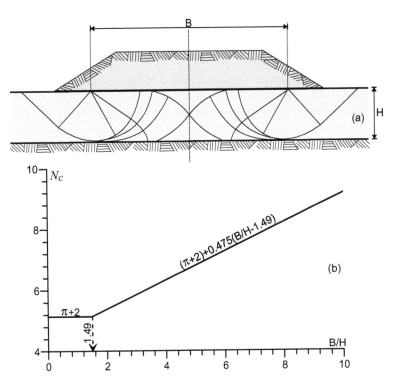

Figure 4.3 Effect of the thickness of a layer of soft soil on the bearing capacity, (a) failure mechanism and (b) bearing capacity coefficient proposed in Ref. [45].

Constitutive mean stress, $\bar{p}$, allows describing the compressibility as

$$de^e = -\bar{k}\ln\left(\bar{p} + \frac{d\bar{p}}{\bar{p}}\right) \quad \text{Elastic compressibility,} \tag{4.22}$$

$$de^{ep} = -\bar{\lambda}\ln\left(\frac{d\bar{p}}{\bar{p}}\right) \quad \text{Elastoplastic compressibility.} \tag{4.23}$$

The compressibility coefficient $\bar{\lambda}$ depends on the effective suction through the following equation:

$$\frac{\bar{\lambda}(\bar{s})}{\lambda(0)} = \bar{r} + (1 - \bar{r})\left[1 + \left(\frac{\bar{s}}{\bar{s}_\lambda}\right)^{1/(1-\bar{\beta})}\right]^{-\bar{\beta}}, \tag{4.24}$$

where $\bar{\lambda}(\bar{s})$ is the coefficient of compressibility for effective suction, $\lambda(0)$ is the saturated coefficient of compressibility, and $\bar{r}$, $\bar{s}_\lambda$ and $\bar{\beta}$ are material parameters.

Moreover, a new loading collapse curve has been defined concerning constitutive stresses in Ref. [4] as

$$\left(\frac{\bar{p}_0}{\bar{p}_c}\right) = \left(\frac{p_0^*}{\bar{p}_c}\right)^{\frac{\lambda(0)-\kappa}{\bar{\lambda}(\bar{s})-\kappa}}, \tag{4.25}$$

where $\bar{p}_c$ is a reference mean constitutive stress.

4.2 EXAMPLE 14: EMBANKMENTS ON SOFT SOILS

This example describes the procedure for analyzing an embankment on soft soil built using the methodology of staged construction. As shown in Figure 4.4, the final level of the embankment must be at 9 m from the natural ground, the platform has 24 m in width, and the slope of the fill is 2/3 (vertical/horizontal).

The embankment should be designed with factors of safety of 1.5 regarding bearing capacity, and 1.25 and 1.5 with respect to the shear strength for the intermediate stages of construction and the end of construction, respectively.

The fill of the embankment has a total unit weight of 22 kN/m³, and the friction angle of the compacted fill is 37°; Table 4.1 presents the properties of the foundation's soil.

Classical analysis of bearing capacity evidence that the construction of the whole embankment produces the failure by shear strength of the foundation; therefore, the embankment should be constructed using the technique of staged construction. The steps for carrying out the analysis for such a technique are as follows:

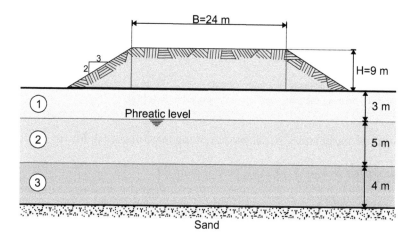

Figure 4.4 Schematic drawing of the embankment of the example.

Table 4.1 Geotechnical properties of the soil of the example

Layer	Depth (m)	γ (kN/m³)	c_{cu} (kPa)	ϕ_{cu}°	e_0	σ_c' (kPa)	C_c	C_r	c_v (10^{-4} cm²/s)	c_h/c_v	E_u (kPa)
	0										
Overconsolidated clay		16	39	14	0.62	100	0.55	0.07	4.5	5	7,000
	3										
Soft compressible clay		14	17	14	0.85	70	0.55	0.07	11.0	5	3,000
	8										
Soft compressible clay		15	26	16	0.73	90	0.4	0.05	6.0	5	4,000
	12										

1   Use the elastic solution for triangular loading to calculate the vertical stress distribution below the axis of symmetry of the embankment.

2   Calculate immediate and consolidation settlements for different fill heights. For this analysis, assume that the undrained Young's modulus remains constant in each layer and neglects settlement due to secondary compression. This settlement analysis allows choosing the height of the fill to be built to reach the required level of the embankment at the end of construction.

3   For the final fill height, evaluate the vertical stress distribution below the embankment.

4   Considering the generalized failure regarding bearing capacity, and the shear strength properties of the soil for consolidated undrained conditions, analyze the possibility of constructing the embankment in several stages, waiting for the 80% of consolidation before placing the subsequent stage.

5   Calculate the dissipation of the excess pore water pressure considering bidimensional consolidation, and evaluate the settlement and the increase of shear strength before placing the subsequent stage.

6   Evaluate the increase in shear strength resulting from the two-dimensional analysis.

7   Using the increase of the shear strength obtained in the previous step, verify the safety factor for each stage of construction by using the finite element and the limit equilibrium methods.

8   Verify the performance of a triangular grid of sand drains, which is installed to accelerate consolidation (sand drains 0.3 m in diameter and 3 m apart in a triangular arrangement). For this grid of vertical drains, calculate the required consolidation time before placing each stage.

4.2.1   Stress distribution beneath the symmetry axis of the embankment

Section 4.1.1 describes a solution that provides the stress distribution in an elastic half-space produced by a triangular distribution of stresses. More precisely, Equation 4.3 permits to compute the distribution of the vertical stress beneath the embankment by subtracting the stress produced by two symmetrical triangular loads, as shown in Figure 4.5.

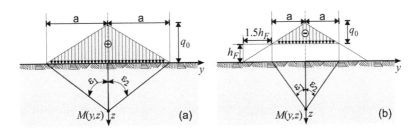

Figure 4.5 Geometric variables for computing the vertical stress below the symmetry axis of two triangular loads, (a) variables for the whole triangular load, and (b) variables for the upper triangular load.

According to Equation 4.3, the vertical stress produced by a triangular load is

$$\sigma_z = \frac{q_0}{\pi a}[a(\varepsilon_1 + \varepsilon_2) + y(\varepsilon_1 - \varepsilon_2)], \tag{4.26}$$

where a is the half base of the triangular load, q_0 is the maximum stress at the apex of the triangle, and ε_1 and ε_2 are the angles between the point of interest and the corners of the triangular load.

Since in the symmetry axis $\varepsilon_1 = \varepsilon_2 = \arctan(a/z)$, Equation 4.26 becomes

$$\sigma_z = \frac{2q_0}{\pi} \arctan\left(\frac{a}{z}\right).$$

The half-width of the base and the stress at the apex of each triangular load, shown in Figure 4.5, are

Triangular load A (full triangle) :
$a = 12 + 1.5h_F$, and

$$q_0 = a\frac{2}{3}\gamma_F = (8 + h_F)\gamma_F.$$

Triangular load B (upper triangle) :
$a = 12$, and

$$q_0 = a\frac{2}{3}\gamma_F = 8\gamma_F.$$

Subtracting both solutions for stress distribution leads to the following expression for the vertical stress beneath the symmetry axis of the embankment:

$$\sigma_z = \frac{2(8 + h_F)\gamma_F}{\pi} \arctan\left(\frac{12 + 1.5h_F}{z}\right) - \frac{16\gamma_F}{\pi} \arctan\left(\frac{12}{z}\right).$$

This expression leads to the distributions of the vertical stress along the symmetry axis for different heights of the fill shown in Figure 4.6. This vertical stress is crucial information for computing immediate and consolidation settlements.

4.2.2   Immediate and consolidation settlements

After computing vertical stresses, Equation 4.5 permits to compute the immediate settlement. In this example, the compressible soil is divided into 12 sublayers of one meter of thickness. Then, using the undrained Young's modulus of each layer presented in Table 4.1, the immediate settlement becomes

$$\rho_i = \sum_{i=1}^{3} \frac{\Delta\sigma_{z_i}}{7,000} + \sum_{i=4}^{8} \frac{\Delta\sigma_{z_i}}{3,000} + \sum_{i=9}^{12} \frac{\Delta\sigma_{z_i}}{4,000}.$$

Table 4.2 presents the results of the evaluation of the immediate settlement, ρ_i, for different heights of the fill.

Regarding consolidation, Equation 4.6 allows evaluating such settlements, ρ_c. Then, using the data of the example, the consolidation settlement becomes

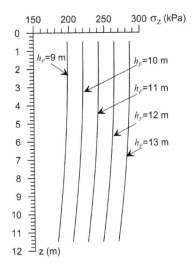

Figure 4.6 Vertical stress along the symmetry axis of the embankment for different heights of the fill.

Table 4.2 Immediate settlements obtained for different heights of the fill

		h_F (m)								
		10		11		12		13		14	
z (m)	E_u (kPa)	$\Delta\sigma_z$ (kPa)	ε_i	$\Delta\sigma_z$ (kPa)	ε_i	$\Delta\sigma_z$ (kPa)	ε_i	$\Delta\sigma_z$ (kPa)	ε_i	$\Delta\sigma_z$ (kPa)	ε_i
0.5	7,000	220.0	0.031	242.0	0.035	264.0	0.038	286.0	0.041	308.0	0.044
1.5	7,000	219.9	0.031	241.9	0.035	263.9	0.038	285.9	0.041	307.9	0.044
2.5	7,000	219.7	0.031	241.7	0.035	263.7	0.038	285.7	0.041	307.7	0.044
3.5	3,000	219.3	0.073	241.3	0.080	263.3	0.088	285.3	0.095	307.2	0.102
4.5	3,000	218.6	0.073	240.5	0.080	262.5	0.087	284.5	0.095	306.4	0.102
5.5	3,000	217.5	0.072	239.4	0.080	261.4	0.087	283.3	0.094	305.3	0.102
6.5	3,000	216.1	0.072	238.0	0.079	259.9	0.087	281.8	0.094	303.7	0.101
7.5	3,000	214.3	0.071	236.1	0.079	258.0	0.086	279.8	0.093	301.7	0.101
8.5	4,000	212.2	0.053	233.9	0.058	255.7	0.064	277.5	0.069	299.4	0.075
9.5	4,000	209.7	0.052	231.4	0.058	253.1	0.063	274.9	0.069	296.7	0.074
10.5	4,000	207.0	0.052	228.6	0.057	250.2	0.063	271.9	0.068	293.6	0.073
11.5	4,000	204.1	0.051	225.6	0.056	247.1	0.062	268.7	0.067	290.3	0.073
Settlement ρ_i (m)			0.664		0.732		0.800		0.867		0.935

$$
\rho_c = \sum_{i=1}^{3} \left[0.043 \log \frac{100}{\sigma'_{v0_i}} + 0.34 \log \frac{\sigma'_{v0_i} + \Delta\sigma_{z_i}}{100} \right]
$$

$$
+ \sum_{i=4}^{8} \left[0.038 \log \frac{70}{\sigma'_{v0_i}} + 0.297 \log \frac{\sigma'_{v0_i} + \Delta\sigma_{z_i}}{70} \right]
$$

$$
+ \sum_{i=8}^{12} \left[0.029 \log \frac{90}{\sigma'_{v0_i}} + 0.231 \log \frac{\sigma'_{v0_i} + \Delta\sigma_{z_i}}{90} \right],
$$

where σ'_{v0_i} is the initial effective vertical stress at mid-depth of each sublayer.

Table 4.3 Consolidation settlements obtained for different heights of the fill

z	γ	$\dfrac{C_c}{1+e_0}$	$\dfrac{C_r}{1+e_0}$	σ'_c	σ'_{v0}	h_F (m) 10	11	12	13	14
(m)	(kN/m³)			(kPa)	(kPa)	ε_c	ε_c	ε_c	ε_c	ε_c
0.5	16	0.340	0.043	100	8	0.169	0.182	0.195	0.206	0.217
1.5	16	0.340	0.043	100	24	0.158	0.171	0.183	0.194	0.204
2.5	16	0.340	0.043	100	40	0.158	0.170	0.181	0.191	0.201
3.5	14	0.297	0.038	70	50	0.179	0.190	0.199	0.208	0.216
4.5	14	0.297	0.038	70	54	0.180	0.190	0.199	0.208	0.216
5.5	14	0.297	0.038	70	58	0.180	0.190	0.199	0.208	0.216
6.5	14	0.297	0.038	70	62	0.180	0.190	0.199	0.207	0.215
7.5	14	0.297	0.038	70	66	0.180	0.190	0.199	0.207	0.215
8.5	15	0.231	0.029	90	70.5	0.118	0.125	0.132	0.139	0.145
9.5	15	0.231	0.029	90	73	0.118	0.125	0.132	0.138	0.144
10.5	15	0.231	0.029	90	75.5	0.117	0.124	0.131	0.138	0.144
11.5	15	0.231	0.029	90	78	0.117	0.124	0.131	0.137	0.143
		Settlement ρ_c (m)				1.854	1.971	2.080	2.181	2.277

Table 4.3 presents the results of the evaluation of the consolidation settlement for different heights of the fill.

The level of the embankment when finalize the immediate and consolidation settlements is calculated from the constructed height minus the settlements (*i.e.*, $h_F - \rho_i - \rho_c$). Therefore, reaching a final level of the embankment of 9 m requires placing a fill whose height exceeds the height when considering only the embankment level. Figure 4.7 allows evaluating the fill height required to reach a final level of 9 m. This evaluation leads to a required height of the fill of 12 m.

4.2.3 Vertical stress distribution under the embankment for the final height of the fill.

Using two triangular loads, as shown in Figure 4.8, allows evaluating vertical stresses produced by the embankment of 12 m height.

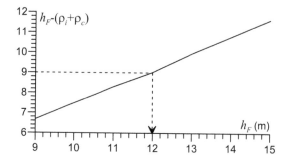

Figure 4.7 Final level of the embankment depending on the height of the fill.

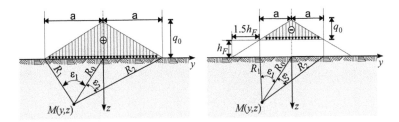

Figure 4.8 Geometric variables to calculate the vertical stress under two triangular loads at any point (y, z).

Angles ε_1 and ε_2 required in Equation 4.26 are

$$\varepsilon_1 = \arctan \frac{a+y}{z} - \arctan \frac{y}{z} \quad \text{and} \quad \varepsilon_2 = \arctan \frac{a-y}{z} + \arctan \frac{y}{z}.$$

Since the base and the apex stress for the triangular load A, which correspond to the whole triangle, are $a = 12 + 1.5h_F = 30$ m, $q_0 = a\frac{2}{3}\gamma_F = (8 + 12)22 = 440$ kPa, the angles become

$$\varepsilon_{1_A} = \arctan \frac{30+y}{z} - \arctan \frac{y}{z} \quad \text{and} \quad \varepsilon_{2_A} = \arctan \frac{30-y}{z} + \arctan \frac{y}{z}.$$

On the other hand, the width of the base and the apex for the triangular load B, which is the upper triangle, are $a = 12$ m, $q_0 = 12\frac{2}{3}\gamma_F = 8 \cdot 22 = 176$ kPa, while the angles ε_{1_B} and ε_{2_B} are

$$\varepsilon_{1_B} = \arctan \frac{12+y}{z} - \arctan \frac{y}{z} \quad \text{and} \quad \varepsilon_{2_B} = \arctan \frac{12-y}{z} + \arctan \frac{y}{z}.$$

Finally, subtracting the stresses produced by both triangular loadings leads to the vertical stress produced by the trapezoidal loading of the embankment as

$$\sigma_z = \sigma_{z_A} - \sigma_B.$$

Figure 4.9 shows the distribution of vertical stresses under the 12 m high embankment. It is important to note that since the slope of the embankment creates a progressive transition between the zero applied stress on the surface and the maximum stress below the full height of the embankment, the increase in vertical stress appears primarily below the embankment with little lateral extension. Therefore, the increase in shear strength, produced by consolidation, occurs mainly below the central part of the embankment.

4.2.4 Evaluation of the bearing capacity for the staged construction

The first stage of the embankment is constructed when the foundation soil is in uncon-solidated conditions; in other words, without undergoing any increase in effective ver-tical stress. For such conditions, Equation 4.15 leads to the following undrained shear strength:

$$c_{uu_1} = c_{uu}(t = 0) = \frac{\sin \phi_{cu}}{1 - \sin \phi_{cu}} \left(\frac{c_{cu}}{\tan \phi_{cu}} \right).$$

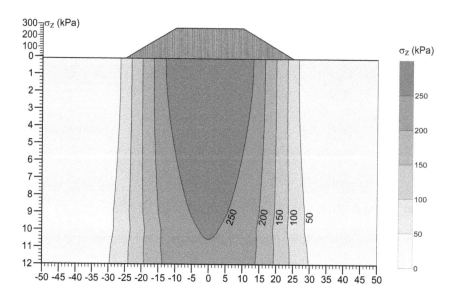

Figure 4.9 Distribution of vertical stress beneath the embankment of 12 m in height.

Table 4.4 shows the values of the undrained shear strength of each sublayer of the foundation's soil. Afterward, Equations 4.16 and 4.17, that were developed by Mandel and Salençon for a homogeneous layer of limited thickness, permit to evaluate the bearing capacity. However, in the case of stratified soils, the shear strength is nonhomogeneous; for this case, using the mean shear strength could be a reasonable approximation providing that the failure mechanism mobilizes the shear strength of the whole layer. Then, from Table 4.4, the mean shear strength for the first stage of construction stage is $c_{uu_1} = 33.0$ kPa.

Table 4.4 Shear strength and safety factors for the first stage of construction, shaded cells indicate the height of the embankment corresponding to $F_s \approx 1.5$.

z (m)	c_{cu} (kPa)	ϕ_{cu}°	$\Delta\sigma_z$	c_{uu} (kPa)		h_F (m)	B (m)	B/H	N_c	F_s
0.5	39	14	0.0	49.9						
1.5	39	14	0.0	49.9		5.0	52.5	4.38	6.51	1.956
2.5	39	14	0.0	49.9		5.2	52.2	4.35	6.50	1.878
3.5	17	14	0.0	21.8		5.4	51.9	4.33	6.49	1.805
4.5	17	14	0.0	21.8	c_{uu_1}	5.6	51.6	4.30	6.48	1.737
5.5	17	14	0.0	21.8	33.0 kPa	5.8	51.3	4.28	6.46	1.647
6.5	17	14	0.0	21.8	$\Longrightarrow$	6.0	51.0	4.25	6.45	1.615
7.5	17	14	0.0	21.8		6.2	50.7	4.23	6.44	1.560
8.5	26	16	0.0	34.5		6.4	50.4	4.20	6.43	1.509
9.5	26	16	0.0	34.5		6.6	50.1	4.18	6.42	1.460
10.5	26	16	0.0	34.5		6.8	49.8	4.15	6.41	1.415
11.5	26	16	0.0	34.5		7.0	49.5	4.13	6.39	1.372
	$\overline{c_{uu}} =$			33.0						

Furthermore, Table 4.4 also shows the safety factor obtained using Equation 4.17 computed using the undrained mean shear strength as follows:

$$F_s = \frac{(\pi + 2) + 0.475(B/12 - 1.49)}{h_F \cdot 22} \cdot 33.0 \quad \text{where} \quad B = 60 - 1.5h_F.$$

As shown in Table 4.4, the first stage of construction can progress up to a fill of 6.4 m in height with a safety factor for bearing capacity >1.5.

After placing the 6.4 m high fill, it produces an increase in total vertical stress that can be evaluated using Equation 4.26 following the same procedure as that used in the first point of this example. In fact, for the 6.4 m high embankment, the increase in vertical stress is

$$a_A = 30\ m,$$
$$a_B = 30 - 1.5 \cdot 6.4 = 20.4\ \text{m},$$
$$q_{0_A} = 30/1.5 * 22 = 440\ \text{kPa},$$
$$q_{0_B} = 20.4/1.5 * 22 = 299.2\ \text{kPa},$$
$$\Delta\sigma_{v_1} = \frac{2 \cdot 440}{\pi}\arctan\left(\frac{30}{z}\right) - \frac{2 \cdot 299.2}{\pi}\arctan\left(\frac{20.4}{z}\right).$$

Therefore, the increased undrained shear strength in each sublayer, for the 80% of consolidation, becomes

$$c_{uu} = \frac{\sin\phi_{cu}}{1 - \sin\phi_{cu}}\left(\frac{c_{cu}}{\tan\phi_{cu}} + 0.8\Delta\sigma_{v_1}\right).$$

Table 4.5 shows the undrained shear strength of each sublayer, achieved after waiting for the 80% of consolidation. For this degree of consolidation, the mean shear strength along the middle axis of the embankment is $\overline{c_{uu}} = 71.3$ kPa. However, as shown in Figure 4.9, the increase in vertical stress occurs mainly in the soil beneath the embankment. Moreover, it is presumable that the shear strength of the soil outside the zone of influence

Table 4.5 Shear strength and safety factors for the second stage of construction, shaded cells indicate the height of the embankment corresponding to $F_s \approx 1.5$.

z (m)	c_{cu} (kPa)	ϕ_{cu}°	$\Delta\sigma_z$	c_{uu} (kPa)		h_F (m)	B (m)	B/H	N_c	F_s
0.5	39	14	140.8	85.9						
1.5	39	14	140.8	85.9		9.0	46.5	3.88	6.27	1.647
2.5	39	14	140.7	85.8		9.2	46.2	3.85	6.26	1.608
3.5	17	14	140.6	57.7		9.4	45.9	3.83	6.25	1.571
4.5	17	14	140.4	57.6	c_{uu_2}	9.6	45.6	3.80	6.24	1.535
5.5	17	14	140.2	57.5	52.0 kPa	9.8	45.3	3.78	6.23	1.501
6.5	17	14	139.8	57.4	$\Longrightarrow$	10.0	45.0	3.75	6.22	1.468
7.5	17	14	139.3	57.3		10.2	44.7	3.73	6.20	1.436
8.5	26	16	138.7	76.7		10.4	44.4	3.70	6.19	1.406
9.5	26	16	137.9	76.5		10.6	44.1	3.68	6.18	1.377
10.5	26	16	137.0	76.2		10.8	43.8	3.65	6.17	1.349
11.5	26	16	136.0	75.9		11.0	43.5	3.63	6.16	1.322
	$\overline{c_{uu}} =$			71.3						

of the embankment remains unchanged (*i.e.*, $c_{uu} = 33.0$). Furthermore, since the failure mechanism for generalized bearing capacity, shown in Figure 4.3a, involves a portion of unconsolidated soil, a reasonable assumption for calculating the safety factor is to use the average shear strength of both areas, consolidated and unconsolidated. Then, after the first stage, the average shear strength is $c_{uu_2} = (71.3 + 33.0)/2 \approx 52.0$ kPa.

Table 4.5 shows the safety factor for different heights of the fill calculated after increasing the shear strength achieved after 80% of consolidation of the first stage. This table shows that the second stage can progress up to an embankment of 9.8 m in height maintaining a safety factor greater than 1.5.

Another relevant point for staged construction is the time required to reach the proper degree of consolidation. Evaluating this time requires calculating the average consolidation coefficient of the entire layer, which is evaluated using Equation 4.11 as follows:

$$\bar{c_v} = \frac{(300 + 500 + 400)^2}{\left(\frac{300}{\sqrt{4.5 \cdot 10^{-4}}} + \frac{500}{\sqrt{1.1 \cdot 10^{-3}}} + \frac{400}{\sqrt{6.0 \cdot 10^{-4}}}\right)^2} = 6.94 \cdot 10^{-4} \ cm^2/s = 6 \cdot 10^{-3} \ m^2/day.$$

Moreover, from Equation 4.9, the time factor for the 80% of consolidation is

$$T_v = \left(\frac{0.5 U_v^6}{1 - U_v^6}\right)^{1/3} = \left(\frac{0.5 \cdot 0.80^6}{1 - 0.80^6}\right)^{1/3} = 0.562.$$

Therefore, the time required to reach 80% consolidation in the layer that drains through both sides (*i.e.*, 6 m drainage length) is

$$T_v = \frac{c_v t}{h^2} = 0.562 \quad \longrightarrow \quad t = \frac{T_v h^2}{c_v} = \frac{0.562 \cdot 6^2}{6 \cdot 10^{-3}} = 3373 \ days.$$

It is clear that waiting 3373 days (*i.e.*, 9.3 years) before placing the later stage is not practical. This time could be shorter if 2D consolidation is considered because usually the consolidation coefficient in the horizontal direction is higher than in the vertical direction; the next point evaluates this possibility. However, a drastic reduction in consolidation time is possible by employing radial consolidation using vertical drains, a possibility which is analyzed in the last point of this example.

Subsequently, the fill of 9.8 m, placed in the second stage of construction, produces the following increase in total vertical stress:

$$a_A = 30 \ m,$$
$$a_B = 30 - 1.5 \cdot 9.8 = 15.3 \ m,$$
$$q_{0_A} = 30/1.5 \cdot 22 = 440 \ kPa,$$
$$q_{0_B} = 15.3/1.5 \cdot 22 = 224.4 \ kPa,$$
$$\Delta\sigma_{v_2} = \frac{2 \cdot 440}{\pi} \arctan\left(\frac{30}{z}\right) - \frac{2 \cdot 224.4}{\pi} \arctan\left(\frac{15.3}{z}\right).$$

However, after waiting 3,373 days for the second construction stage, the consolidation time of the first stage is $2 \times 3,373 = 6,746$ days, which corresponds to the following time factor and degree of consolidation:

$$T_v = \frac{c_v t}{h^2} = \frac{6 \cdot 10^{-3} \cdot 6746}{6^2} = 1.124 \quad \longrightarrow \quad U_v = \left(\frac{1.124^3}{0.5 + 1.124^3}\right)^{1/6} = 0.95.$$

Therefore, the increased shear strength after the second stage of consolidation becomes

$$c_{uu} = \frac{\sin\phi_{cu}}{1 - \sin\phi_{cu}}\left[\frac{c_{cu}}{\tan\phi_{cu}} + 0.95\Delta\sigma_{v_1} + 0.8(\Delta\sigma_{v_2} - \Delta\sigma_{v_1})\right].$$

Table 4.6 shows that the embankment of the total height of 12 m is attainable after waiting for the second stage of consolidation. The mean shear strength at the middle axis is $c_{uu} = 97.8$ kPa, and considering the unconsolidated region, the average shear strength is $c_{uu_3} = 0.5(97.8 + 33.0) = 65.4$ kPa.

Figure 4.10 summarizes the results of the safety factor for different embankment heights and each stage of construction.

4.2.5 Evaluation of the bidimensional consolidation

Evaluating two-dimensional consolidation requires solving the 2D PDE, that describes the consolidation process, using a numerical method such as the finite difference method, which linearizes the PDE using a grid.

Table 4.6 Shear strength and safety factors for the third stage of construction, shaded cells indicate the height of the embankment corresponding to $F_s \approx 1.5$.

z (m)	c_{cu} (kPa)	ϕ_{cu}°	$\Delta\sigma_z$	c_{uu} (kPa)		h_F (m)	B (m)	B/H	N_c	F_s
0.5	39	14	215.6	112.1						
1.5	39	14	215.6	112.1		10.0	39.0	3.25	5.98	1.777
2.5	39	14	215.4	112.1		10.2	39.3	3.28	5.99	1.746
3.5	17	14	215.2	83.9		10.4	39.6	3.30	6.00	1.715
4.5	17	14	214.8	83.8	c_{uu_3}	10.6	39.9	3.33	6.01	1.686
5.5	17	14	214.1	83.6	65.4 kPa	10.8	40.2	3.35	6.03	1.658
6.5	17	14	213.2	83.3	$\Longrightarrow$	11.0	40.5	3.38	6.04	1.632
7.5	17	14	212.1	83.0		11.2	40.8	3.40	6.05	1.606
8.5	26	16	210.7	107.1		11.4	41.1	3.43	6.06	1.580
9.5	26	16	209.1	106.5		11.6	41.4	3.43	6.06	1.556
10.5	26	16	207.2	105.9		11.8	41.7	3.48	6.08	1.533
11.5	26	16	205.1	105.2		12.0	42.0	3.50	6.10	1.510
	$\overline{c_{uu}} =$			97.8						

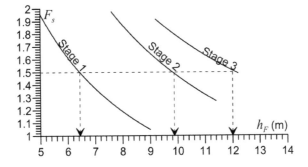

Figure 4.10 Safety factor for different heights of the embankment and each construction stage.

Since the embankment of this example has a symmetry axis, the consolidation can be computed on the half of the embankment using a rectangular grid of 12 m depth (the depth of the soft layer) and 50 m length. This rectangular model uses distances between nodes of $\Delta x = 1$ m and $\Delta z = 0.5$ m, as shown in Figure 4.11. Also, solving the differential equation requires initial and boundary conditions. As shown in Figure 4.11, boundary conditions of this example are

- Zero water flow at $x = 0$, or $j = 1$ (because it is the axis of symmetry), and at $x = 50$ m, or $j = 51$ which is considered a distance far from the fill and, therefore, the flow of water is not affected by the presence of the embankment.
- Since the layer undergoing consolidation is drained through the top and the bottom, the excess of pore water pressure is zero for $z = 0$ and $z = 12$ m (i.e., $u_{wc(1,j)} = 0$, $u_{wc(25,j)} = 0$).

Regarding the initial conditions, the Terzaghi theory of consolidation suggests that the initial excess of pore water pressure at time $t = 0$ is equal to the increase in vertical stress $u_{wc} = \Delta\sigma_v$.

The first step required for computing the bidimensional consolidation is evaluating the increase of the vertical stress in each point, $\Delta\sigma_v(i, j)$. As described previously, this evaluation is performed using the method of the two triangular loads shown in Figure 4.8. The geometrical characteristics corresponding to the triangular load A (the whole triangle) and the triangular load B (top triangle) for each loading stage are

$$\textit{First stage}: \quad a_A = 30.0 \text{ m} \quad a_B = 20.4 \text{ m} \quad q_{0_A} = 440.0 \text{ kPa} \quad q_{0_B} = 299.2 \text{ kPa},$$
$$\textit{Second stage}: a_A = 20.4 \text{ m} \quad a_B = 15.3 \text{ m} \quad q_{0_A} = 299.2 \text{ kPa} \quad q_{0_B} = 224.4 \text{ kPa},$$
$$\textit{Third stage}: \quad a_A = 15.3 \text{ m} \quad a_B = 12.0 \text{ m} \quad q_{0_A} = 224.4 \text{ kPa} \quad q_{0_B} = 176.0 \text{ kPa},$$

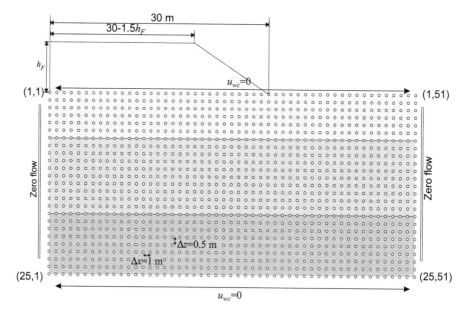

Figure 4.11 Rectangular grid to calculate 2D consolidation below the embankment.

$$\varepsilon_{1_{A,B}}(i,j) = \arctan\frac{a_{A,B} + (j-1)\Delta x}{(i-1)\Delta z} - \arctan\frac{(j-1)\Delta x}{(i-1)\Delta z},$$

$$\varepsilon_{2_{A,B}}(i,j) = \arctan\frac{a_{A,B} - (j-1)\Delta x}{(i-1)\Delta z} + \arctan\frac{(j-1)\Delta x}{(i-1)\Delta z}.$$

Therefore, the increase in vertical stress is

$$\Delta\sigma_v(i,j) = \frac{q_{0_A}}{\pi a_A}\left[a_A(\varepsilon_{1_A}(i,j) + \varepsilon_{2_A}(i,j) + (j-1)\Delta x(\varepsilon_{1_A}(i,j) - \varepsilon_{2_A}(i,j)\right]$$
$$- \frac{q_{0_B}}{\pi a_B}\left[a_B(\varepsilon_{1_B}(i,j) + \varepsilon_{2_B}(i,j) + (j-1)\Delta x(\varepsilon_{1_B}(i,j) - \varepsilon_{2_B}(i,j)\right].$$

The dissipation of the excess of pore water pressure is given by Equation 4.8 which can be transformed into a set of linear equations using the following approximations of the partial derivatives in time and space:

Discretization in space:

$$\frac{\partial^2 u_{wc}}{\partial x^2} = \frac{u_{wc_{i+1,j}}^t - 2u_{wc_{i,j}}^t + u_{wc_{i-1,j}}^t}{\Delta x^2}, \tag{4.27}$$

$$\frac{\partial^2 u_{wc}}{\partial z^2} = \frac{u_{wc_{i+1,j}}^t - 2u_{wc_{i,j}}^t + u_{wc_{i-1,j}}^t}{\Delta z^2}. \tag{4.28}$$

Discretization in time:

$$\frac{\partial u_{wc}}{\partial t} = \frac{u_{wc_{i,j}}^{t+\Delta t} - u_{wc_{i,j}}^t}{\Delta t}. \tag{4.29}$$

Using the approximations of the derivatives given in Equations 4.27, 4.28, and 4.29, the 2D consolidation equation becomes

$$u_{wc}^{t+\Delta t}(i,j) = \frac{c_h(i,j)\Delta t}{\Delta x^2}\left[u_{wc_{i,j+1}}^t + u_{wc_{i,j-1}}^t\right] + \frac{c_v(i,j)\Delta t}{\Delta z^2}\left[u_{wc_{i-1,j}}^t + u_{wc_{i+1,j}}^t\right]$$
$$+ \left[1 - \frac{2c_v(i,j)\Delta t}{\Delta z^2} - \frac{2c_h(i,j)\Delta t}{\Delta x^2}\right]u_{wc_{i,j}}^t + \frac{\sigma_v(i,j)^{t+\Delta t} - \sigma_v(i,j)^t}{\Delta t}$$

As described in Chapter 2, this equation corresponds to an explicit scheme of solution. Therefore, to avoid instability problems, the time lapse Δt must be

$$\Delta t < \frac{1}{2}\frac{\min(\Delta x^2, \Delta z^2)}{\max(c_v, c_h)}. \tag{4.30}$$

The vertical coefficients of consolidation of each layer are $c_v(i = 1, ..., 6, j) = 4.5 \cdot 10^{-4}$ cm²/s, $c_v(i = 8, ..., 16, j) = 1.1 \cdot 10^{-3}$ cm²/s, $c_v(i = 18, ..., 25, j) = 6.0 \cdot 10^{-4}$ cm²/s; and for the horizontal direction, $c_h/c_v = 5$ and then $c_h(i = 1, ..., 6, j) = 2.25 \cdot 10^{-3}$ cm²/s, $c_h(i = 8, ..., 16, j) = 5.5 \cdot 10^{-3}$ cm²/s, $c_h(i = 18, ..., 25, j) = 3.0 \cdot 10^{-3}$ cm²/s.

Rows $i = 7$ and $i = 17$ are interfaces between materials. Therefore, the mean coefficient of consolidation in the vertical direction is computed using Equation 4.11. In contrast, in the horizontal direction, the mean consolidation coefficient comes from the arithmetic mean since it is a parallel system. Then, the consolidation coefficients at the interfaces are

$$\overline{c_v}(7,j) = \frac{4}{\left(\frac{1}{\sqrt{4.5 \cdot 10^{-4}}} + \frac{1}{\sqrt{1.1 \cdot 10^{-3}}}\right)^2} = 6.7 \cdot 10^{-4} \; cm^2/s,$$

$$\overline{c_h}(7,j) = \frac{1}{2}(2.25 \cdot 10^{-3} + 5.5 \cdot 10^{-3}) = 3.88 \cdot 10^{-3} \ cm^2/s,$$

$$\overline{c_v}(17,j) = \frac{4}{\left(\frac{1}{\sqrt{1.1 \cdot 10^{-3}}} + \frac{1}{\sqrt{6.0 \cdot 10^{-4}}}\right)^2} = 7.94 \cdot 10^{-4} \ cm^2/s,$$

$$\overline{c_h}(17,j) = \frac{1}{2}(5.5 \cdot 10^{-3} + 3.0 \cdot 10^{-3}) = 4.25 \cdot 10^{-3} \ cm^2/s.$$

For the explicit solution, the stability condition is given in Equation 4.30 which leads to

$$\min\left(\frac{1}{2}\frac{100^2}{1.1 \cdot 10^{-3}}, \frac{1}{2}\frac{50^2}{5.5 \cdot 10^{-3}}\right), \ \text{then}$$

$$\Delta t < 2.27 \cdot 10^5 \ s \ \rightarrow \ \Delta t < 2.63 \ \text{days}.$$

Since the lateral limits have zero water flow, Equation 4.30 must be modified assuming imaginary node columns that have zero water pressure gradients (as already explained in Chapter 2). This modification leads to the following equation that applies to the left boundary:

$$
\begin{aligned}
u_{wc}^{t+\Delta t}(i = 2..24, 1) = {} & \frac{c_h(i,1)\Delta t}{\Delta x^2}\left[2u_{wc_{i,2}}^t\right] + \frac{c_v(i,1)\Delta t}{\Delta z^2}\left[u_{wc_{i-1,1}}^t + u_{wc_{i+1,1}}^t\right] \\
& + \left[1 - \frac{2c_v(i,1)\Delta t}{\Delta z^2} - \frac{2c_h(i,1)\Delta t}{\Delta x^2}\right]u_{wc_{i,1}}^t \\
& + \frac{\sigma_v(i,1)^{t+\Delta t} - \sigma_v(i,1)^t}{\Delta t},
\end{aligned}
$$

and for the right boundary

$$
\begin{aligned}
u_{wc}^{t+\Delta t}(i = 2..24, 51) = {} & \frac{c_h(i,51)\Delta t}{\Delta x^2}\left[2u_{wc_{i,50}}^t\right] + \frac{c_v(i,51)\Delta t}{\Delta z^2}\left[u_{wc_{i-1,51}}^t + u_{wc_{i+1,51}}^t\right] \\
& + \left[1 - \frac{2c_v(i,51)\Delta t}{\Delta z^2} - \frac{2c_h(i,51)\Delta t}{\Delta x^2}\right]u_{wc_{i,51}}^t \\
& + \frac{\sigma_v(i,51)^{t+\Delta t} - \sigma_v(i,51)^t}{\Delta t}.
\end{aligned}
$$

Regarding the top and bottom boundaries, the pore water pressure is zero, and therefore,

$$u_{wc}^{t+\Delta t}(1,j) = 0,$$
$$u_{wc}^{t+\Delta t}(25,j) = 0.$$

The MATLAB script, provided with this book, allows solving the 2D consolidation problem using the set of equations described above. Also, this script verifies the time required to reach a degree of consolidation of 80% and places the next stage of the embankment. Figure 4.12 shows the evolution of water pressure during the consolidation process of each construction stage. Furthermore, Figure 4.13 shows the evolution of the degree of consolidation over time, comparing the cases of 1D and 2D. Although the 2D consolidation analysis shows a faster process, the reduction in the consolidation time is not significant. Therefore, it is necessary to consider another option to accelerate the consolidation process, which is the installation of vertical drains.

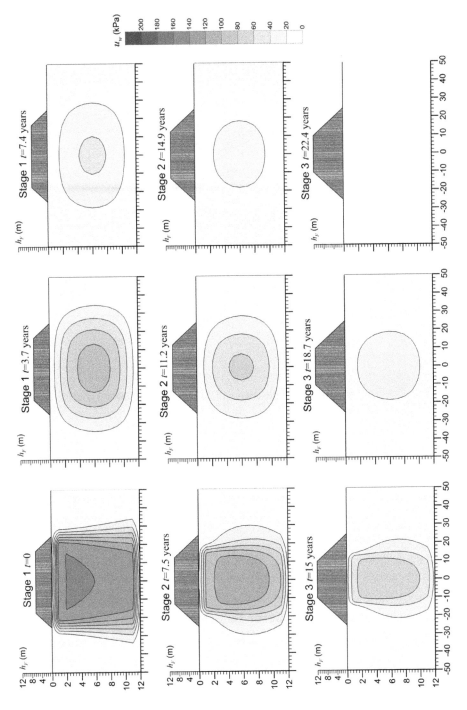

Figure 4.12 Evolution of the pore pressure for the 2D consolidation of the three stages.

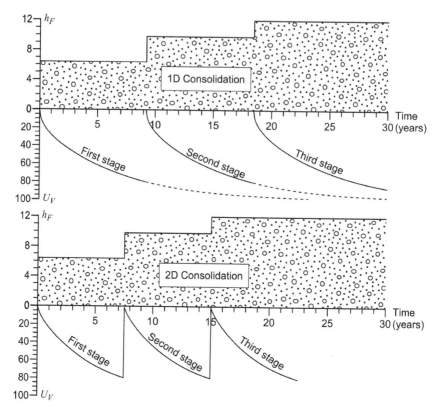

Figure 4.13 Comparison of the evolution of the degree of consolidation for the 1D and 2D analyses.

4.2.6   Evolution of the undrained shear strength considering the 2D consolidation

The results of the evolution of the pore water pressure, analyzed in 2D, permit to apply Equation 4.15 to compute the increase in the shear strength of the soil during the consolidation process. Such an increase is also computed by the MATLAB script leading to the results shown in Figure 4.14. These results permit to propose the sectorization of shear strength shown in Figure 4.15, which is crucial to analyze the stability of the embankment regarding shear strength explained in the next point of this example.

4.2.7   Evaluation of the safety factor before placing each stage

The analysis of the generalized bearing capacity presented in the fourth point of this example assumes that the failure process involves all the layers and, for this reason, considers the average undrained shear strength. However, the 2D consolidation analysis shows that the increase in shear strength in the intermediate layer is less than in the upper and lower layers because drainage of the intermediate layer advances slowly. Therefore, it is presumable that the failure mechanism corresponding to the lowest safety factor crosses this intermediate layer because it has a lower shear strength.

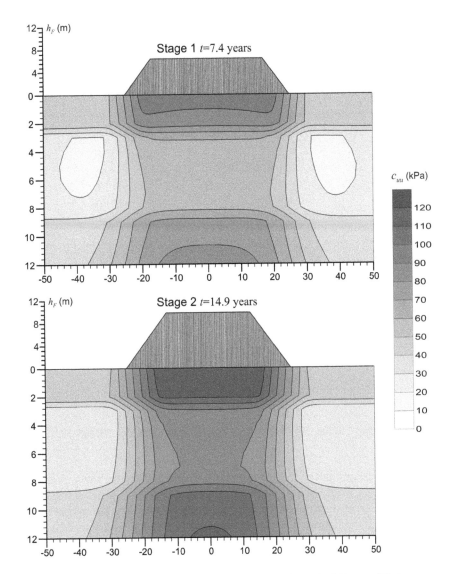

Figure 4.14 Increase of the undrained shear strength resulting from the 2D consolidation.

Two methodologies are possible to analyze the stability regarding shear strength:

- The Limit Equilibrium Method assumes circular or noncircular failure surfaces; it places the maximum shear strength along these surfaces and searches for the surface that has the lower safety factor.
- The Finite Element Method (FEM) also permits approaching the failure. For this purpose, the method reduces the shear strength dividing it by a safety factor (in the case of this example c_{uu}/F_s). Afterward, the method computes the displacement field of the embankment and the foundation. This process is repeated for different assumptions of F_s searching the value that produces displacements that increase to infinity.

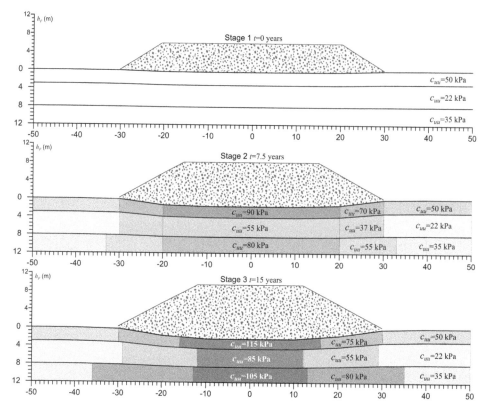

Figure 4.15 Sectorization of the undrained shear strength of the deformed soil layers.

These two methods, applied to the sectorization of shear strength shown in Figure 4.15, lead to the results of safety factors shown in Figure 4.16. It is clear from this figure that safety factors are lower than the obtained when analyzing the generalized bearing capacity. As explained, it occurs because the failure mechanisms for the Limit Equilibrium or the FEM privilege the failure through the middle layer.

Moreover, there are some differences between the safety factors obtained with the analysis of Limit Equilibrium or FEM (Plaxis®). These discrepancies appear because the analysis of the Limit Equilibrium in this example uses circular surfaces, and the results of the FEM show that the displacement field could differ from a circular movement, mainly for stage 3. Nevertheless, all safety factors exceed or are very close to the required values of 1.25 for the intermediate stages and 1.5 for the last stage of construction.

4.2.8   Analysis of the radial drainage

Similarly to the analysis of the 2D consolidation, the Finite Difference Method also allows solving the radial consolidation. In this case, the discretization in space is applied to a half transverse section of the cylinder undergoing radial consolidation, as shown in Figure 4.17. This plane forms a rectangular section, which is divided into a rectangular grid with Δr and Δz which are the distances between nodes.

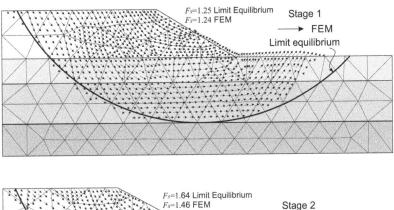

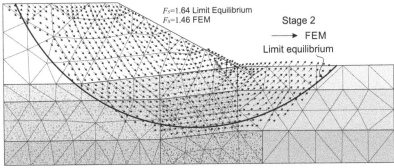

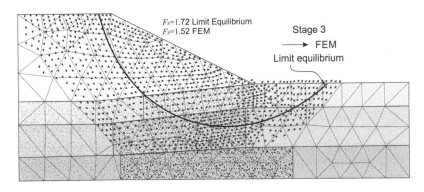

Figure 4.16 Stability of the three stages of the embankment computed using the Finite Element Method and Limit Equilibrium.

The solution of the diffusion problem requires knowing the initial conditions of the computed variable (u_{wc} in the case of consolidation) and a particular treatment of the boundary conditions.

The components of the partial differential Equation 4.12 are substituted by their approximations in the form of Finite Differences as follows:

Discretization in space:

$$\frac{\partial u_{wc}}{\partial r} = \frac{u^t_{wc_{i,j+1}} - u^t_{wc_{i,j-1}}}{2\Delta r},$$

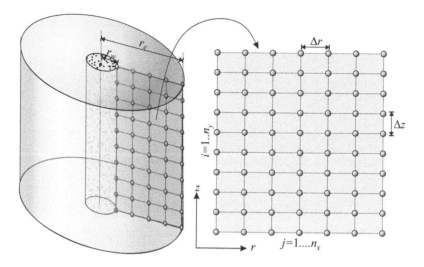

Figure 4.17 Discretization in space around a vertical drain to apply the Finite Difference Method.

$$\frac{\partial^2 u_{wc}}{\partial r^2} = \frac{u^t_{wc_{i,j+1}} - 2u^t_{wc_{i,j}} + u^t_{wc_{i,j+1}}}{\Delta r^2},$$

$$\frac{\partial u_{wc}}{\partial z} = \frac{u^t_{wc_{i+1,}} - u^t_{wc_{i,j}}}{\Delta z},$$

$$\frac{\partial^2 u_{wc}}{\partial z^2} = \frac{u^t_{wc_{i+1,j}} - 2u^t_{wc_{i,j}} + u^t_{wc_{i-1,j}}}{\Delta z^2}.$$

Discretization in time:

$$\frac{\partial u_{wc}}{\partial t} = \frac{u^{t+\Delta t}_{wc_{i,j}} - u^t_{wc_{i,j}}}{\Delta t}.$$

Then, using the explicit resolution method, the recursive equation that permits to compute the pore water pressure in each node at time $t + \Delta t$ knowing the pore water pressure in all nodes at time t is

$$u^{t+\Delta t}_{wc(i,j)} = \frac{c_h \Delta t}{\Delta r^2} \left[\left(1 - \frac{1}{2(j-1+\rho)} \right) u^t_{wc_{i,j-1}} + \left(1 + \frac{1}{2(j-1+\rho)} \right) u^t_{wc_{i,j+1}} \right]$$

$$+ \frac{c_v \Delta t}{\Delta z^2} \left[u^t_{wc_{i-1,j}} + u^t_{wc_{i+1,j}} \right] + \left[1 - \frac{2c_v \Delta t}{\Delta z^2} - \frac{2c_h \Delta t}{\Delta r^2} \right] u^t_{wc_{i,j}}, \qquad (4.31)$$

where $r = r_w + (j-1)\Delta r$ and $\rho = r_w/\Delta r$.

Boundary conditions for solving Equation 4.31 are

- Zero flow of water on the external radius of the cylinder.
- Zero pore water pressure in the external radius of the vertical drain.
- Zero pore water pressure on the top and bottom surfaces.

The interfaces between adjacent sublayers represent a discontinuity in the consolidation coefficient. One option to solve this problem is to use an average consolidation coefficient, as was done for the 2D consolidation. Another more rigorous option is to use a supplemental condition that describes the continuity of the water flow. For a horizontal interface placed in row k of the grid, this continuity equation emerges by writing Darcy's law in the interface, as follows:

$$k_{w_{k-1}} \frac{u_{wc_{k,j}} - u_{wc_{k-1,j}}}{\Delta z_{k-1}} = k_{w_{k+1}} \frac{u_{wc_{k+1,j}} - u_{wc_{k,j}}}{\Delta z_{k+1}}.$$

Finally, the initial conditions for the pore water pressure are considered similar to those of the 2D consolidation analysis.

This book also provides a MATLAB script that allows solving the radial consolidation of this example. Figure 4.18 shows the results of the evolution of the pore water pressure obtained with this script. These results show that the third stage can be placed six months after the staged construction begins. It is a significant reduction in time compared to the 15 years required without installing vertical drains. Finally, it can be concluded that the technology of vertical drains makes possible the construction of the embankment using the staged construction.

4.3 EXAMPLE 15: ANALYSIS OF THE COLLAPSE OF EMBANKMENTS UNDER SOAKING USING THE BBM

This example, which is based on the data presented in Ref. [3], describes the procedure for computing the volumetric strain of compacted soils submitted to wetting. The volumetric strain is calculated depending on the compaction density, water content, and overburden stress. The soil is a low-plasticity material and has the following characteristics: liquid limit $w_L = 34\%$, plasticity index $PI = 15\%$, and density of solid particles $\frac{\gamma_s}{\gamma_w} = 2.73$. Proctor compaction characteristics are

- Standard Proctor (SP): $\gamma_{d_{max}} = 18.1\ kN/m^3\ w_{opt} = 15\%$
- Modified Proctor (MP): $\gamma_{d_{max}} = 20.2\ kN/m^3\ w_{opt} = 10\%$

Moreover, the parameters of the BBM and the water retention curve for the soil of the example are given in Table 4.7. The objective of the example is to calculate the volumetric strain of compacted soils submitted to soaking for different overburden stresses, up to 5 MPa. The effects of the compaction water content and compaction density are analyzed using three different water contents (5%, 10%, and 15%), and two different densities corresponding to the 88% of the maximum unit weight of the standard and modified Proctor tests. As explained in Refs. [10,14], the coefficient of horizontal stresses at rest, K_0, depends on the water content, suction, vertical stress, and loading path. However, for simplicity, in this example the values of K_0 are assumed as constant depending on the stage of compaction as follows: $K_{0_{ic}} = 0.25$ on compaction, $\frac{K_{0_{pc}}}{K_{0_{ic}}} = 0.4$ for the post-compaction stage, and $K_{0_{sat}} = 0.4$ when soaking.

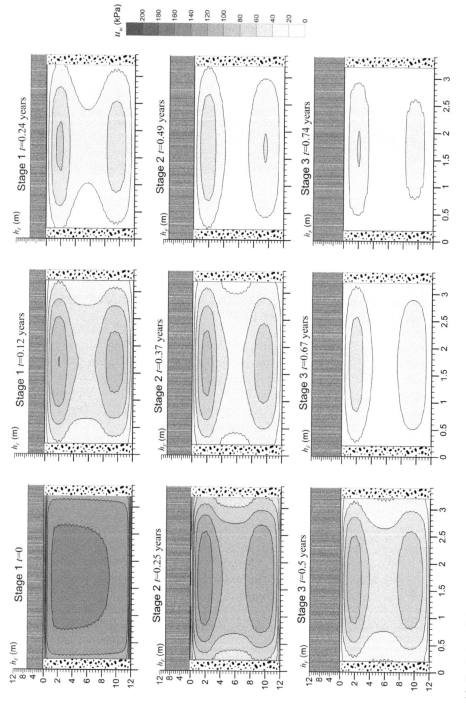

Figure 4.18 Evolution of the pore pressure for the radial consolidation of the three stages.

Table 4.7 Parameters of the BBM and the WRC for the soil of the example

Parameter	Meaning	Value	Parameter	Meaning	Value
$N(0)$	Compressibility	1.9	r	Compressibility	0.4
$\lambda(0)$	in saturated	0.09	β	in unsaturated	0.8 MPa^{-1}
κ	state	0.009	p^c	state	5 kPa
κ_s	Compressibility	0.001	ϕ_{wr}		0.008 kPa^{-1}
λ_s	due to suction		ψ_{wr}	Water	2.5
M	Critical state line	1.2	n_{wr}	retention	2.0
k_c	Tensile strength	0.8	m_{wr}	curve	0.5
			k_{swr}		0.00025 kPa^{-1}

The BBM, which is summarized in Section 3.1.1, can be used to compute volumetric strains of compacted soils in a soaked embankment. Obtaining the volumetric strain after soaking using the BBM requires analyzing the evolution of the different state variables along the different stages of soil compaction and soaking. Therefore, strain calculation requires the following steps:

1 Compute of the specific volume, and unit weight, obtained by oedometric compaction at different vertical stresses σ_{vc} and constant water content. Afterward, evaluate the vertical stresses corresponding to the required unit weights (i.e., $0.88\gamma_{d_{max}}^{SP}$, and $0.88\gamma_{d_{max}}^{MP}$).

2 Compute the unloading process down to zero vertical stress.

3 Compute the reloading process at different vertical stresses, from zero to $\sigma_{v_f} = 5$ MPa.

4 Compute the soaking process under constant vertical stress σ_{v_f}.

The above points are developed first, algebraically, and then the results of the analysis are applied to the soil whose parameters are given in Table 4.7.

4.3.1 Simulation of the oedometric compaction

The first step consists in simulating the process of oedometric compaction of the embankment using the BBM. This stage of the computation requires calculating the evolution of specific volume for a given vertical stress, σ_{vic}, and then assessing the vertical stress that corresponds to the required relative compaction density (i.e., $0.88\gamma_{d_{max}}$ of the standard and modified Proctor tests).

Compaction interpreted within the framework of the BBM is a hardening process represented by a growing of the elliptical yield surface as the compaction stress increases, as shown in Figure 4.19b. The state variables that represent the soil at the end of the loading process are p_{ic}, s_{ic}, q_{ic}.

Figures 4.19c and d represent the stress state on compaction in the p, q and p, s planes. These planes can be used to identify the positions of two overconsolidation stresses that corresponds to the saturated and unsaturated states (i.e., $p_{0_{ic}}^*$ and $p_{0_{ic}}$).

Moreover, compaction reduces the void space, and when assuming constant water content, it leads to an increase in the saturation degree. Consequently, as the void ratio decreases and the saturation degree increases, the matric suction of the soil changes

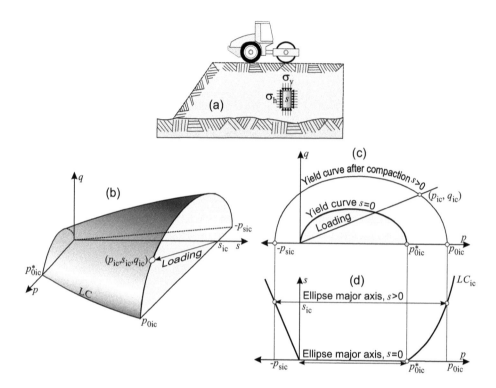

Figure 4.19 Stress state of an embankment during compaction according to the BBM [10], (a) principal stresses during compaction, (b) stress path during loading in the (p, q, s) plane, and (c) yield curves in the (p, q) and (p, s) planes.

following the set of water retention curves given by Equation 3.11. This equation indicates that the matric suction for different void ratios, e, and saturation degrees, S_r, is represented by

$$s_{ic} = \frac{1}{\phi_{wr}} \left(S_{ric}^{-1/m_{wr}} - 1 \right)^{1/n_{wr}} e^{-\psi_{wr}}.$$

However, for a constant water content w_0, the saturation degree is $S_{ric} = \frac{w_0}{e_{ic}} \frac{\gamma_s}{\gamma_w}$, and then

$$s_{ic} = \frac{1}{\phi_{wr}} \left[\left(\frac{w_0}{e_{ic}} \frac{\gamma_s}{\gamma_w} \right)^{-1/m_{wr}} - 1 \right]^{1/n_{wr}} e_{ic}^{-\psi_{wr}}.$$

Also, since the void ratio e_{ic} in terms of specific volume v_{ic} is $e_{ic} = v_{ic} - 1$, the evolution of suction during compaction at constant water content becomes

$$s_{ic} = \frac{1}{\phi_{wr}} \left[\left(\frac{w_0}{v_{ic} - 1} \frac{\gamma_s}{\gamma_w} \right)^{-1/m_{wr}} - 1 \right]^{1/n_{wr}} (v_{ic} - 1)^{-\psi_{wr}}. \tag{4.32}$$

Regarding the overconsolidation stress, Equation 3.28 permits its calculation as

$$p_0 = p \left[1 + \frac{\alpha^2}{M^2 \left(1 + \frac{p_s}{p} \right)} \right]. \tag{4.33}$$

However, considering that compaction occurs under a vertical stress σ_{vic}, and assuming oedometric conditions, the horizontal stress is $\sigma_{hic} = K_{0ic}\sigma_{vic}$. Therefore, the mean and deviator net stresses are

$$p_{ic} = \frac{\sigma_{vic} + 2\sigma_{hic}}{3} = \frac{\sigma_{vic}(1 + 2K_{0ic})}{3}, \text{ and}$$

$$q_{ic} = \sigma_{vic} - \sigma_{hic} = \sigma_{vic}(1 - K_{0ic}).$$

Therefore, the slope of the compaction path in the p, q plane is

$$\alpha_{ic} = \frac{q_{ic}}{p_{ic}} = \frac{3(1 - K_{0ic})}{1 + 2K_{0ic}}.$$

Then, the overconsolidation stress for the unsaturated state, p_{0ic}, is

$$p_{0ic} = \frac{1 + 2K_{0ic}}{3} \left\{ 1 + \frac{9(1 - K_{0ic})^2}{M^2(1 + 2K_{0ic})^2 \left[1 + \frac{3p_s}{(1+2K_{0ic})\sigma_{vic}} \right]} \right\} \sigma_{vic}. \tag{4.34}$$

Moreover, in the original BBM, the tensile strength p_s resulting from a suction s_{ic} is $p_s = k_c s_{ic}$, and then Equation 4.34 becomes

$$p_{0ic} = \frac{1 + 2K_{0ic}}{3} \left\{ 1 + \frac{9(1 - K_{0ic})^2}{M^2(1 + 2K_{0ic})^2 \left[1 + \frac{3k_c s_{ic}}{(1+2K_{0ic})\sigma_{vic}} \right]} \right\} \sigma_{vic}. \tag{4.35}$$

Regarding the evolution of the specific volume that results in an increase in the mean net stress p_{0ic}, Equations 3.3, 3.4, and 3.8 lead to

$$v_{ic} = N(0) - \kappa_s \ln \frac{s_{ic} + p_{atm}}{p_{atm}} - \lambda(0) \left[(1 - r)e^{-\beta s_{ic}} + r \right] \ln \frac{p_{0ic}}{p^c}. \tag{4.36}$$

To sum up, the evolution of suction and specific volume, or void ratio, of a soil submitted to oedometric compaction at a constant water content results from the solution of the system of nonlinear equations given in Table 4.8.

Computing the overconsolidation mean stress for the saturated state, p_{0ic}^*, is possible using the equation of the LC curve after knowing the suction and the specific volume, leading to

$$p_{0ic}^* = p^c \left(\frac{p_0(\sigma_{vic})}{p^c} \right)^{\frac{\lambda(s_{ic})-\kappa}{\lambda(0)-\kappa}}. \tag{4.37}$$

Table 4.8 System of nonlinear equations for computing suction and specific volume on loading

Equation	Expression
4.32	$s_{ic} = \frac{1}{\phi_{wr}} \left[\left(\frac{w_0}{v_{ic}-1} \frac{\gamma_s}{\gamma_w} \right)^{-1/m_{wr}} - 1 \right]^{1/n_{wr}} (v_{ic} - 1)^{-\psi_{wr}}$
4.35	$p_{0ic} = \frac{1+2K_{0ic}}{3} \left\{ 1 + \frac{9(1-K_{0ic})^2}{M^2(1+2K_{0ic})^2 \left[1+\frac{3k_c s}{(1+2K_{0ic})\sigma_{vic}} \right]} \right\} \sigma_{vic}$
4.36	$v_{ic} = N(0) - \kappa_s \ln \frac{s_{ic}+p_{atm}}{p_{atm}} - \lambda(0) \left[(1-r)e^{-\beta s_{ic}} + r \right] \ln \frac{p_{0ic}}{p^c}$

After the algebraic derivation of the compaction equations applied to the BBM, the next step is to use the values of the parameters given in Table 4.7. Therefore, the equations that allow calculating the specific volume for a given vertical stress are

Saturation degree,

$$S_{r_{ic}} = \frac{2.73w_0}{v_{ic} - 1}. \tag{4.38}$$

Suction in kPa,

$$s_{ic} = 125(v_{ic} - 1)^{-2.5}\left[S_{r_{ic}}^{-2} - 1\right]^{1/2}. \tag{4.39}$$

Overconsolidation stress in unsaturated state,

$$p_{0_{ic}} = 0.5\left(1 + \frac{1.5625}{1 + 1.6\frac{s_{ic}}{\sigma_{v_{ic}}}}\right)\sigma_{v_{ic}}. \tag{4.40}$$

Specific volume,

$$v_{ic} = 1.9 - 0.001\ln\frac{s_{ic} + 101.3}{101.3} - 0.09\left(0.6e^{-0.0008s_{ic}} + 0.4\right)\ln\frac{p_{0_{ic}}}{5}. \tag{4.41}$$

The set of nonlinear Equations 4.38–4.41 can be easily solved using the following iterative procedure:

1   Choose a value of vertical stress, $\sigma_{v_{ic}}$.
2   Assume a value of specific volume v_{ic}.
3   Compute the suction, s_{ic}, using Equation 4.39.
4   Compute $p_{0_{ic}}$ using Equation 4.40.
5   Compute v_{ic} using Equation 4.41.
6   Return to step 2 until convergence.
7   Choose another value of vertical stress.

Finally, the equation of the LC curve allows computing the overconsolidation stress for the saturated state, $p_{0_{ic}}^*$, as

$$p_{0_{ic}}^* = 5\left(0.2p_{0_{ic}}\right)^{\frac{1}{3}(2e^{-0.0008s_{ic}}+1)} \quad \text{kPa} \tag{4.42}$$

Table 4.9 presents the results of specific volume and stresses achieved during the loading stage for different water contents.

Figure 4.20 shows the suction paths achieved during compression with different water contents. In this figure, it is noticeable that for low water content, the suction path progress toward higher suctions for incremental compression, one reason for this increase in suction is the reduction in pore sizes as the void ratio, or specific volume, decreases. On the other hand, for the water content of 15%, the suction initially increases and then decreases for high degrees of saturation, reaching saturation (i.e., zero suction) for compression above 2,000 kPa.

Figure 4.21 illustrates the three oedometric compression curves obtained for different constant water contents. These curves initiate with a linear trend; however, the curve for 5% water content finalizes with a slight concave shape that results on the increase of

Table 4.9 Results of the loading stage

σ_{vic} (kPa)	$w_0 = 5\%$				$w_0 = 10\%$				$w_0 = 15\%$			
	v_{ic}	s_{ic} (kPa)	p_{0ic} (kPa)	p^*_{0ic} (kPa)	v_{ic}	s_{ic} (kPa)	p_{0ic} (kPa)	p^*_{0ic} (kPa)	v_{ic}	s_{ic} (kPa)	p_{0ic} (kPa)	p^*_{0ic} (kPa)
10	1.897	1065.4	5.0	5.0	1.897	513.6	5.1	5.1	1.896	320.0	5.1	5.1
100	1.768	1338.4	53.5	18.9	1.734	675.6	56.6	28.8	1.712	415.4	60.2	37.7
200	1.732	1436.9	112.5	27.2	1.687	737.9	122.6	47.4	1.656	448.9	134.0	69.2
300	1.711	1498.7	176.1	33.5	1.660	778.1	195.5	63.0	1.623	468.1	217.0	99.0
400	1.696	1545.2	243.5	38.8	1.640	808.2	273.8	76.8	1.599	480.7	306.9	127.8
500	1.685	1582.9	314.4	43.3	1.626	832.3	356.6	89.5	1.580	489.3	402.2	155.9
600	1.676	1614.0	388.4	47.4	1.614	852.6	443.2	101.1	1.565	495.2	502.2	183.7
700	1.668	1641.7	465.1	51.1	1.604	870.1	533.0	111.9	1.552	499.2	605.4	211.3
800	1.661	1665.8	544.3	54.5	1.596	885.5	625.6	122.1	1.540	501.4	712.1	238.9
900	1.656	1687.3	625.8	57.6	1.588	899.2	720.6	131.7	1.530	502.2	821.5	266.6
1,000	1.651	1706.6	709.4	60.6	1.581	911.9	817.7	140.7	1.521	501.8	933.3	294.7
2,000	1.617	1842.3	1631.6	83.3	1.540	995.6	1869.8	213.8	1.338	0.0	2562.5	2562.5
3,000	1.599	1925.7	2656.3	99.2	1.518	1044.9	3005.0	268.1	1.302	0.0	3843.8	3843.8
4,000	1.586	1986.2	3741.4	111.7	1.501	1082.9	4182.3	310.6	1.276	0.0	5125.0	5125.0
5,000	1.576	2033.6	4866.3	122.1	1.491	1106.9	5384.5	349.6	1.256	0.0	6406.3	6406.3

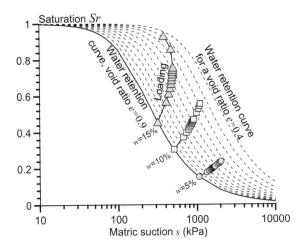

Figure 4.20 Suction paths achieved during compaction at the three different constant water contents ($w_0 = 5\%$, $w_0 = 10\%$, $w_0 = 15\%$).

suction during compression, while the curve for 15% water content ends with a convex shape due to the reduction in suction occurring at high compression stresses and reaches the saturated compression line for high vertical stresses.

Assessing vertical strains during soaking for the two different compaction densities (88% of the standard and modified Proctor tests), which is the central question of this example, requires computing the vertical stresses that correspond to the respective specific volumes. Using a phase diagram, the specific volume of a soil is $v = \frac{\gamma_s}{\gamma_d}$, and so the specific volume for the required compaction densities are

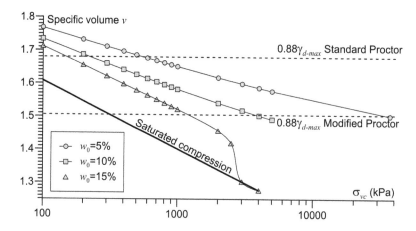

Figure 4.21 Compression curves achieved during the oedometric compaction at the three different constant water contents ($w_0 = 5\%$, $w_0 = 10\%$, $w_0 = 15\%$).

Table 4.10 Stresses required to reach the specific volumes corresponding to 88% of γ_{dmax} of the standard and modified Proctor tests

Compaction degree	v_{ic}	$w_0 = 5\%$			$w_0 = 10\%$			$w_0 = 15\%$		
		$\sigma_{v_{ic}}$ (kPa)	s_{ic} (kPa)	$p^*_{0_{ic}}$ (kPa)	$\sigma_{v_{ic}}$ (kPa)	s_{ic} (kPa)	$p^*_{0_{ic}}$ (kPa)	$\sigma_{v_{ic}}$ (kPa)	s_{ic} (kPa)	$p^*_{0_{ic}}$ (kPa)
$0.88\gamma_{dmax}$ SP	1.680	550	1599.2	45.4	220	747.2	50.7	150	434.9	53.7
$0.88\gamma_{dmax}$ MP	1.505	33,500	2456.8	231.9	3,800	1073.8	303.9	1,190	497.8	350.0

$$v_{0.88SP} = \frac{2.73 \cdot 9.8}{0.88 \cdot 18.1} = 1.680 \quad \text{SP, and}$$

$$v_{0.88MP} = \frac{2.73 \cdot 9.8}{0.88 \cdot 20.2} = 1.505 \quad \text{MP.}$$

The procedure described above, which applies to any vertical stress value, allows calculating the vertical stress values for each density, as well as the overconsolidation stress in the saturated state, which is useful information for calculating the collapse strain after soaking. Table 4.10 shows those results.

4.3.2   Post compaction

After applying the compaction stress, the compactor passes, and the soil follows an unloading path to zero vertical stress. This stress condition is known as the post-compaction state and is represented by the point p_{pc}, s_{pc}, q_{pc} in Figure 4.22.

As a result of the reduction in vertical stress, the post-compaction stress state implies changes in the specific volume, saturation, and suction. Assuming at rest conditions, the horizontal stress is $\sigma_{h_{pc}} = K_{0_{pc}} \sigma_{v_{ic}}$, while the vertical stress decreases to zero, then mean net stress after unloading becomes

$$p_{pc} = \frac{2}{3} K_{0_{pc}} \sigma_{v_{ic}}.$$

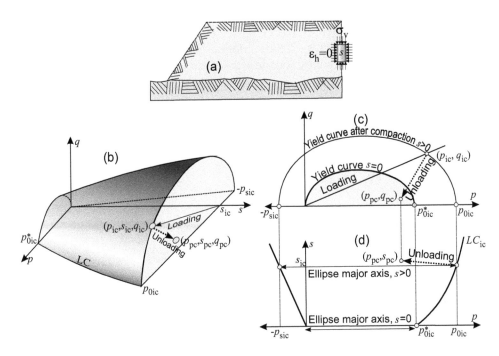

Figure 4.22 Post compaction stress state of an embankment according to the BBM [10], (a) stress state during unloading, (b) stress path during loading and then unloading in the (p, q, s) plane, and (c) yield curves after loading and unloading in the (p, q) and (p, s) planes.

The soil undergoes an increase in the specific volume due to the elastic unloading from $p_{ic} = \frac{1+2K_{0ic}}{3}\sigma_{vic}$, to $p_{pc} = \frac{2}{3}K_{0pc}\sigma_{vic}$. Equation 3.5 allows computing this increase in specific volume as

$$v_{pc} = v_{ic} - \kappa \ln \frac{2K_{0pc}}{1+2K_{0ic}}.$$

Then, for the data of this example, the change in specific volume on unloading is

$$\Delta v = v_{pc} - v_{ic} = 0.009 \ln \left(\frac{1 + 2 \cdot 0.25}{2 \cdot 0.25 \cdot 0.4} \right) = 1.81 \cdot 10^{-2}.$$

Moreover, the increase in specific volume, at constant water content, leads to the following decrease of the saturation degree:

$$\Delta S_r = w_0 \frac{\gamma_s}{\gamma_w} \left(\frac{1}{v_{pc}-1} - \frac{1}{v_{ic}-1} \right).$$

For each compaction energy, the decrease in saturation degrees for the data of the example are

$$\Delta S_r = 2.73 w_0 \left(\frac{1}{0.680 + 1.81 \cdot 10^{-2}} - \frac{1}{0.680} \right) = -0.1044 w_0 \quad \text{For SP, and}$$

$$\Delta S_r = 2.73 w_0 \left(\frac{1}{0.505 + 1.81 \cdot 10^{-2}} - \frac{1}{0.505} \right) = -0.1873 w_0 \quad \text{For MP.}$$

Since the saturation degree decreases, the suction-saturation path follows the curve given by Equation 3.12. Then, the matric suction after unloading, s_{pc}, becomes

$$s_{pc} = s_{ic} - \frac{\Delta S_r}{k_{SWR}}.$$

So, using the data of the example, changes in suction become

$$\Delta s = 417.13 w_0 \text{ kPa} \quad \text{For SP, and}$$
$$\Delta s = 749.57 w_0 \text{ kPa} \quad \text{For MP.}$$

According to the previous analysis, the changes in specific volume, saturation degree, and suction, as well as their final values at the end of the post-compaction state are presented in Tables 4.11 and 4.12.

4.3.3 Reloading

The construction of the embankment continues by placing subsequent compacted layers, which produces an increase of the overburden stress to (p_{fc}, q_{fc}). However, depending on the magnitude of the overburden stress, the yield surface may grow or remain constant:

- For a low overburden stresses, the increase in vertical stress does not increase the size of the initial yield curve. This situation appears when the stress state (p_{fc}, q_{fc}) falls within the area limited by the initial yield curve reached on the compaction stage (i.e., $p_{fc} < p_{ic}$), as shown in Figure 4.23.
- In contrast, for a high overburden stress, that occurs when (p_{fc}, q_{fc}) is higher than the compaction stress (p_{ic}, q_{ic}), the elliptic yield curve of the soil grows and produces additional settlement during construction. Simultaneously, the loading collapse curve moves toward a new position denoted by LC_{fc} in Figure 4.24.

Table 4.11 Changes in the specific volume, saturation degree, and suction at the post-compaction state

Compaction degree	Δv	$w_0 = 5\%$		$w_0 = 10\%$		$w_0 = 15\%$	
		ΔS_r	Δs (kPa)	ΔS_r	Δs (kPa)	ΔS_r	Δs (kPa)
$0.88 \gamma_{d max}$ SP	$1.81 \cdot 10^{-2}$	$-5.22 \cdot 10^{-3}$	20.9	$-1.04 \cdot 10^{-2}$	41.8	$-1.57 \cdot 10^{-2}$	62.6
$0.88 \gamma_{d max}$ MP	$1.81 \cdot 10^{-2}$	$-9.37 \cdot 10^{-3}$	37.5	$-1.87 \cdot 10^{-2}$	74.9	$-2.81 \cdot 10^{-2}$	112.4

Table 4.12 Specific volume, saturation degree, and suction at the end of the post-compaction state

Compaction degree	v_{pc}	$w_0 = 5\%$		$w_0 = 10\%$		$w_0 = 15\%$	
		S_{rpc}	s_{pc} (kPa)	S_{rpc}	s_{pc} (kPa)	S_{rpc}	s_{pc} (kPa)
$0.88 \gamma_{d max}$ SP	1.6981	0.196	1620.0	0.391	788.8	0.587	497.6
$0.88 \gamma_{d max}$ MP	1.5231	0.261	2494.3	0.522	1148.9	0.783	610.2

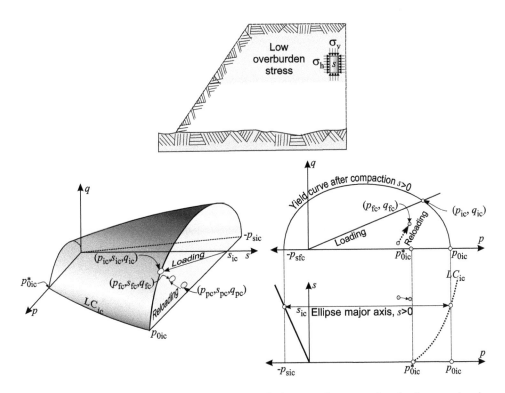

Figure 4.23 Stress state of an embankment at the final stage of construction for low overburden stresses [10], (a) stress state during reloading, (b) stress path during loading, unloading and reloading in the (p, q, s) plane, and (c) yield curves after loading unloading and reloading in the (p, q) and (p, s) planes.

The stress path on reloading up to a vertical stress $\sigma_{v_{fc}}$ follows initially the elastic domain in both planes: (v, p) and (s, S_r). Afterward, when the vertical stress reaches the initial vertical stress, the loading path continues with the initial trend, increasing the plastic strains. The following analysis permits to obtain the values of specific volume and suction for the two domains of behavior (*i.e.*, elastic and elastoplastic).

4.3.3.1 Elastic domain

When the final vertical stress, $\sigma_{v_{fc}}$, is lower than the compaction vertical stress, $\sigma_{v_{ic}}$, the soil remains within the elastic domain. In this case, the horizontal stress remains constant at the same value than on unloading (*i.e.*, $\sigma_{h_{cf}} = K_{0_{pc}}\sigma_{v_{ic}}$), and so the mean net stress is

$$p_{fc} = \frac{1}{3}(\sigma_{v_{fc}} + 2K_{0_{pc}}\sigma_{v_{ic}}),$$

and the specific volume after reloading is

$$v_{fc} = v_{pc} - \kappa \ln \frac{\sigma_{v_{fc}} + 2K_{0_{pc}}\sigma_{v_{ic}}}{2K_{0_{pc}}\sigma_{v_{ic}}}. \tag{4.43}$$

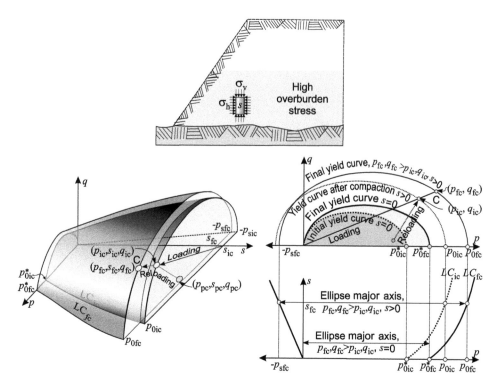

Figure 4.24 Stress state of an embankment at the end of the construction for high overburden stresses [10], (a) stress state during reloading, (b) stress path during loading, unloading and reloading in the (p, q, s) plane, and (c) yield curves after loading unloading and reloading in the (p, q) and (p, s) planes.

As a result, the reduction of specific volume at constant water content leads to the following increase in the saturation degree:

$$\Delta S_r = w_0 \frac{\gamma_s}{\gamma_w} \left(\frac{1}{v_{fc} - 1} - \frac{1}{v_{pc} - 1} \right). \tag{4.44}$$

Moreover, the suction decreases to

$$s_{fc} = s_{pc} - \frac{\Delta S_r}{k_{sWR}}. \tag{4.45}$$

Computing the evolution of specific volume, saturation and suction along the reloading elastic path is possible including in Equations 4.43–4.45 the data of the example and the results of the post-compaction process, as follows:

Specific volume:

$$v_{fc} = 1.6981 - 0.009 \ln \left(5 \frac{\sigma_{vfc}}{\sigma_{vic}} + 2.5 \right) \quad \text{For SP, and}$$

$$v_{fc} = 1.5231 - 0.009 \ln \left(5 \frac{\sigma_{vfc}}{\sigma_{vic}} + 2.5 \right) \quad \text{For MP.}$$

Change in saturation degree $\Delta S_r = S_{r_{fc}} - S_{r_{pc}}$:

$$\Delta S_r = 2.73 w_0 \left(\frac{1}{v_{fc} - 1} - 1.4324 \right) \quad \text{For SP, and}$$

$$\Delta S_r = 2.73 w_0 \left(\frac{1}{v_{fc} - 1} - 1.9116 \right) \quad \text{For MP.}$$

Change in suction $\Delta s = s_{fc} - s_{pc}$:

$$\Delta s = -4,000 \Delta S_r \, kPa \quad \text{For SP and MP.}$$

As shown in Figures 4.25a and b, the overconsolidation stresses for unsaturated and saturated states remain at the same position than during compaction, so $p_{0_{fc}} = p_{0_{ic}}$, and $p^*_{0_{fc}} = p^*_{0_{ic}}$.

4.3.3.2 Elastoplastic domain

For the elastoplastic reloading, the hydromechanical paths in the (s, S_r) and (p, v) planes continue along the same paths than on the initial loading. Therefore, the set of equations given in Table 4.8 for the initial loading also allows computing the soil variables for reloading, using $\sigma_{v_{fc}}$ instead of $\sigma_{v_{ic}}$ as the maximum stress.

Similarly, computing the overconsolidation mean stress for the unsaturated and saturated states achieved after reloading $p_{0_{fc}}$ and $p^*_{0_{fc}}$ is possible using Equations 4.40 and 4.42 (see Figure 4.25c).

4.3.4 Soaking

As described in Caicedo [10], during soaking, changes of specific volume depend on the level of stresses reached on reloading, as follows:

- For low stress levels (Figure 4.26a), the saturated elliptic yield curve at the end of construction remains in the same position as in the final stage of compaction. Therefore, the saturated overconsolidation stresses at saturation are also equal: $p^*_{0_{ic}} = p^*_{0_{fc}}$.

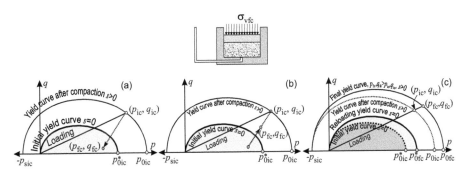

Figure 4.25 Projection in the (p, q, s=0) plane of the elliptic yield curves at the end of compaction, three cases are possible: (a) the stress state (p, q) falls into the zone limited by the initial saturated yield curve, (b) the stress state (p, q) falls into the area limited by the initial saturated yield curve and the projection of the unsaturated yield curve after compaction, and (c) the stress state (p, q) falls outside the projection of the unsaturated yield curve after compaction.

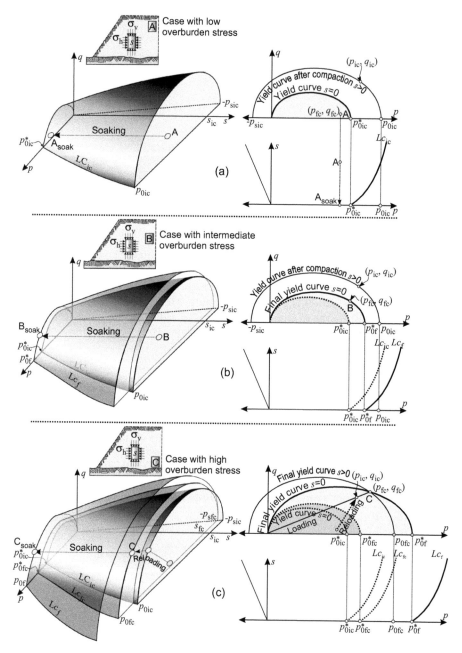

Figure 4.26 Evolution of the yield curves of the BBM model after soaking, three cases are possible [10]: (a) the stress state (p, q) falls into the zone limited by the initial saturated yield curve, and therefore the yield surface does not moves, (b) the stress state (p, q) falls into the area limited by the initial saturated yield curve and the projection of the unsaturated yield curve after compaction; therefore, the LC curve and the elliptic surface move towards higher stresses, and (c) the stress state (p, q) falls outside the projection of the initial unsaturated yield curve after compaction, and therefore the LC curve and the elliptical yield surface grows according to the position of the stress after soaking.

- On the other hand, when the projection of the final stress state (p_{fc}, q_{fc}) in the saturated plane is outside the area delimited by the initial saturated yield curve, and the embankment becomes soaked, the saturated yield curve grows, as shown in Figure 4.26b. As a result, the growth of the saturated yield curve produces plastic strains, and the soil collapses.

- This phenomenon also occurs at high overburden stresses, as shown in Figure 4.26c. In this case, the saturated yield curve also grows producing plastic strains due to collapse.

In other words, when the soil soaks, the stress state in the plane (p, q) remains at the same position than on reloading, but the suction decreases to zero. Therefore, depending on the magnitude of the stresses and the value of the saturated overconsolidation stress, the saturated yield curve could grow to produce plastic strains (*i.e.*, collapse) or remains in the same position as on reloading. Figures 4.26 and 4.27 illustrate the three possible cases for the stress state.

Strains due to soaking, Δv, occur in two directions: first, strains due to swelling, Δv_s, which appear due to reduction in suction, and collapse strains, Δv_c, which appear because of the increase in saturated overconsolidation stress from $p_{0_{fc}}^*$ to $p_{0_f}^*$, and then

$$\Delta v = \Delta v_c + \Delta v_s.$$

Collapse occurs because of the growing of the saturated yield curve, which is represented by the ellipse on the plane $(p, q, s = 0)$. This ellipse passes through two points: the origin at $p = 0$ and $q = 0$, and the point representing the final stress at p_{fc}, q_{fc}. Likewise, the relationship $\alpha = q/p$ between the deviator and the mean net stress passes from α_{ic} in the final stage of compaction going to α_{fc} in the final stage of construction finalizing in α_f on soaking. Then, Equation 4.33, applied in the saturated plane (*i.e.*, for $s = 0$ and $p_{sic} = 0$), leads to the new saturated overconsolidation stress $p_{0_f}^*$, which is

$$p_{0f}^* = p_{fc}\left[1 + \frac{\alpha_f^2}{M^2}\right]. \tag{4.46}$$

After computing p_{0f}^*, the change in specific volume due to collapse is

$$\Delta v_c = \lambda(0) \ln \frac{p_{0fc}^*}{p_{0f}^*}.$$

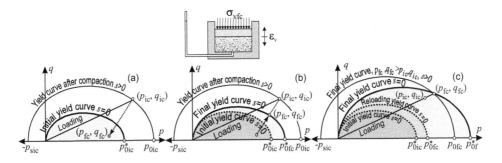

Figure 4.27 Growing of the saturated elliptic yield curve after soaking according to the three cases depicted in Figure 4.26, which are: (a) the saturated yield surface does not move, and (b, c) the saturated yield surface grows according to the stress state after soaking.

Using the data of the example, the coefficient of horizontal stresses at rest K_0 increases up to 0.4 on soaking. So, the relationship between deviator and mean stresses, α_f, is

$$\alpha_f = \frac{q_f}{p_f} = \frac{3\sigma_{vf}(1 - K_{0\text{sat}})}{\sigma_{vf}(1 + 2K_{0\text{sat}})} = \frac{3\sigma_{vf}(1 - 0.4)}{\sigma_{vf}(1 + 2 \cdot 0.4)} = 1.$$

Afterward, Equation 4.46 leads to the following equation that provides the overconsolidation stress on soaking

$$p_{0f}^* = p_f \left[1 + \frac{1}{1.2^2}\right] = 1.694 p_f = 1.0167\sigma_{vf}.$$

Therefore, the change in specific volume due to collapse becomes

$$\Delta v_c = 0 \qquad \text{if} \quad p_{0f}^* < p_{0fc}^*, \text{ or}$$

$$\Delta v_c = -0.09 \ln \left(\frac{p_{0f}^*}{p_{0fc}^*}\right) \quad \text{if} \quad p_{0f}^* \geq p_{0fc}^*.$$

Regarding swelling, Equation 3.6 can be used to compute the change in specific volume as

$$\Delta v_s = \kappa_s \left[\ln \left(\frac{s_{fc} + p_{\text{atm}}}{p_{\text{atm}}}\right)\right].$$

When using the data of the example, the change in specific volume due to expansion is

$$\Delta v_s = \frac{1}{1,000} \left[\ln \left(\frac{s_{fc} + 101.3}{101.3}\right)\right].$$

The vertical strain, ε_v, results by computing the rate between the change in the specific volume and its initial value. Therefore, assuming positive values in compression, ε_v becomes

$$\varepsilon_v = -\frac{\Delta v}{v} = -\frac{\Delta v_c + \Delta v_s}{v}.$$

Finally, the total vertical displacement of the embankment, Δz, results from the integration of the vertical strain as follows:

$$\Delta z = \int \varepsilon_z dz.$$

Figure 4.28 presents the results of vertical strain for the different water contents and densities. Also, Tables 4.13–4.18 present the values of specific volume, suction, and saturated overconsolidation stresses (p_{0f}^* and p_{0fc}^*) that allow computing the changes in specific volume due to expansion and collapse as well as the vertical strains.

4.3.5 Concluding remarks

Results provide pieces of evidence of the relevance of compaction density and water content on the volumetric behavior of compacted soils in embankments, mainly when they soak.

- Compaction at low water contents (*i.e.*, 5% in this example) produce significant collapse strains on soaking. The magnitude of the collapse decreases when increasing

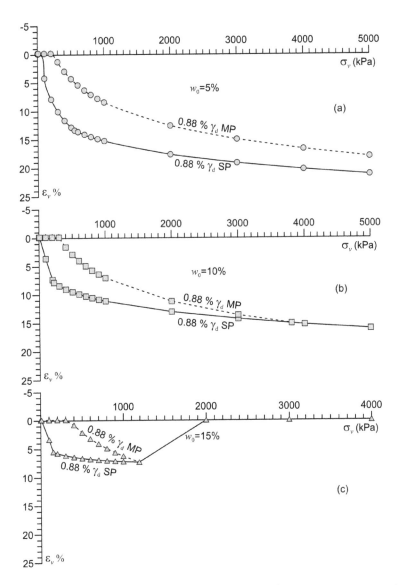

Figure 4.28 Results of the vertical strain after soaking for the three water contents and the two compaction energies.

the relative density; however, this reduction is noticeable mainly at low vertical stresses. In fact, high vertical stress mitigates the effect of the high overconsolidation stress achieved during compaction.

- Intermediate water contents (*i.e.*, 10% in this example) produce less collapsing strains than dry compaction. The effect of the compaction density is also noticeable, but it disappears for high vertical stresses (*i.e.*, when the vertical stress exceeds the overconsolidation vertical stress achieved during compaction).

- High water contents (*i.e.*, 15% in this example) produce low collapsing strains. This effect is particularly clear for high densities and low vertical stresses.

Table 4.13 Volumetric strain due to soaking for a soil compacted at $0.88\gamma_{d\max}$ of the standard Proctor and 5% water content, shaded cells highlights the domain of elastic behavior.

σ_{vf} (kPa)	v_{fc}	s_{fc} (kPa)	p^*_{0f} (kPa)	p^*_{0fc} (kPa)	Behavior Loading	Behavior Soaking	Δv_c	Δv_s	ε_v (%)
0	1.698	1620.0	0.0	45.4	Elastic	Elastic	0.000	0.003	−0.17
10	1.690	1610.3	10.2	45.4	Elastic	Elastic	0.000	0.003	−0.17
100	1.687	1607.4	101.7	45.4	Elastic	Elastoplastic	−0.073	0.003	4.14
200	1.685	1605.0	203.3	45.4	Elastic	Elastoplastic	−0.135	0.003	7.84
300	1.683	1602.9	305.0	45.4	Elastic	Elastoplastic	−0.171	0.003	10.02
400	1.682	1601.3	406.7	45.4	Elastic	Elastoplastic	−0.197	0.003	11.57
500	1.681	1599.8	508.3	45.4	Elastic	Elastoplastic	−0.217	0.003	12.77
600	1.676	1614.4	610.0	47.4	Elastoplastic	Elastoplastic	−0.230	0.003	13.56
700	1.668	1641.7	711.7	51.1	Elastoplastic	Elastoplastic	−0.237	0.003	14.04
800	1.661	1665.8	813.3	54.5	Elastoplastic	Elastoplastic	−0.243	0.003	14.47
900	1.656	1687.3	915.0	57.6	Elastoplastic	Elastoplastic	−0.249	0.003	14.86
1,000	1.651	1706.6	1016.7	60.6	Elastoplastic	Elastoplastic	−0.254	0.003	15.20
2,000	1.617	1842.3	2033.3	83.3	Elastoplastic	Elastoplastic	−0.288	0.003	17.60
3,000	1.599	1925.7	3050.0	99.2	Elastoplastic	Elastoplastic	−0.308	0.003	19.10
4,000	1.586	1986.2	4066.7	111.7	Elastoplastic	Elastoplastic	−0.324	0.003	20.21
5,000	1.576	2033.6	5083.4	122.1	Elastoplastic	Elastoplastic	−0.336	0.003	21.10

Table 4.14 Volumetric strain due to soaking for a soil compacted at $0.88\gamma_{d\max}$ of the modified Proctor and 5% water content, shaded cells highlights the domain of elastic behavior.

σ_{vf} (kPa)	v_{fc}	s_{fc} (kPa)	p^*_{0f} (kPa)	p^*_{0fc} (kPa)	Behavior Loading	Behavior Soaking	Δv_c	Δv_s	ε_v (%)
0	1.523	2494.3	0.0	238.8	Elastic	Elastic	0.000	0.003	−0.21
10	1.515	2477.5	10.2	238.8	Elastic	Elastic	0.000	0.003	−0.21
100	1.515	2477.4	101.7	238.8	Elastic	Elastic	0.000	0.003	−0.21
200	1.515	2477.3	203.3	238.8	Elastic	Elastic	0.000	0.003	−0.21
300	1.515	2477.2	305.0	238.8	Elastic	Elastoplastic	−0.022	0.003	1.24
400	1.515	2477.1	406.7	238.8	Elastic	Elastoplastic	−0.048	0.003	2.95
500	1.515	2476.9	508.3	238.8	Elastic	Elastoplastic	−0.068	0.003	4.28
600	1.515	2476.8	610.0	238.8	Elastic	Elastoplastic	−0.084	0.003	5.36
700	1.514	2476.7	711.7	238.8	Elastic	Elastoplastic	−0.098	0.003	6.28
800	1.514	2476.6	813.3	238.8	Elastic	Elastoplastic	−0.110	0.003	7.07
900	1.514	2476.5	915.0	238.8	Elastic	Elastoplastic	−0.121	0.003	7.77
1,000	1.514	2476.4	1016.7	238.8	Elastic	Elastoplastic	−0.130	0.003	8.40
2,000	1.514	2475.4	2033.3	238.8	Elastic	Elastoplastic	−0.193	0.003	12.52
3,000	1.513	2474.4	3050.0	238.8	Elastic	Elastoplastic	−0.229	0.003	14.94
4,000	1.513	2473.5	4066.7	238.8	Elastic	Elastoplastic	−0.255	0.003	16.65
5,000	1.513	2472.6	5083.4	238.8	Elastic	Elastoplastic	−0.275	0.003	17.98

Table 4.15 Volumetric strain due to soaking for a soil compacted at $0.88\gamma_{d\max}$ of the standard Proctor and 10% water content, shaded cells highlights the domain of elastic behavior.

σ_{vf} (kPa)	v_{fc}	s_{fc} (kPa)	$p^*_{0_f}$ (kPa)	$p^*_{0_{fc}}$ (kPa)	Behavior Loading	Behavior Soaking	Δv_c	Δv_s	ε_v (%)
0	1.699	788.8	0.0	50.7	Elastic	Elastic	0.000	0.002	−0.13
10	1.689	768.2	10.2	50.7	Elastic	Elastic	0.000	0.002	−0.13
100	1.684	756.6	101.7	50.7	Elastic	Elastoplastic	−0.063	0.002	3.59
200	1.681	748.4	203.3	50.7	Elastic	Elastoplastic	−0.125	0.002	7.31
300	1.660	778.1	305.0	63.0	Elastoplastic	Elastoplastic	−0.142	0.002	8.42
400	1.640	808.2	406.7	76.8	Elastoplastic	Elastoplastic	−0.150	0.002	9.01
500	1.626	832.3	508.3	89.5	Elastoplastic	Elastoplastic	−0.156	0.002	9.48
600	1.614	852.6	610.0	101.1	Elastoplastic	Elastoplastic	−0.162	0.002	9.89
700	1.604	870.1	711.7	111.9	Elastoplastic	Elastoplastic	−0.167	0.002	10.24
800	1.596	885.5	813.3	122.1	Elastoplastic	Elastoplastic	−0.171	0.002	10.56
900	1.588	899.2	915.0	131.7	Elastoplastic	Elastoplastic	−0.174	0.002	10.84
1,000	1.581	911.9	1016.7	140.7	Elastoplastic	Elastoplastic	−0.178	0.002	11.11
2,000	1.540	995.6	2033.3	213.8	Elastoplastic	Elastoplastic	−0.203	0.002	13.01
3,000	1.518	1044.9	3050.0	268.1	Elastoplastic	Elastoplastic	−0.219	0.002	14.26
4,000	1.501	1082.9	4066.7	310.6	Elastoplastic	Elastoplastic	−0.231	0.002	15.26
5,000	1.491	1106.9	5083.4	349.6	Elastoplastic	Elastoplastic	−0.241	0.002	16.00

Table 4.16 Volumetric strain due to soaking for a soil compacted at $0.88\gamma_{d\max}$ of the modified Proctor and 10% water content, shaded cells highlights the domain of elastic behavior.

σ_{vf} (kPa)	v_{fc}	s_{fc} (kPa)	$p^*_{0_f}$ (kPa)	$p^*_{0_{fc}}$ (kPa)	Behavior Loading	Behavior Soaking	Δv_c	Δv_s	ε_v (%)
0	1.523	1148.9	0.0	303.9	Elastic	Elastic	0.000	0.003	−0.17
10	1.515	1115.1	10.2	303.9	Elastic	Elastic	0.000	0.002	−0.16
100	1.514	1113.4	101.7	303.9	Elastic	Elastic	0.000	0.002	−0.16
200	1.514	1111.6	203.3	303.9	Elastic	Elastic	0.000	0.002	−0.16
300	1.514	1109.9	305.0	303.9	Elastic	Elastic	0.000	0.002	−0.14
400	1.513	1108.2	406.7	303.9	Elastic	Elastoplastic	−0.026	0.002	1.57
500	1.513	1106.6	508.3	303.9	Elastic	Elastoplastic	−0.046	0.002	2.90
600	1.512	1105.1	610.0	303.9	Elastic	Elastoplastic	−0.063	0.002	3.98
700	1.512	1103.6	711.7	303.9	Elastic	Elastoplastic	−0.077	0.002	4.90
800	1.512	1102.2	813.3	303.9	Elastic	Elastoplastic	−0.089	0.002	5.70
900	1.511	1100.9	915.0	303.9	Elastic	Elastoplastic	−0.099	0.002	6.40
1,000	1.511	1099.5	1016.7	303.9	Elastic	Elastoplastic	−0.109	0.002	7.03
2,000	1.508	1088.3	2033.3	303.9	Elastic	Elastoplastic	−0.171	0.002	11.18
3,000	1.506	1079.6	3050.0	303.9	Elastic	Elastoplastic	−0.208	0.002	13.62
4,000	1.501	1082.9	4066.7	310.6	Elastoplastic	Elastoplastic	−0.231	0.002	15.26
5,000	1.491	1106.9	5083.4	349.6	Elastoplastic	Elastoplastic	−0.241	0.002	16.00

Table 4.17 Volumetric strain due to soaking for a soil compacted at $0.88\gamma_{d\max}$ of the standard Proctor and 15% water content, shaded cells highlights the domain of elastic behavior.

σ_{vf} (kPa)	v_{fc}	s_{fc} (kPa)	$p_{0_f}^*$ (kPa)	$p_{0_{fc}}^*$ (kPa)	Behavior Loading	Behavior Soaking	Δv_c	Δv_s	ε_v (%)
0	1.698	497.6	0.0	53.7	Elastic	Elastic	0.000	0.002	−0.10
10	1.689	465.6	10.2	53.7	Elastic	Elastic	0.000	0.002	−0.10
100	1.682	442.9	101.7	53.7	Elastic	Elastoplastic	−0.057	0.002	3.31
200	1.656	448.9	203.3	69.2	Elastoplastic	Elastoplastic	−0.097	0.002	5.76
300	1.623	468.1	305.0	99.0	Elastoplastic	Elastoplastic	−0.101	0.002	6.14
400	1.599	480.7	406.7	127.8	Elastoplastic	Elastoplastic	−0.104	0.002	6.41
500	1.580	489.3	508.3	155.9	Elastoplastic	Elastoplastic	−0.106	0.002	6.62
600	1.565	495.2	610.0	183.7	Elastoplastic	Elastoplastic	−0.108	0.002	6.79
700	1.552	499.2	711.7	211.3	Elastoplastic	Elastoplastic	−0.109	0.002	6.93
800	1.540	501.4	813.3	238.9	Elastoplastic	Elastoplastic	−0.110	0.002	7.04
900	1.530	502.2	915.0	266.6	Elastoplastic	Elastoplastic	−0.111	0.002	7.14
1,000	1.521	501.8	1016.7	294.7	Elastoplastic	Elastoplastic	−0.111	0.002	7.21
2,000	1.338	0.0	2033.3	2562.5	Elastoplastic	Elastoplastic	0.000	0.000	0.00
3,000	1.302	0.0	3050.0	3843.8	Elastoplastic	Elastoplastic	0.000	0.000	0.00
4,000	1.276	0.0	4066.7	5125.0	Elastoplastic	Elastoplastic	0.000	0.000	0.00
5,000	1.256	0.0	5083.4	6406.3	Elastoplastic	Elastoplastic	0.000	0.000	0.00

Table 4.18 Volumetric strain due to soaking for a soil compacted at $0.88\gamma_{d\max}$ of the modified Proctor and 15% water content, shaded cells highlights the domain of elastic behavior.

σ_{vf} (kPa)	v_{fc}	s_{fc} (kPa)	$p_{0_f}^*$ (kPa)	$p_{0_{fc}}^*$ (kPa)	Behavior Loading	Behavior Soaking	Δv_c	Δv_s	ε_v (%)
0	1.523	610.2	0.0	350.0	Elastic	Elastic	0.000	0.002	−0.13
10	1.515	559.0	10.2	350.0	Elastic	Elastic	0.000	0.002	−0.12
100	1.513	551.2	101.7	350.0	Elastic	Elastic	0.000	0.002	−0.12
200	1.512	543.7	203.3	350.0	Elastic	Elastic	0.000	0.002	−0.12
300	1.511	537.0	305.0	350.0	Elastic	Elastic	0.000	0.002	−0.12
400	1.510	531.1	406.7	350.0	Elastic	Elastoplastic	−0.014	0.002	0.77
500	1.509	525.6	508.3	350.0	Elastic	Elastoplastic	−0.034	0.002	2.10
600	1.509	520.6	610.0	350.0	Elastic	Elastoplastic	−0.050	0.002	3.19
700	1.508	516.1	711.7	350.0	Elastic	Elastoplastic	−0.064	0.002	4.12
800	1.507	511.8	813.3	350.0	Elastic	Elastoplastic	−0.076	0.002	4.92
900	1.507	507.8	915.0	350.0	Elastic	Elastoplastic	−0.086	0.002	5.62
1,000	1.506	504.1	1016.7	350.0	Elastic	Elastoplastic	−0.096	0.002	6.25
2,000	1.338	0.0	2033.3	2562.5	Elastoplastic	Elastoplastic	0.000	0.000	0.00
3,000	1.302	0.0	3050.0	3843.8	Elastoplastic	Elastoplastic	0.000	0.000	0.00
4,000	1.276	0.0	4066.7	5125.0	Elastoplastic	Elastoplastic	0.000	0.000	0.00
5,000	1.256	0.0	5083.4	6406.3	Elastoplastic	Elastoplastic	0.000	0.000	0.00

4.4   EXAMPLE 16: EFFECT OF THE SOIL'S MICROSTRUCTURE IN THE COLLAPSE OF EMBANKMENTS

The example below describes the procedure to evaluate the effect of the microstructure on the volumetric strain of two compacted soils submitted to wetting. Both soils are compacted at the same density, but on the dry and wet sides of the Proctor compaction curve, they are denoted as DD for the soil compacted on the dry side, and WD for the soil compacted on the wet side, as shown in Figure 4.29. Afterward, the soil compacted on the wet side is dried to reach the same suction as the soil compacted on the dry side. This procedure creates soils having the same density and suction but having different microstructures.

Once soils have the same density and suction, they are submitted to oedometric compression up to a net mean stress of 1.5 MPa under constant suction pressure (i.e., 1 MPa), and finally, both soils are soaked. This example describes the procedure to calculate the specific volume of both soils to demonstrate the effect of the microstructure.

Table 4.19 shows the parameters of the BBM, including micromechanical features, which are based on the data given in Ref. [4]. Note that all parameters are the same for both soils, and only the microstructural variable ξ_m is different. Moreover, the initial void's ratio of both soils is e = 0.538.

Computing the vertical strain due to collapse requires using the microstructural-based BBM, which was described in Section 4.1.7. The following procedure allows this calculation:

1   Use the data given in Table 4.19 to establish the initial conditions of both soils.

2   Calculate the oedometric compression curve for both soils up to a net stress of 1.5 MPa with a constant suction of 1 MPa.

3   For the saturated state, calculate the oedometric compression curve for both soils up to a net stress of 1.5 MPa, and calculate the change in specific volume due to soaking and the respective vertical strain.

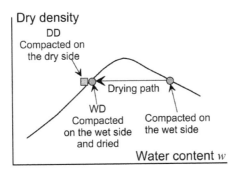

Figure 4.29 Position of the soils of the example in the Proctor's plane.

Table 4.19 Parameters of the WRC and the BBM, including the microstructural features

Parameter	Meaning	DD	WD
ξ_m	Microstructural	0.36	0.41
n_{sm}	variables	3	
$\overline{\lambda}(0)$		0.07	
κ		0.006	
$\overline{\kappa}_s$	Compressibility	0.0005	
$\overline{r}$	in saturated and	0.8	
$\overline{\beta}$	unsaturated states	0.8	
$\overline{s}_\lambda$		0.25 MPa	
$\overline{p}_0^*$		0.45 MPa	
$\overline{p}^c$		0.001 MPa	
m_{WR}		0.35	
n_{WR}	Water retention	1.54	
ϕ_{WR}	curve	5.45 MPa	
ψ_{WR}		0.6	
e_0	Initial	0.538	
p_0^*	conditions	0.45 MPa	
s_0		1.0 MPa	

Source: Based on the data given in Ref. [4].

4.4.1 Initial conditions

Using the data given in Table 4.19, Equation 3.11 allows computing the initial saturation degree of both soils for the initial void ratio of $e = 0.538$ and suction of $s = 1$ MPa, which remains constant throughout the compression stage.

$$S_r = \left\{ \frac{1}{1 + \left[5.45 \cdot 1 \cdot 0.538^{0.6}\right]^{1.54}} \right\}^{0.35} = 0.469.$$

The saturation degree is the same for both soils. However, the effective saturation degree is different because of the differences in the microstructural variable, ξ_m. Then, using Equation 4.19 each effective saturation degree is

$$\overline{S}_r(DD) = \frac{0.469 - 0.36}{1 - 0.36} + \frac{1}{3} \ln \left[1 + e^{-3\frac{0.469-0.36}{1-0.36}}\right] = 0.327, \text{ and}$$

$$\overline{S}_r(WD) = \frac{0.469 - 0.41}{1 - 0.41} + \frac{1}{3} \ln \left[1 + e^{-3\frac{0.469-0.41}{1-0.41}}\right] = 0.285.$$

Therefore, using Equation 4.21, the effective suctions of each soil is

$$\overline{s}_0(DD) = \overline{S}_r s = 0.327 \cdot 1 = 0.327 \text{ MPa, and}$$
$$\overline{s}_0(WD) = \overline{S}_r s = 0.285 \cdot 1 = 0.285 \text{ MPa}.$$

Moreover, from Equation 4.24, the compressibility coefficients $\overline{\lambda}(\overline{s})$ for the respective effective suctions are

$$\overline{\lambda}(\overline{s}_{DD}) = 0.07 \left\{ 0.8 + (1 - 0.8) \left[1 + \left(\frac{0.327}{0.25}\right)^5\right]^{-0.8} \right\} = 0.060, \text{ and}$$

$$\bar{\lambda}(\bar{s}_{WD}) = 0.07 \left\{ 0.8 + (1 - 0.8) \left[1 + \left(\frac{0.285}{0.25} \right)^5 \right]^{-0.8} \right\} = 0.062.$$

Finally, Equation 4.25, which corresponds to the curve LC considering the microstructural features, allows calculating the overconsolidation stresses of each soil as

$$\bar{p}_0(DD) = 0.001 \, (450)^{\frac{0.064}{0.060-0.006}} = 1.403 \text{ MPa, and}$$

$$\bar{p}_0(WD) = 0.001 \, (450)^{\frac{0.064}{0.062-0.006}} = 1.088 \text{ MPa.}$$

4.4.2   Oedometric compression

During compression, the specific volume decreases, first in the elastic domain, and then in the elastoplastic one. As usual, the limit between these two domains depends on the overconsolidation stress, which, when considering the microstructural effect, depends on the constitutive stress.

4.4.2.1   Elastic compression

The Equation 3.12 indicates that the elastic loading of soils, when carried out at constant suction, also implies a constant saturation degree. As a result, the effective suction remains constant during loading. Also, since the initial total stress is zero, the constitutive stress at the beginning of the loading process is $\bar{p} = \bar{s}_0$. Therefore, Equation 4.22 that allows computing the changes in void ratio during the elastic compression becomes

$$\Delta e^e = -\bar{k} \ln \left(\frac{\bar{p}_0}{\bar{s}_0} \right).$$

Resulting in

$$\Delta e^e(DD) = -0.006 \ln \left(\frac{1.403}{0.327} \right) = -0.008738, \text{ and}$$

$$\Delta e^e(WD) = -0.006 \ln \left(\frac{1.088}{0.285} \right) = -0.008038.$$

Then, the values of void ratio at the elastic limit of both soils are

$$e_Y(DD) = 0.538 - 0.008726 = 0.52926, \text{ and}$$
$$e_Y(WD) = 0.538 - 0.008049 = 0.52996.$$

Finally, from Equation 4.20, the net mean stress for the elastic limit is $p_0 = \bar{p}_0 - \bar{S}_r s$ resulting in

$$p_0(DD) = 1.403 - 0.327 = 1.076 \text{ MPa, and}$$
$$p_0(WD) = 1.09 - 0.285 = 0.803 \text{ MPa.}$$

4.4.2.2   Elastoplastic compression

Elastoplastic compression produces changes in all the state variables: *i.e.*, void ratio, degree of saturation, effective suction, and constitutive stress. Furthermore, at the beginning of the elastoplastic compression, the values of these variables correspond to the values reached at the end of the elastic loading, which were calculated previously.

Since all the state variables evolve simultaneously, calculating their changes requires using an incremental iterative process which consists in choosing a small increment of net stress, Δp, and then computing the evolution of the soil variables as follows:

1   Begin with initial values of the specific volume and the mean net stress.
2   Increase the net stress,

$$p_{i+1} = p_i + \Delta p.$$

3   Compute the saturation degree,

$$S_{r_{i+1}} = \left\{ \frac{1}{1 + \left[5.45 \cdot 1 \cdot (v_i - 1)^{0.6} \right]^{1.54}} \right\}^{0.35}.$$

4   Compute the effective suction,

$$\bar{s}_{i+1} = \overline{S}_{r_{i+1}} \cdot 1 = \frac{S_{r_{i+1}} - \xi_m}{1 - \xi_m} + \frac{1}{3} \ln \left[1 + e^{-3 \frac{S_{r_{i+1}} - \xi_m}{1 - \xi_m}} \right].$$

5   Compute the compressibility coefficient,

$$\overline{\lambda}(\bar{s}_{i+1}) = 0.07 \left\{ 0.8 + (1 - 0.8) \left[1 + \left(\frac{\bar{s}_{i+1}}{0.25} \right)^5 \right]^{-0.8} \right\}.$$

6   Compute the constitutive stress,

$$\bar{p}_{i+1} = p_{i+1} + \bar{s}_{i+1}.$$

7   Compute the change in void ratio,

$$\Delta e^{ep} = -\overline{\lambda}(\bar{s}_{i+1}) \ln \left(\frac{\bar{p}_{i+1}}{\bar{p}_i} \right).$$

8   Compute the specific volume in step $i + 1$:

$$v_{i+1} = v_i + \Delta e^{ep}$$

9   Return to step 2.

Tables 4.20 and 4.21 present the results of compression in the elastic and elastoplastic domains for both soils (DD and WD). Also, Figure 4.30 shows the loading paths in the planes $\bar{p} - \bar{s}$ and $\varepsilon_v - p_{net}$. As observed in Figure 4.30c and d, for a mean net stress of 1.5 MPa, the DD soil exhibits a stiffer behavior compared with the WD soil because of the difference in microstructure.

4.4.3   Saturated oedometric compression

The saturated compression up to a net stress of 1.5 MPa also occurs in two domains of behavior:

Table 4.20   Volumetric strain on compression for the soil compacted on the dry side, shaded cells highlight the domain of elastic behavior.

p_{net} (kPa)	v	$\bar{s}$ (kPa)	$\bar{p}$	$\bar{\lambda}(\bar{s})$	Behavior	ε_V (%)
0	1.538	327.4	327.4	0.05996	Elastic	
1,075	1.529	327.4	1402.6	0.05996	Elastic	0.57
1,100	1.528	329.8	1429.8	0.05987	Elastoplastic	0.63
1,200	1.524	330.8	1530.8	0.05983	Elastoplastic	0.90
1,300	1.520	331.8	1631.8	0.05979	Elastoplastic	1.15
1,400	1.517	332.7	1732.7	0.05976	Elastoplastic	1.39
1,500	1.513	333.6	1833.6	0.05973	Elastoplastic	1.61

Table 4.21   Volumetric strain on compression for the soil compacted on the wet side and then dried, shaded cells highlight the domain of elastic behavior.

p_{net} (kPa)	v	$\bar{s}$ (kPa)	$\bar{p}$	$\bar{\lambda}(\bar{s})$	Behavior	ε_V (%)
0	1.538	285.2	285.2	0.06192	Elastic	
802	1.530	285.2	1087.6	0.06192	Elastic	0.53
900	1.525	288.4	1188.4	0.06175	Elastoplastic	0.86
1,000	1.520	289.7	1289.7	0.06168	Elastoplastic	1.19
1,100	1.515	291.0	1391.0	0.06161	Elastoplastic	1.50
1,200	1.511	292.1	1492.1	0.06155	Elastoplastic	1.78
1,300	1.507	293.2	1593.2	0.06150	Elastoplastic	2.05
1,400	1.503	294.2	1694.2	0.06144	Elastoplastic	2.29
1,500	1.499	295.2	1795.2	0.06139	Elastoplastic	2.53

- A first stage consists of decreasing suction from 1 MPa to zero, which produces elastic expansion.
- The process follows with the elastic loading up to the overconsolidation stress, for the saturated state, of 0.45 MPa.
- Afterward, the elastoplastic compression in the saturated state progress up to 1.5 MPa.

The change in specific volume due to the saturation and the subsequent compression in the elastic domain is

$$\Delta e^e = \bar{\kappa}_s \ln \left(\frac{\bar{s}_0 + p_{atm}}{p_{atm}} \right) - \kappa \ln \left(\frac{p_0^*}{\bar{s}_0} \right).$$

Therefore, for the DD and the WD soils, the change in void ratio is

$$\Delta e^e(DD) = 0.0005 \ln \left(\frac{0.327 + 0.101}{0.101} \right) - 0.006 \ln \left(\frac{0.45}{0.327} \right) = -0.00119, \text{ and}$$

$$\Delta e^e(WD) = 0.0005 \ln \left(\frac{0.285 + 0.101}{0.101} \right) - 0.006 \ln \left(\frac{0.45}{0.285} \right) = -0.00207.$$

For both soils, the saturated loading from the overconsolidation stress of 0.45 MPa to the stress of 1.5 MPa produces the following change in void ratio:

$$\Delta e^{ep} = -0.07 \ln \left(\frac{1.5}{0.45} \right) = -0.08428.$$

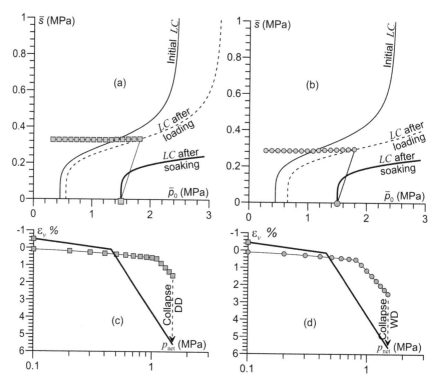

Figure 4.30 Results that show the effect of the soil's microstructure on the collapse of compacted soils, based on the data given in Ref. [4], (a, b) hydraulic stress path in the $(\bar{p}_0, \bar{s})$ plane for the DD and the WD soils, and (c, d) evolution of the volumetric strain due to collapse after soaking for the DD and WD soils.

This change implies that the specific volume for each soil after compression in the saturated state is

$$v_{sat}(DD) = 1.538 - 0.00119 - 0.08428 = 1.45253, \text{ and}$$
$$v_{sat}(WD) = 1.538 - 0.00207 - 0.08428 = 1.45165.$$

Finally, the collapse due to soaking results from the difference between the specific volumes achieved during oedometric compression in the unsaturated and saturated states as

$$\varepsilon_v(DD) = \frac{1.51325 - 1.45253}{1.51325} = 0.04012 \quad \rightarrow \quad \epsilon_v(DD) = 4.01\%, \text{ and}$$
$$\varepsilon_v(WD) = \frac{1.49913 - 1.45253}{1.49913} = 0.03167 \quad \rightarrow \quad \epsilon_v(WD) = 3.17\%.$$

The results show that the ratio between the collapsing vertical strains is $4.01/3.17 = 1.265$, which implies that the DD soil has a collapsing strain 26.5% higher than the WD soil. This difference depends only on the soil's microstructure because all other soil's characteristics are the same for both soils.

Chapter 5

Mechanical behavior of road materials

5.1 RELEVANT EQUATIONS

Various models allow the evolution of the elastic modulus to be described as a function of stress, density, or suction. This chapter only briefly describes the models used in the examples below.

5.1.1 Models describing the resilient modulus

The simplest model is known as the $k - \theta$ model; it was proposed in 1963 by Dunlap [27]. This model relates the resilient Young's modulus E_r to the bulk stress θ as

$$E_r = k_1 p_a \left(\frac{\theta}{p_a} \right)^{k_2},$$
(5.1)

where k_1 and k_2 are fitting coefficients, and p_a is a reference pressure (usually the atmospheric pressure, or 100 kPa).

The bulk modulus θ is given by the sum of the principal stresses σ_1 and σ_3, and it is related to the mean stress p as follows:

$$\theta = \sigma_1 + \sigma_2 + \sigma_3 = 3p, \quad \text{which becomes}$$
(5.2)

$$\theta = \sigma_1 + 2\sigma_3 \quad \text{when} \quad \sigma_2 = \sigma_3.$$
(5.3)

This model generally adopts a value for the Poisson ratio of $v = 0.3$.

Uzan's model [61] uses the octahedral shear stress, τ_{oct}, in a model which is also known as the $k_1 - k_3$ model, or the universal model, whose equation is as follows:

$$E_r = k_1 p_a \left(\frac{\theta}{p_a} \right)^{k_2} \left(\frac{\tau_{\text{oct}}}{p_a} + 1 \right)^{k_3}.$$
(5.4)

The following equations give the octahedral shear stress τ_{oct}:

$$\tau_{\text{oct}} = \frac{1}{3} \left[(\sigma_1 - \sigma_2)^2 + (\sigma_1 - \sigma_3)^2 + (\sigma_2 - \sigma_3)^2 \right]^{0.5}, \quad \text{becoming}$$
(5.5)

$$\tau_{\text{oct}} = \frac{\sqrt{2}}{3} q \quad \text{when} \quad \sigma_2 = \sigma_3,$$
(5.6)

where q is the deviator stress given by $q = \sigma_1 - \sigma_3$.

Coefficients $k_1, k_2,$ and k_3 are obtained from a regression analysis. These coefficients have the following characteristics [10]:

- Coefficient k_1 involves elastic characteristics of grains, the density of the material, and Poisson's ratio of the arrangement of particles; k_1 is always positive.
- Coefficient k_2 characterizes the behavior at the contact between particles, its value is $k_2 = 1/3$ for contacts between spherical particles and $k_2 = 0.5$ for flat contacts, and this coefficient decreases when the bonding between particles is strong.
- Coefficient k_3 is zero for small strains, and it could be negative, indicating a reduction of Young's resilient modulus when the shear stress increases.

Tatsuoka et al. [35,37,56,58] proposed an equation that involves the effect of the density of the material by considering its void ratio e. Also, they demonstrated that Young's resilient modulus is a function of the stress applied in the direction of loading. Thus, they proposed to relate the vertical resilient modulus with the vertical stress σ_v and the void ratio e as follows:

$$E_{rv} = k_{E_v} p_a \frac{(2.17 - e)^2}{1 + e} \left(\frac{\sigma_v}{p_a}\right)^n, \tag{5.7}$$

$\sigma_v = q + \sigma_3$ for axisymmetric conditions.

The coefficient n has a meaning similar to that of k_2 in the other models.

5.1.2   Models describing the resilient Young's modulus and Poisson's ratio

The model proposed by Boyce [7] describes the resilient coefficient of volumetric compressibility K_r, and the resilient shear modulus, G_r, as

$$K_r = \frac{\Delta p_c}{\Delta \varepsilon_{vr}} \quad \text{and} \quad G_r = \frac{\Delta q_c}{3\Delta \varepsilon_{qr}}, \tag{5.8}$$

where Δq_c and Δp_c are the deviator and mean cyclic stresses, and $\Delta \varepsilon_{vr}$ and $\Delta \varepsilon_{qr}$ are the resilient volumetric and shear strains defined as follows:

$$\Delta \varepsilon_{vr} = \Delta \varepsilon_{1r} + 2\Delta \varepsilon_{3r} = \frac{\Delta p_c}{K_r}, \quad \text{and} \quad \Delta \varepsilon_{qr} = \frac{2}{3}(\Delta \varepsilon_{1r} - \Delta \varepsilon_{3r}) = \frac{\Delta q_c}{3G_r}. \tag{5.9}$$

Moduli K_r and G_r are stress-dependent according to the following relationships:

$$K_r = \frac{\left(\frac{p_c}{p_a}\right)^{1-n}}{\frac{1}{K_a} - \frac{\beta}{K_a}\left(\frac{q_c}{p_c}\right)^2} \quad \text{and} \quad G_r = G_a \left(\frac{p_c}{p_a}\right)^{1-n}, \tag{5.10}$$

where K_a, G_a, and n are fitting coefficients, while β is a variable relating these coefficients as

$$\beta = (1 - n)\frac{K_a}{6G_a}. \tag{5.11}$$

Volumetric and shear resilient strains become

$$\varepsilon_{vr} = \frac{1}{K_a} p_a^{1-n} p_c^n \left[1 - \beta \left(\frac{q_c}{p_c}\right)^2\right], \quad \text{and} \quad \varepsilon_{qr} = \frac{1}{3G_a} p_a^{1-n} p_c^n \left(\frac{q_c}{p_c}\right). \tag{5.12}$$

The relationships linking elastic constants can be used to obtain the resilient Young's modulus and the resilient Poisson's ratio as follows:

$$E_r = \frac{9 G_a \left(\frac{p_c}{p_a}\right)^{1-n}}{3 + \frac{G_a}{K_a}\left[1 - \beta \left(\frac{q_c}{p_c}\right)^2\right]}, \quad \text{and} \tag{5.13}$$

$$v_r = \frac{\frac{3}{2} - \frac{G_a}{K_a}\left[1 - \beta \left(\frac{q_c}{p_c}\right)^2\right]}{3 + \frac{G_a}{K_a}\left[1 - \beta \left(\frac{q_c}{p_c}\right)^2\right]}. \tag{5.14}$$

Boyce's model describes volumetric and shear behavior of road materials reasonably well, except for dilatant strains. For this particular case, a coefficient of anisotropy was included in the original model [38].

5.1.3 Effect of water in the resilient Young's modulus

The Mechanistic-Empirical Pavement Design Guide (MEPDG 2004) [5] recommends to use Uzan's universal Equation 5.4 to estimate the resilient modulus at optimum moisture content, and then correct the value of $E_{r-\text{opt}}$ using the following equation:

$$\log \frac{E_r}{E_{r-\text{opt}}} = a + \frac{b - a}{1 + \exp\left[\ln\left(-\frac{b}{a}\right) + k_m(S_r - S_{r-\text{opt}})\right]}, \tag{5.15}$$

where S_r is the degree of saturation and $S_{r-\text{opt}}$ is the degree of saturation at the optimum water content (both are in decimals). The suggested values of the parameters are

- $a = -0.5934$, $b = 0.4$, and $k_m = 6.1324$ for fine-grained soils, and
- $a = -0.3123$, $b = 0.3$, and $k_m = 6.8157$ for coarse-grained soils.

However, a large number of studies have shown that the relationship between resilient modulus and moisture content is highly dependent on the type of soil and must explicitly include the effect of suction pressure.

An advantageous way to consider the effect of water was proposed in Ref. [11]; they suggested to use the following linear model:

$$E_r = E_{0\sigma_v}(s) + A_{\sigma_v}\sigma_v. \tag{5.16}$$

This model has two constants: $E_{0\sigma_v}$, which represents the value of Young's resilient modulus for zero vertical stress and depends on the suction pressure and A_{σ_v} which is the slope relating how Young's modulus grows as the vertical stress increases.

The effect of water can also be modeled using the principle of effective stress. When using this principle, the mean and deviator stresses become

$$p'_c = (p_c - u_a) + \chi s, \quad \text{and} \quad q'_c = q_c. \tag{5.17}$$

The coefficient χ was proposed initially by Bishop, but other relationships were proposed more recently. One of them is the following equation proposed by Khalili and Khabbaz [42]:

$$\chi = \left(\frac{s}{s_b}\right)^{-0.55} \quad \text{for } s \geq s_b, \text{ and}$$

$$\chi = 1 \quad \text{for } s < s_b, \tag{5.18}$$

where s_b is the air entry suction pressure.

5.2 EXAMPLE 17: ADJUSTMENT OF THE MEASURED RESILIENT YOUNG'S MODULUS USING DIFFERENT MODELS

The following example describes some methodologies that can be used to adjust the experimental results of the measurement of the resilient elastic properties of road materials.

Table 5.1 shows the results of the measurement of the resilient Young's modulus and resilient Poisson's ratio carried out in a triaxial apparatus equipped with internal measurement of the axial and radial strains [11]. Tests were carried out following a constant confining pressure (CCP) procedure.

For predicting the resilient elastic constants, the example uses four models: the $k - \theta$ model, Uzan's model, Boyce's model, and a linear model.

Finally, the example compares the performance of each model for predicting the resilient elastic constants.

Table 5.1 Results of the measurement of the resilient Young's moduli and Poisson's ratio (CCP procedure)

σ_3 (kPa)	q (kPa)	E_r (MPa)	ν_r
10	30	144.5	0.20
10	40	159.3	0.22
10	50	149.4	0.22
10	60	171.5	0.24
10	70	202.4	0.26
20	65	200.9	0.19
20	90	205.8	0.22
20	115	220.7	0.25
20	140	242.6	0.26
30	50	172.6	0.16
30	90	225.4	0.19
30	130	269.5	0.22
30	170	289.1	0.24
30	210	325.9	0.26
40	60	249.9	0.17
40	115	281.8	0.19
40	175	308.8	0.23
40	225	335.7	0.25
40	280	384.8	0.25

5.2.1   Fitting the experimental results using the $k-\theta$ model

The $k - \theta$ model relates the bulk stress θ with the resilient Young's modulus E_r through Equation 5.1. The bulk modulus θ is the sum of the principal stresses that, for axisymmetric conditions, is $\theta = \sigma_1 + 2\sigma_3$, which in turn is also three times the mean stress p. Knowing the confining stress σ_3, and the deviator stress q, the mean stress is

$$q = \sigma_1 - \sigma_3 \;\rightarrow\; \sigma_1 = q + \sigma_3, \quad \text{then} \quad p = \frac{\sigma_1 + 2\sigma_3}{3} \;\rightarrow\; p = \frac{q}{3} + \sigma_3.$$

Therefore, the bulk stress θ becomes

$$\theta = q + 3\sigma_3.$$

The reference pressure p_a is usually adopted as $p_a = 100$ kPa.

The fitting parameters k_1 and k_2 of the $k - \theta$ model can be easily obtained relating the values of θ/p_a and E_r and then using a power fitting function, as shown in Figure 5.1.

According to Figure 5.1, the fitting parameters are $k_1 = 1,780$ (using $p_a = 0.1$ MPa to obtain E_r in MPa) and $k_2 = 0.5055$. Therefore, the relationship between the resilient Young's modulus and the bulk stress is

$$E_r = 1780 p_a \left(\frac{\theta}{p_a}\right)^{0.5055}.$$

The mean square error (MSE), that is, the average squared difference between estimated and experimental values, permits to evaluate the performance of the model. The MSE is defined as

$$\text{MSE} = \frac{1}{n_m} \sum_{i=1}^{n_m} \left(E_r^{\text{Exp}} - E_r^{\text{Model}}\right)^2,$$

where n_m is the number of measures, and E_r^{Exp} and E_r^{Model} are the measured and predicted resilient Young's moduli, respectively. Table 5.2 presents a comparison between them leading to a MSE of 218.6 MPa^2.

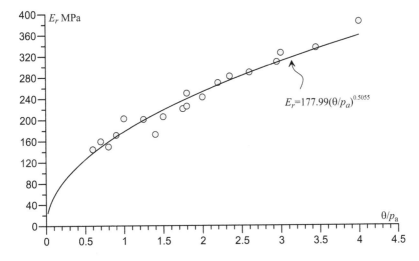

Figure 5.1   Relationship between the resilient Young's modulus E_r and the dimensionless bulk stress θ/p_a to find the fitting parameters of the $k - \theta$ model.

Table 5.2 Comparison between the measured and predicted resilient Young's moduli achieved using the $k - \theta$ model. Shaded columns highlight the experimental and theoretical Young's moduli.

σ_3 (kPa)	q (kPa)	p (kPa)	θ/p_a	Test E_r (MPa)	Model E_r (MPa)
10	30	20.00	0.60	144.51	137.48
10	40	23.33	0.70	159.27	148.63
10	50	26.67	0.80	149.40	159.00
10	60	30.00	0.90	171.48	168.76
10	70	33.33	1.00	202.38	177.99
20	65	41.67	1.25	200.89	199.24
20	90	50.00	1.50	205.78	218.48
20	115	58.33	1.75	220.75	236.18
20	140	66.67	2.00	242.62	252.68
30	50	46.67	1.40	172.65	210.99
30	90	60.00	1.80	225.42	239.57
30	130	73.33	2.20	269.49	265.15
30	170	86.67	2.60	289.13	288.51
30	210	100.00	3.00	325.87	310.16
40	60	60.00	1.80	249.95	239.57
40	115	78.33	2.35	281.80	274.14
40	175	98.33	2.95	308.77	307.53
40	225	115.00	3.45	335.74	332.86
40	280	133.33	4.00	384.80	358.70

5.2.2 Fitting the experimental results using the three parameters model

The three parameters model proposed by Uzan [61] is also known as the universal model and is the model recommended in the MEPDG. This model, described by Equation 5.4, uses the octahedral shear stress, τ_{oct}, and the bulk stress, θ, to fit the experimental results.

A straightforward method to obtain the fitting parameters of Uzan's model is to linearize Equation 5.4 by applying a logarithm in both sides of the equation as follows:

$$\log(E_r) = \log(k_1 p_a) + k_2 \log\left(\frac{\theta}{p_a}\right) + k_3 \log\left(\frac{\tau_{oct}}{p_a} + 1\right).$$

Most spreadsheets and, of course, all specialized packages for data analysis allow adjusting linear equations that have three parameters, but, in this example, it is preferred to make the adjustment using the more classic way of adjusting a linear equation that has two parameters. The method consists in assuming a value of the parameter k_3, and then placing the term of the equation involving the octahedral stress on the right side of the equation as follows:

$$\log(E_r) - k_3 \log\left(\frac{\tau_{oct}}{p_a} + 1\right) = \log(k_1 p_a) + k_2 \log\left(\frac{\theta}{p_a}\right).$$

Subsequently, it is possible to fit the equation using a linear fit, resulting in the intersection of the linear fit with the vertical axis, which is $\log(k_1 p_a)$, and its slope is k_2.

This adjustment process is repeated for different values of the constant k_3, as shown in Figure 5.2.

For each assumed value of the constant k_3, it is possible to compute the MSE and plot the evolution of this error for the different values of k_3, as shown in Figure 5.3, which indicates that the minimum MSE corresponds to a value of $k_3 = 0.3$.

For a constant $k_3 = 0.3$, the linear fit allows to find the other two constants: $k_2 = 0.4039$ and $\log(k_1 p_a) = 2.2194$. Therefore, the resilient modulus becomes

$$E_r = 1657.3 p_a \left(\frac{\theta}{p_a}\right)^{0.4039} \left(\frac{\tau_{oct}}{p_a} + 1\right)^{0.3}.$$

Note that for the data of the example, E_r is given in MPa, and therefore, $p_a = 0.1$ MPa, while θ is given in kPa, and then $p_a = 100$ kPa.

Table 5.3 presents a comparison between the measured and predicted resilient Young's moduli obtained using Uzan's model.

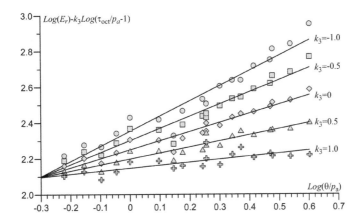

Figure 5.2 Linear fitting of Uzan's model for different values of the parameter k_3.

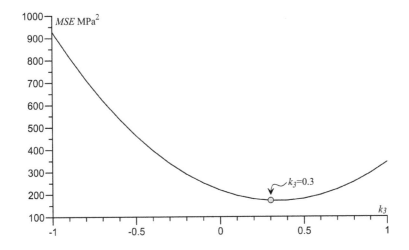

Figure 5.3 Evolution of the mean square error for different values of the parameter k_3.

Table 5.3 Comparison between the resilient Young's moduli measured and computed using Uzan's model. Shaded columns highlight the experimental and theoretical Young's moduli.

σ_3 (kPa)	q (kPa)	p (kPa)	θ/p_a	τ_{oct}/p_{a+1}	Test E_r (MPa)	Model E_r (MPa)
10	30	20.00	0.60	1.141	144.51	140.29
10	40	23.33	0.70	1.189	159.27	151.13
10	50	26.67	0.80	1.236	149.40	161.37
10	60	30.00	0.90	1.283	171.48	171.15
10	70	33.33	1.00	1.330	202.38	180.53
20	65	41.67	1.25	1.306	200.89	196.50
20	90	50.00	1.50	1.424	205.78	217.07
20	115	58.33	1.75	1.542	220.75	236.59
20	140	66.67	2.00	1.660	242.62	255.28
30	50	46.67	1.40	1.236	172.65	202.30
30	90	60.00	1.80	1.424	225.42	233.66
30	130	73.33	2.20	1.613	269.49	263.02
30	170	86.67	2.60	1.801	289.13	290.86
30	210	100.00	3.00	1.990	325.87	317.51
40	60	60.00	1.80	1.283	249.95	226.44
40	115	78.33	2.35	1.542	281.80	266.51
40	175	98.33	2.95	1.825	308.77	307.28
40	225	115.00	3.45	2.061	335.74	339.49
40	280	133.33	4.00	2.320	384.80	373.43

5.2.3 Fitting the experimental results using Boyce's model

Boyce's model, proposed in Ref. [7], permits to fit both the resilient Young's modulus and the resilient Poisson's ratio. However, the model is defined in terms of the resilient coefficient of volumetric compressibility K_r, and the resilient shear modulus, G_r, which are computed from the experimental data given in Table 5.1 as follows:

$$K_r = \frac{E_r}{3(1-2v_r)} \quad \text{and} \quad G_r = \frac{E_r}{2(1+v_r)}.$$

Expressions for the elastic constants of this model were derived in terms of incremental stresses beginning at zero stress. Then, it is better adapted to adjust the resilient modulus obtained in triaxial tests following a variable confining pressure procedure. Nevertheless, as remarked in Ref. [17], the results obtained in variable confining pressure (VCP) or constant confining pressure (CCP) triaxial tests can be described using the same relationship; in other words, the resilient performance is independent of the procedure. This evidence allows using Boyce's model to fit the results of the CCP triaxial test of this example.

The model has three parameters $(G_a, K_a,$ and $1 - n)$ as shown in Equations 5.13 and 5.14. The fitting process can be performed in two phases. The first step consists in fitting a potential equation relating the resilient shear modulus G_r with the dimensionless mean stress p/p_a, as shown in Figure 5.4. This fit leads to the following constants: $G_a = 125.99$ MPa and $1 - n = 0.492$; therefore, the resilient shear modulus G_r becomes

$$G_r = 125.99 \left(\frac{p}{p_a}\right)^{0.492} \quad \text{in MPa.}$$

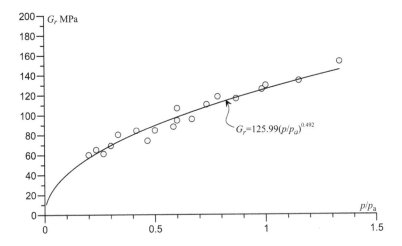

Figure 5.4 Fitting of the resilient shear modulus G_r using Boyce's model.

After obtaining the constants G_a, and $1 - n$, the constant K_a results from adjusting the resilient coefficient of volumetric compressibility K_r. From Equations 5.10, the relationship that gives K_r is

$$K_r = \frac{\left(\frac{p}{p_a}\right)^{1-n}}{\frac{1}{K_a} - \frac{\beta}{K_a}\left(\frac{q}{p}\right)^2},$$

where $\beta = (1 - n)\frac{K_a}{6G_a}$. Then, placing the value of β into the previous equation, the modulus k_r becomes

$$K_r = \frac{\left(\frac{p}{p_a}\right)^{1-n}}{\frac{1}{K_a} - \frac{1-n}{6G_a}\left(\frac{q}{p}\right)^2}.$$

leading to

$$\frac{1}{K_a} - \frac{1-n}{6G_a}\left(\frac{q}{p}\right)^2 = \frac{1}{K_r}\left(\frac{p}{p_a}\right)^{1-n},$$

and then the value of $1/K_a$ becomes

$$\frac{1}{K_a} = \frac{1}{K_r}\left(\frac{p}{p_a}\right)^{1-n} + \frac{1-n}{6G_a}\left(\frac{q}{p}\right)^2.$$

This equation suggests that any measure of K_r leads to the same value of $1/K_a$. Figure 5.5 shows the different values of K_a calculated using the equation above, leading to an average value of K_a, denoted as $\overline{K_a}$, which is $\overline{K_a} = 133.23$ MPa becoming the third parameter of the model.

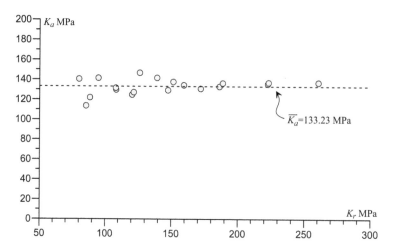

Figure 5.5 Calculation of the mean value of K_a of Boyce's model using the different values of the coefficient of volumetric compressibility K_r.

Finally, considering that $\beta = 0.492 \frac{133.23}{6 \cdot 125.99} = 8.67 \cdot 10^{-2}$, Equation 5.13 leads to the following relationship for obtaining the resilient Young's modulus using Boyce's model

$$E_r = \frac{1133.91 \left(\frac{p}{p_a}\right)^{0.492}}{3 + 0.9457 \left[1 - 8.67 \cdot 10^{-2} \left(\frac{q}{p}\right)^2\right]},$$

and, from Equation 5.14, the resilient Poisson's ratio is

$$\nu_r = \frac{\frac{3}{2} - 0.9457 \left[1 - 8.67 \cdot 10^{-2} \left(\frac{q}{p}\right)^2\right]}{3 + 0.9457 \left[1 - 8.67 \cdot 10^{-2} \left(\frac{q}{p}\right)^2\right]}.$$

Table 5.4 shows a comparison between the experimental and predicted resilient Young's moduli and Poisson's ratios obtained using Boyce's model. Regarding the resilient Young's modulus, the mean square error is MSE= 198.3 MPa², which is only slightly higher than the MSE of Uzan's model that also uses three parameters. However, one of the main benefits of Boyce's model is the possibility of also predicting the evolution of Poisson's ratio depending on stresses.

5.2.4   Fitting the experimental results using the linear model

A linear model is a simplistic approach, but, at low stresses, it is more reliable than the other models. In fact, the $k - \theta$, Uzan, or Boyce models, when used in total stresses, predict zero resilient Young's modulus for zero mean or bulk stresses. This prediction is undoubtedly unrealistic because, due to the capillary forces, compacted materials exhibit a no negligible modulus when no stresses are applied to them.

Table 5.4 Comparison between the measured and computed resilient Young's moduli and Poisson's ratios achieved using Boyce's model. Shaded columns highlight the experimental and theoretical Young's moduli and Poisson's ratio.

σ_3 (kPa)	q (kPa)	p (kPa)	v_r	E_r (MPa)	K_r (MPa)	G_r (MPa)	K_a (MPa)	E_r (MPa)	v_r
				Test				**Model**	
10	30	20.00	0.20	144.51	80.11	60.25	140.49	136.53	0.20
10	40	23.33	0.22	159.27	94.72	65.29	141.41	149.54	0.21
10	50	26.67	0.22	149.40	88.34	61.32	122.02	161.77	0.23
10	60	30.00	0.24	171.48	108.39	69.35	129.78	173.30	0.24
10	70	33.33	0.26	202.38	139.12	80.47	141.70	184.24	0.26
20	65	41.67	0.19	200.89	108.09	84.39	131.63	196.73	0.20
20	90	50.00	0.22	205.78	120.42	84.67	124.79	219.07	0.22
20	115	58.33	0.25	220.75	147.44	88.27	129.32	239.79	0.24
20	140	66.67	0.26	242.62	171.89	95.92	130.95	259.15	0.26
30	50	46.67	0.16	172.65	85.39	74.22	113.70	202.32	0.17
30	90	60.00	0.19	225.42	121.62	94.63	127.24	234.46	0.20
30	130	73.33	0.22	269.49	159.23	110.63	134.46	263.94	0.22
30	170	86.67	0.24	289.13	186.01	116.50	133.06	291.13	0.24
30	210	100.00	0.26	325.87	222.48	129.74	135.76	316.39	0.26
40	60	60.00	0.17	249.95	126.12	106.84	146.68	228.24	0.16
40	115	78.33	0.19	281.80	151.35	118.44	137.70	266.79	0.19
40	175	98.33	0.23	308.77	188.44	125.83	136.52	305.10	0.22
40	225	115.00	0.25	335.74	223.04	134.39	137.08	334.47	0.24
40	280	133.33	0.25	384.80	260.50	153.45	137.10	364.52	0.26
						$\overline{K_a}$	133.23		

The possibility of predicting a not null resilient Young's modulus at low stresses is a convenient characteristic for analyzing the effect of water on road structures having thick layers of unbonded granular materials. This characteristic makes the linear model a useful methodology for analyzing flexible road structures for which the behavior of the granular layers plays a significant role in its performance.

On the other hand, as it was already mentioned, experimental researches carried out by Tatsuoka et al. [35,37,56,58] demonstrated that Young's resilient modulus is a function of the stress applied in the direction of loading. Thus, it is more rational to relate the vertical resilient modulus to the vertical stress σ_v.

Figure 5.6 shows the linear fitting of the resilient Young's modulus with the dimensionless vertical stress σ_v/p_a. This fitting leads to the following equation:

$$E_r = 120.51 + 85.1\frac{\sigma_v}{p_a} \text{ in MPa.}$$

Table 5.5 shows a comparison of the measured and predicted resilient Young's modulus obtained using the linear model, and the mean square error of this model is $\text{MSE} = 256.6 \text{ MPa}^2$.

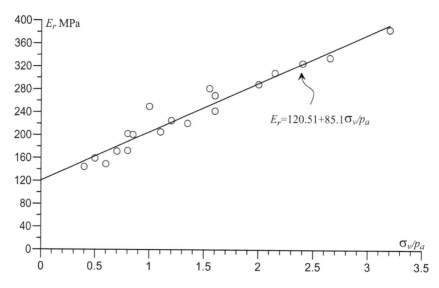

Figure 5.6 Fitting the resilient Young's modulus E_r using a linear model.

Table 5.5 Comparison between experimental and predicted resilient Young's moduli achieved using a linear model. Shaded columns highlight the experimental and theoretical Young's moduli.

				Test	Model
σ_3 (kPa)	q (kPa)	p (kPa)	σ_v/p_a	E_r (MPa)	E_r (MPa)
10	30	20.00	0.4	144.51	154.55
10	40	23.33	0.5	159.27	163.06
10	50	26.67	0.6	149.40	171.57
10	60	30.00	0.7	171.48	180.08
10	70	33.33	0.8	202.38	188.59
20	65	41.67	0.85	200.89	192.85
20	90	50.00	1.1	205.78	214.12
20	115	58.33	1.35	220.75	235.40
20	140	66.67	1.6	242.62	256.67
30	50	46.67	0.8	172.65	188.59
30	90	60.00	1.2	225.42	222.63
30	130	73.33	1.6	269.49	256.67
30	170	86.67	2	289.13	290.71
30	210	100.00	2.4	325.87	324.75
40	60	60.00	1	249.95	205.61
40	115	78.33	1.55	281.80	252.42
40	175	98.33	2.15	308.77	303.48
40	225	115.00	2.65	335.74	346.03
40	280	133.33	3.2	384.80	392.83

5.2.5   Performance of the different models to predict resilient Young's moduli and Poisson's ratios

Figures 5.7 and 5.8 show a comparison between the resilient Young's moduli and Poisson's ratios obtained with the different models and the measured values. It is important to remark that the only model that has the capacity for predicting Poisson's ratio is Boyce's model; for the other models, Poisson's ratio must be assumed; for example, a value of $v_r = 0.3$ is assumed in Figures 5.7b,d, and 5.8d.

Regarding the resilient Young's moduli, all the models lead to a reasonable prediction as it can be observed in Figures 5.7 and 5.8. In fact, the MSE of each model is

- $k - \theta$ model MSE = 218.6 MPa^2.
- Uzan's model MSE = 171.4 MPa^2.
- Boyce's model MSE = 198.3 MPa^2.
- Linear model MSE = 256.6 MPa^2.

In the first view, the differences between the MSE appear to be significant. However, to get a real view of the differences, it is possible to calculate the standard deviation of the prediction, which is the square root of MSE, and also the coefficient of variation of

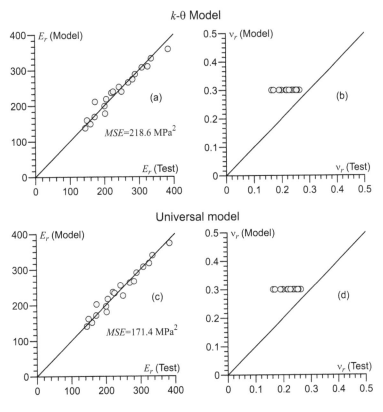

Figure 5.7 Comparison of the measured and predicted Young's moduli and Poisson's ratio: (a and b) $k - \theta$ model; (c and d) Uzan's model.

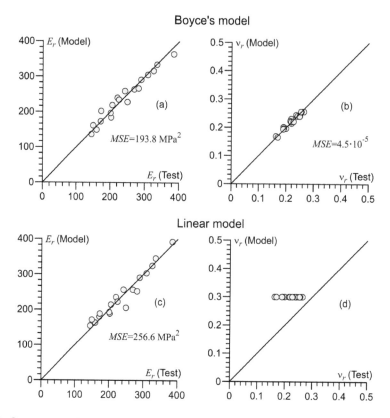

Figure 5.8 Comparison of the measured and predicted Young's moduli and Poisson's ratio: (a and b) Boyce's model; (c and d) linear model.

the prediction COV, which is the standard deviation divided by the mean value. In other words, the coefficient of variation of the resilient Young's modulus is

$$\text{COV}_{E_r} = \frac{\sqrt{\text{MSE}}}{\overline{E_r}},$$

where $\overline{E_r}$ is the mean value of the resilient Young's modulus.

As a result, the coefficients of variation of the predictions for each model are

- $k - \theta$ model $\text{COV}_{E_r} = 6.2\%$.
- Uzan's model $\text{COV}_{E_r} = 5.5\%$.
- Boyce's model $\text{COV}_{E_r} = 5.8\%$.
- Linear model $\text{COV}_{E_r} = 6.7\%$.

This last comparison shows that the difference between the highest and the lowest COV of the predictions is only 1.2%, which is a small difference, suggesting that for the data in this example, all models are equivalent in terms of the predictability of Young's modulus.

5.3   EXAMPLE 18: ASSESSMENT OF THE EFFECT OF THE WATER CONTENT OF THE GRANULAR LAYER ON THE FATIGUE LIFE OF A LOW-TRAFFIC ROAD STRUCTURE

This example describes the methodology for evaluating the effect of the water content of the granular layer on the fatigue life of a low traffic road structure.

As shown in Figure 5.9, the road structure has a 7-cm-thick bituminous layer resting on a 50 cm granular layer which in turn rests on the subgrade.

The density of the bituminous layer is $\rho_b = 2,200 \ kg/m^3$, and its fatigue life can be described with the following power equation $\varepsilon_t = \varepsilon_{ref} N^{-0.25}$, where ε_t is the tensile strain at the bottom of the bituminous layer, ε_{ref} is a reference tensile strain, and N is the number of load repetitions. The resilient Young's modulus of the bituminous layer is 4,000 MPa and its Poisson's ratio $\nu = 0.4$.

The granular layer has a certain amount of fines, i.e., percentage of particles passing the sieve #200 $P_{200} = 12.4\%$, and its plasticity characteristics are liquid limit, $w_L = 26\%$ and plasticity index $PI = 14.25\%$.

Other characteristics of the granular layer are

- *Density of grains $\rho_s = 2.64 \ g/cm^3$.*
- *Dry density $\rho_d = 1,980 \ kg/m^3$.*
- *Optimum water content (modified Proctor test) $w_{opt} = 7\%$.*
- *Coefficient of at rest horizontal stress $K_0 = 0.7$.*

Table 5.6 shows the different couples of the degree of saturation-suction representing the water retention curve of the granular material which was measured in the laboratory.

Resilient Young's moduli of the granular layer were measured at three water contents $w_{opt} - 1\%$, $w_{opt} - 2.4\%$, and $w_{opt} - 3.8\%$. Table 5.7 shows the measured resilient Young's moduli for the different stress paths applied in the laboratory (using the constant confinement pressure procedure) and the different water contents. At the same time, Poisson's ratio of the granular material must be assumed as $\nu = 0.3$.

Regarding the subgrade, its resilient Young's modulus is 100 MPa and its Poisson's ratio $\nu = 0.3$.

The load over the road structure is a circular load with a radius of $a = 0.15 \ m$ and a uniform pressure of 670 kPa.

The purpose of the example is to evaluate the ratio between the number of load repetitions until failure due to fatigue of the bituminous layer when the granular layer has a water content w, N_w, and the number of load repetitions when the layer has the optimum water content, N_{opt}. The steps to achieve this purpose are as follows:

1   Fit the results of the water retention curve using the van Genuchten equation.
2   Evaluate the best methodology for describing the evolution of resilient Young's moduli for the different water contents. Three methodologies must be evaluated: the methodology recommended in the MEPDG, a linear model with two state variables (total vertical stress and suction), and a model based on effective stress.

3   Using the best methodology obtained in the precedent step, evaluate the tensile strain at the bottom of the bituminous layer for five water contents ($w_{opt} - 3\%$, $w_{opt} - 2\%$, $w_{opt} - 1\%$, w_{opt}, and $w_{opt} + 1\%$). Based on the results of tensile stress, calculate the number of load repetitions up to bituminous layer fatigue's failure and its ratio to the number of load repetitions for optimal water content.

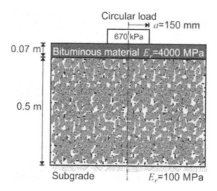

Figure 5.9 Road structure for analyzing the effect of the water content of the granular layer on the tensile strain at the bottom of the bituminous layer.

Table 5.6 Water retention curve of the material of the example

S_r	0.93	0.91	0.87	0.78	0.67	0.47	0.37
s (kPa)	3	5	10	20	30	100	140.0

Table 5.7 Resilient Young's moduli measured on samples having different water contents

σ_3 (kPa)	p (kPa)	q (kPa)	$w_{opt} - 1.0\%$ E_r (MPa)	$w_{opt} - 2.4\%$ E_r (MPa)	$w_{opt} - 3.8\%$ E_r (MPa)
10	20.00	30	138.28	203.91	358.20
10	23.33	40	140.63	215.63	367.97
10	26.67	50	152.34	201.56	363.28
10	30.00	60	171.09	199.22	358.59
10	33.33	70	154.69	208.59	370.31
20	41.67	65	150.00	212.30	410.16
20	50.00	90	157.03	243.75	414.84
20	58.33	115	175.78	243.75	473.44
20	66.67	140	168.75	250.20	440.63
30	46.67	50	150.10	222.66	421.88
30	60.00	90	170.50	239.06	420.50
30	73.33	130	180.47	255.47	466.41
30	86.67	170	199.22	267.19	447.66
30	100.00	210	236.72	285.94	478.13
40	60.00	60	157.03	234.38	461.72
40	78.33	115	196.88	276.56	503.91
40	98.33	175	213.28	292.97	496.88
40	115.00	225	234.38	325.78	562.50
40	133.33	280	248.44	328.13	534.38

5.3.1   Fitting the experimental measures of suction using the van Genuchten equation

Several equations allow adjusting the experimental measurements of the pairs of the degree of saturation and suction or water content and suction. Among them, the van Genuchten equation, given by Equation 5.19, remains one of the most widely used:

$$S_r = \frac{1}{[1 + (as)^n]^m},$$
(5.19)

where a is related to the inverse of the air entry value and n is a shape factor that depends on the pore size distribution of the material. Usually, the value of m is related to n as follows: $m = 1 - 1/n$. However, considering m as an independent variable permits to achieve a better fit of the experimental results.

To find out the values of the constants n, m, and a, it is possible to linearize Equation 5.19. First, it is possible to rearrange the terms of the equation as follows:

$$\left(\frac{1}{S_r}\right)^{1/m} - 1 = (as)^n,$$

and then it is possible to extract the logarithm of both sides of the equation, which leads to

$$\log\left[\left(\frac{1}{S_r}\right)^{1/m} - 1\right] = \log\left[(as)^n\right], \text{ and then}$$

$$\log\left[\left(\frac{1}{S_r}\right)^{1/m} - 1\right] = n\log(as).$$

Finally, a linear form of the original expression of the van Genuchten equation is

$$\log\left[\left(\frac{1}{S_r}\right)^{1/m} - 1\right] = n\log a + n\log s.$$
(5.20)

Equation 5.20 suggests that it is possible to relate the values of $\log\left[(1/S_r)^{1/m} - 1\right]$ and $\log s$ using a linear equation. The slope of such a linear relationship gives the value of n, and the intercept with the vertical axis gives the value of $n\log a$. However, as Equation 5.20 has three unknowns, the method to find the best fit consists of choosing a value of m, then performing the linear fit and computing the square of the coefficient of correlation R^2. The value of m leading to the highest value of R^2 produces the best fit of the equation.

Figure 5.10a shows the linear relationship applied to the experimental results of the example, while Figure 5.10b shows the evolution of the square of the correlation coefficient. This last figure indicates that for $m = 3$, the value of R^2 is 0.992, and then it remains constant for higher values of m.

When using $m = 3$, the linear equation that allows adjusting the experimental results is

$$\log\left[\left(\frac{1}{S_r}\right)^{1/3} - 1\right] = 0.7387\log s - 2.0016, \quad R^2 = 0.992.$$

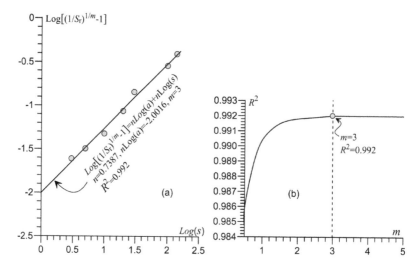

Figure 5.10 (a) Linear relationship for fitting the van Genuchten equation and (b) R^2 for different values of the coefficient m.

Therefore, the values of the fitting parameters n and a are $n = 0.7387$, and $n \log a = -2.0016$, and then $a = 10^{-2.0016/0.7387} = 1.9515 \cdot 10^{-3}$ kPa^{-1}.

Finally, the water retention curve for the material of the example is represented by the following equation:

$$S_r = \frac{1}{\left[1 + \left(1.9515 \cdot 10^{-3} s\right)^{0.7387}\right]^3}.$$

Figure 5.11 shows the very good agreement between the experimental results and the van Genuchten equation.

It is also useful to calculate the suction s for a particular value of the degree of saturation S_r; this is possible by algebraically rearranging Equation 5.19 that leads to

$$s = \frac{1}{a}\left[\left(\frac{1}{S_r}\right)^{1/m} - 1\right]^{1/n}. \tag{5.21}$$

Considering the numerical values of the fitting parameters this equation becomes

$$s = \frac{1}{1.9515 \cdot 10^{-3}}\left[\left(\frac{1}{S_r}\right)^{1/3} - 1\right]^{1/0.7387}. \tag{5.22}$$

The phase's diagram, represented in Figure 5.12, permits to compute the degree of saturation for a particular value of water content following the analysis below:

Volume solid particles $V_s = \dfrac{\rho_d}{\rho_s}$,

Volume voids $V_v = 1 - \dfrac{\rho_d}{\rho_s}$,

Volume of water $\qquad V_w = \dfrac{w\rho_d}{\rho_w},$

Degree of saturation $\qquad S_r = \dfrac{V_w}{V_v} = \dfrac{\frac{w\rho_d}{\rho_w}}{1 - \frac{\rho_d}{\rho_s}}.$

This procedure permits to compute the degree of saturation of each sample for which the resilient Young's moduli were tested at different water content, and then, Equation 5.22 permits to compute their suction pressure. Table 5.8 shows the results of the degree of saturation and suction of the three samples.

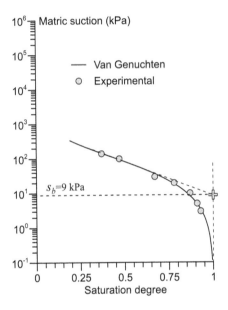

Figure 5.11 Experimental measurements of suction fitted using the van Genuchten equation.

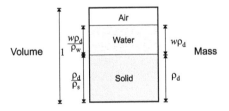

Figure 5.12 Phase's diagram for computing the degree of saturation of the soil.

Table 5.8 Degree of saturation and suction computed using the van Genuchten equation for the three values of the water content of the tested samples

w%	S_r		s (kPa)
$w_{opt}- 1.0$	6.0	0.4752	92.1
$w_{opt}- 2.4$	4.6	0.3642	148.3
$w_{opt}- 3.8$	3.2	0.2534	245.2

5.3.2 Evaluation of the models that describe the effect of the water content on the resilient Young's modulus

5.3.2.1 Models recommended in the MEPD

The MEPDG recommends to use Uzan's model to fit the experimental measures of the resilient Young's modulus and then, correcting this fit for the effect of water content by applying Equation 5.15. Therefore, the MEPDG requires the measure of the resilient Young's modulus on only one sample at one water content. For this purpose, this example uses the measure corresponding to the water content closer to the optimum value *i.e.,* $w_{opt} - 1\%$.

The algebraic procedure to linearize Uzan's equation is described in Example 5.2, and this procedure leads to the following equation:

$$\log(E_r) - k_3 \log\left(\frac{\tau_{oct}}{p_a} + 1\right) = \log(k_1 p_a) + k_2 \log\left(\frac{\theta}{p_a}\right).$$

This equation suggests that fitting the values of $\log(E_r) - k_3 \log\left(\frac{\tau_{oct}}{p_a} + 1\right)$ and $\log\left(\frac{\theta}{p_a}\right)$ using a linear equation, as shown in Figure 5.13a, permits to find out the values of k_1 and k_2. Also, as explained in the Example 5.2, the value of k_3 is obtained searching the minimum value of the *MSE* of the linear fitting. It is important to note that finding the minimum *MSE* is equivalent to finding the maximum value of R^2 as performed in this example to fit the van Genuchten equation.

As shown in Figure 5.13b, a value of $k_3 = 0.7$ gives the minimum *MSE* and, therefore, the adjustment using this value of k_3 allows finding the other two model constants that are $\log(k_1 p_a) = 2.1095 \rightarrow k_1 p_a = 128.68$ and $k_2 = 0.05183$.

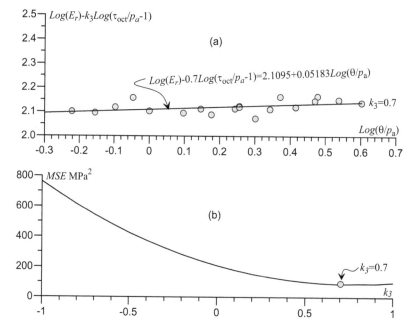

Figure 5.13 (a) Linear equation for fitting Uzan's model using the result of the resilient Young's modulus corresponding to a water content of $w_{opt} - 1\%$, (b) values of *MSE* for different values of k_3.

As a result, the expression for the resilient Young's modulus when using Uzan's model becomes

$$E_r = 1286.8 p_a \left(\frac{\theta}{p_a}\right)^{0.05183} \left(\frac{\tau_{oct}}{p_a} + 1\right)^{0.7}.$$

On the other hand, Equation 5.15, recommended in the MEPDG to take into account the effect of water, becomes

for fine graded materials,

$$\log \frac{E_r}{E_{r_{opt}}} = \frac{0.9934}{1 + exp\left[0.3944 + 6.1324(S_r - S_{r-opt})\right]} - 0.5934, \text{ or}$$

for coarse graded materials,

$$\log \frac{E_r}{E_{r_{opt}}} = \frac{0.6123}{1 + exp\left[0.0402 + 6.8157(S_r - S_{r-opt})\right]} - 0.3123,$$

where $S_r - S_{r-opt}$ is the difference, in decimals, between the actual degree of saturation and the degree of saturation at the optimum water content. Figure 5.14 shows the shape of both correction curves.

The following equation combines the correction factors to obtain the resilient Young's modulus for any water content E_{r_w} based on the already-adjusted resilient Young's modulus that corresponds to a water content of $w_{opt} - 1$, $E_r(w_{opt} - 1)$:

$$\frac{E_{r_w}}{E_r(w_{opt} - 1)} = \frac{E_{r_w}}{E_{r_{opt}}} \cdot \frac{E_{r_{opt}}}{E_r(w_{opt} - 1)}.$$

Table 5.9 presents the correction factors for the three water contents of the example. Although the granular material of the example is coarse, it has a nonnegligible amount of fines. Therefore, this example uses the correction factors given by both functions (*i.e.*, for fine and coarse materials).

Tables 5.10 and 5.11 compare the experimental results of the resilient Young's moduli for the three water contents and the prediction of Uzan's model and both correction factors recommended in the MEPDG. Moreover, Figure 5.15 compares the accuracy of

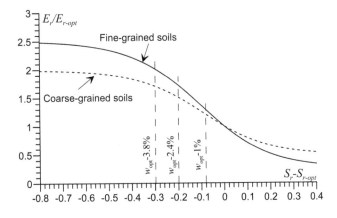

Figure 5.14 Correction factors recommended in the MEPDG.

Table 5.9 Correction factor recommended in the MEPDG depending on the degree of saturation

w%	S_r	$S_r - S_{ropt}$	Fine		Coarse	
			$\dfrac{E_r}{E_{ropt}}$	$\dfrac{E_r}{E_{r(wopt-1)}}$	$\dfrac{E_r}{E_{ropt}}$	$\dfrac{E_r}{E_{r(wopt-1)}}$
$w_{opt} - 1.0$	6.0	0.4752	−0.079	1.28	1.00	1.20	1.00
$w_{opt} - 2.4$	4.6	0.3642	−0.190	1.69	1.31	1.49	1.24
$w_{opt} - 3.8$	3.2	0.2534	−0.300	2.02	1.57	1.71	1.42

Table 5.10 Comparison of the resilient Young's moduli for the three water contents achieved using the correction factor for coarse materials

σ_3	q	p	$\dfrac{\theta}{P_a}$	$\dfrac{\tau_{oct}}{P_a} + 1$	E_r Test (MPa) w_{opt}			E_r Model (MPa) w_{opt}		
(kPa)	(kPa)	(kPa)			−1%	−2.4%	−3.8%	−1%	−2.4%	−3.8%
10	30	20.0	0.60	1.141	138.3	203.9	358.2	137.5	170.02	195.3
10	40	23.3	0.70	1.189	140.6	215.6	368.0	142.6	176.31	202.5
10	50	26.7	0.80	1.236	152.3	201.6	363.3	147.5	182.43	209.5
10	60	30.0	0.90	1.283	171.1	199.2	358.6	152.4	188.42	216.4
10	70	33.3	1.00	1.330	154.7	208.6	370.3	157.1	194.30	223.2
20	65	41.7	1.25	1.306	150.0	212.3	410.2	157.0	194.12	223.0
20	90	50.0	1.50	1.424	157.0	243.8	414.8	168.3	208.17	239.1
20	115	58.3	1.75	1.542	175.8	243.8	473.4	179.4	221.85	254.8
20	140	66.7	2.00	1.660	168.8	250.2	440.6	190.2	235.21	270.2
30	50	46.7	1.40	1.236	150.1	222.7	421.9	151.8	187.80	215.7
30	90	60.0	1.80	1.424	170.5	239.1	420.5	169.9	210.15	241.4
30	130	73.3	2.20	1.613	180.5	255.5	466.4	187.3	231.65	266.1
30	170	86.7	2.60	1.801	199.2	267.2	447.7	204.1	252.47	290.0
30	210	100.0	3.00	1.990	236.7	285.9	478.1	220.5	272.70	313.2
40	60	60.0	1.80	1.283	157.0	234.4	461.7	157.9	195.31	224.3
40	115	78.3	2.35	1.542	196.9	276.6	503.9	182.1	225.26	258.7
40	175	98.3	2.95	1.825	213.3	293.0	496.9	207.4	256.45	294.6
40	225	115.0	3.45	2.061	234.4	325.8	562.5	227.6	281.48	323.3
40	280	133.3	4.00	2.320	248.4	328.1	534.4	249.2	308.18	354.0

the predictions; it is clear that the prediction is reasonably well for the water content close to the value of water content used for the adjusting Uzan's model. However, when the water content of the material differs in some points from it, the accuracy of the prediction is very low. In fact, when considering the whole measured points, the coefficients of variation, COV_{E_r}, are poor; *i.e.*, 51% or 41% when using the functions for coarse or fine materials, respectively.

5.3.2.2 *Model with two state variables: vertical total stress and suction*

Equation 5.16 represents the model with two state variables, vertical stress and suction, that was proposed in Ref. [11]. Each state variable controls the parameters of the model as follows: $E_{0\sigma_v}$ represents the value of Young's resilient modulus for zero vertical stress and depends on the suction pressure, and the dependence of the resilient Young's modulus upon the vertical stress is controlled by the constant A_{σ_v}.

Table 5.11 Comparison of the resilient Young's moduli for the three water contents achieved using the correction factor for fine materials

σ_3 (kPa)	q (kPa)	p (kPa)	$\frac{\theta}{P_a}$	$\frac{\tau_{oct}}{P_a}+1$	E_r Test (MPa) w_{opt} −1%	−2.4%	−3.8%	E_r Model (MPa) w_{opt} −1%	−2,4%	−3.8%
10	30	20.0	0.60	1.141	138.3	203.9	358.2	137.5	180.7	215.7
10	40	23.3	0.70	1.189	140.6	215.6	368.0	142.6	187.4	223.7
10	50	26.7	0.80	1.236	152.3	201.6	363.3	147.5	193.9	231.5
10	60	30.0	0.90	1.283	171.1	199.2	358.6	152.4	200.3	239.1
10	70	33.3	1.00	1.330	154.7	208.6	370.3	157.1	206.5	246.5
20	65	41.7	1.25	1.306	150.0	212.3	410.2	157.0	206.3	246.3
20	90	50.0	1.50	1.424	157.0	243.8	414.8	168.3	221.2	264.1
20	115	58.3	1.75	1.542	175.8	243.8	473.4	179.4	235.8	281.5
20	140	66.7	2.00	1.660	168.8	250.2	440.6	190.2	250.0	298.4
30	50	46.7	1.40	1.236	150.1	222.7	421.9	151.8	199.6	238.3
30	90	60.0	1.80	1.424	170.5	239.1	420.5	169.9	223.3	266.6
30	130	73.3	2.20	1.613	180.5	255.5	466.4	187.3	246.2	293.9
30	170	86.7	2.60	1.801	199.2	267.2	447.7	204.1	268.3	320.3
30	210	100.0	3.00	1.990	236.7	285.9	478.1	220.5	289.8	346.0
40	60	60.0	1.80	1.283	157.0	234.4	461.7	157.9	207.6	247.8
40	115	78.3	2.35	1.542	196.9	276.6	503.9	182.1	239.4	285.8
40	175	98.3	2.95	1.825	213.3	293.0	496.9	207.4	272.5	325.4
40	225	115.0	3.45	2.061	234.4	325.8	562.5	227.6	299.2	357.1
40	280	133.3	4.00	2.320	248.4	328.1	534.4	249.2	327.5	391.0

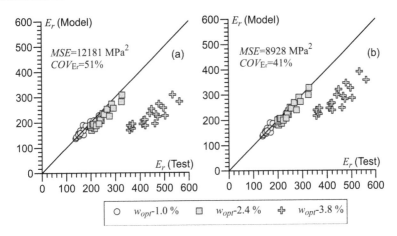

Figure 5.15 Comparison of the measured and predicted Young's moduli obtained for different water contents using the recommendations of the MEPDG: (a) using the correction curve for coarse materials and (b) using the correction curve for fine materials.

The procedure to fit the experimental results using this model begins using linear equations for each test carried out at different water contents, as shown in Figure 5.16. The intersections of the linear fittings with the vertical axis permit to obtain the resilient Young's modulus at zero stress, denoted as $E_{0\sigma_v}$.

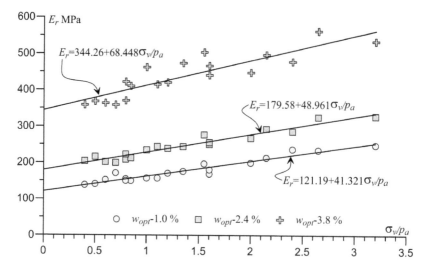

Figure 5.16 Linear functions to adjust the resilient Young's moduli and the vertical total stress.

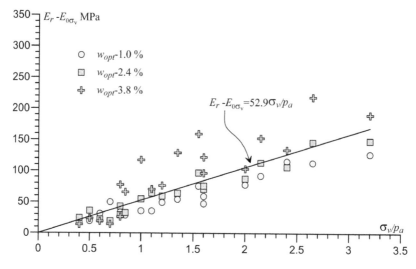

Figure 5.17 Linear function to find the slope that gives the increase in resilient Young's moduli depending on the total vertical stress.

As the fitting lines for each water content are roughly parallel, the increase of the resilient Young's modulus can be represented with a unique slope denoted as A_{σ_v}. To find out this slope, it is possible to draw all the results in a graph relating $E_r - E_{0\sigma_v}$ and σ_v/p_a, and then to adjust the points with a linear equation through zero, as shown in Figure 5.17.

Afterward, it is possible to relate $E_r - E_{0\sigma_v}$ to the matric suction of each sample, as shown in Figure 5.18.

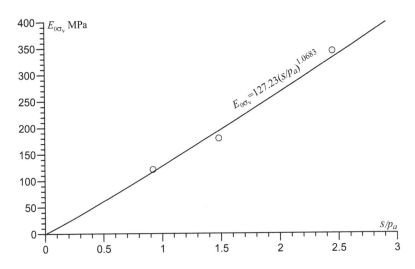

Figure 5.18 Power function for fitting the resilient Young's moduli at zero total vertical stress.

As a result, the equation giving the resilient Young's modulus depending on the vertical stress and suction is

$$E_r(\sigma_v, s) = 52.9\frac{\sigma_v}{p_a} + 127.23\left(\frac{s}{p_a}\right)^{1.0683}.$$

Table 5.12 shows a comparison of the experimental measures for the three water contents and the results of the model that uses two-state variables. Furthermore, Figure 5.19 permits to appreciate the performance of the model, which, for the experimental results of this example, is quite good since the coefficient of variation of the resilient Young's modulus is 10.7%.

5.3.2.3 Model based on effective stress

Using an effective stress approach allows the effect of total stress and suction to be included in a single expression. The mean effective stress p' is $p' = p + \chi s$, where p is the mean total stress, χ is a coefficient representing the volume of pores filled with water, and s the matrix suction. Although the samples tested in this example have low suction pressures and low degrees of saturation, it seems that a value of $\chi = 1$ allows all the measurements of the resistant Young's modulus to be represented with a single function, as shown in Figure 5.20. This value of χ, at first glance, appears to be in contradiction with the low degrees of saturation of the samples; however, one explanation of this high value of χ could be the presence of fine particles in the coarse grain contacts that create saturated lumps between them.

Figure 5.20 shows that the following potential expression describes reasonably well the relationship between resilient Young's modulus and mean effective stress:

$$E_r(p') = 1050.8 p_a\left(\frac{p'}{p_a}\right)^{1.2205}. \tag{5.23}$$

Finally, Table 5.13 allows comparing the measured values of the resilient Young's moduli with the prediction obtained using Equation 5.23. Moreover, Figure 5.21 permits to

Table 5.12 Comparison of the measured and predicted Young's moduli achieved using a model with two independent state variables: vertical stress and suction

σ_3	q	p		E_r Test (MPa)			E_r Model (MPa)		
				s (kPa)			s (kPa)		
(kPa)	(kPa)	(kPa)	$\frac{\sigma_v}{P_a}$	92.1	148.3	245.2	92.1	148.3	245.2
10	30	20.00	0.40	138.3	203.9	358.2	137.7	215.0	352.9
10	40	23.33	0.50	140.6	215.6	368.0	143.0	220.3	358.2
10	50	26.67	0.60	152.3	201.6	363.3	148.3	225.6	363.5
10	60	30.00	0.70	171.1	199.2	358.6	153.6	230.9	368.7
10	70	33.33	0.80	154.7	208.6	370.3	158.9	236.2	374.0
20	65	41.67	0.85	150.0	212.3	410.2	161.5	238.8	376.7
20	90	50.00	1.10	157.0	243.8	414.8	174.7	252.0	389.9
20	115	58.33	1.35	175.8	243.8	473.4	188.0	265.3	403.1
20	140	66.67	1.60	168.8	250.2	440.6	201.2	278.5	416.4
30	50	46.67	0.80	150.1	222.7	421.9	158.9	236.2	374.0
30	90	60.00	1.20	170.5	239.1	420.5	180.0	257.3	395.2
30	130	73.33	1.60	180.5	255.5	466.4	201.2	278.5	416.4
30	170	86.67	2.00	199.2	267.2	447.7	222.4	299.6	437.5
30	210	100.00	2.40	236.7	285.9	478.1	243.5	320.8	458.7
40	60	60.00	1.00	157.0	234.4	461.7	169.4	246.7	384.6
40	115	78.33	1.55	196.9	276.6	503.9	198.5	275.8	413.7
40	175	98.33	2.15	213.3	293.0	496.9	230.3	307.6	445.5
40	225	115.00	2.65	234.4	325.8	562.5	256.7	334.0	471.9
40	280	133.33	3.20	248.4	328.1	534.4	285.8	363.1	501.0

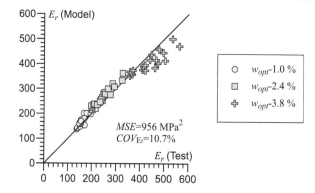

Figure 5.19 Comparison of the measured and predicted Young's moduli obtained for different water contents using a model with two state variables: vertical total stress and suction.

appreciate the good agreement between experimental and predicted values obtained with the effective stress approach. In fact, using this approach, the coefficient of variation of the resilient Young's modulus decreases to 8.4%.

5.3.2.4 Comparison of models' performance

Comparing the performance of each model is possible by analyzing their coefficients of variation of the resilient Young's modulus COV_{E_r}, and such COVs are

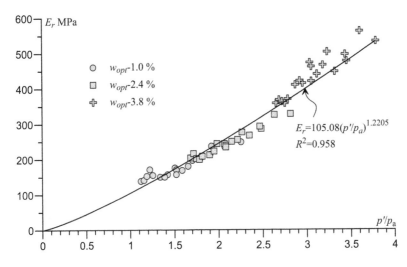

E_r MPa

- w_{opt}-1.0 %
- w_{opt}-2.4 %
- w_{opt}-3.8 %

$E_r = 105.08(p'/p_a)^{1.2205}$
$R^2 = 0.958$

p'/p_a

Figure 5.20 Power function for fitting the resilient Young's moduli using an effective stress approach.

- $COV_{E_r} = 51\%$ for Uzan's model corrected with the MEPDG function for coarse materials.
- $COV_{E_r} = 41\%$ for Uzan's model corrected with the MEPDG function for fine materials.

Table 5.13 Comparison between the measured and computed values of the resilient Young's moduli achieved using a model in effective stress

σ_3 (kPa)	q (kPa)	p (kPa)	E_r Test (MPa)			E_r Model (MPa)		
			s (kPa)			s (kPa)		
			92.1	148.3	245.2	92.1	148.3	245.2
10	30	20.00	138.3	203.9	358.2	120.8	198.4	345.6
10	40	23.33	140.6	215.6	368.0	125.2	203.2	350.9
10	50	26.67	152.3	201.6	363.3	129.6	208.0	356.2
10	60	30.00	171.1	199.2	358.6	134.1	212.8	361.5
10	70	33.33	154.7	208.6	370.3	138.6	217.7	366.9
20	65	41.67	150.0	212.3	410.2	149.9	230.0	380.3
20	90	50.00	157.0	243.8	414.8	161.4	242.3	393.9
20	115	58.33	175.8	243.8	473.4	173.0	254.8	407.5
20	140	66.67	168.8	250.2	440.6	184.7	267.4	421.2
30	50	46.67	150.1	222.7	421.9	156.8	237.4	388.4
30	90	60.00	170.5	239.1	420.5	175.3	257.3	410.2
30	130	73.33	180.5	255.5	466.4	194.3	277.6	432.2
30	170	86.67	199.2	267.2	447.7	213.5	298.1	454.4
30	210	100.00	236.7	285.9	478.1	233.1	318.8	476.7
40	60	60.00	157.0	234.4	461.7	175.3	257.3	410.2
40	115	78.33	196.9	276.6	503.9	201.4	285.2	440.5
40	175	98.33	213.3	293.0	496.9	230.7	316.2	473.9
40	225	115.00	234.4	325.8	562.5	255.5	342.5	502.1
40	280	133.33	248.4	328.1	534.4	283.4	371.8	533.5

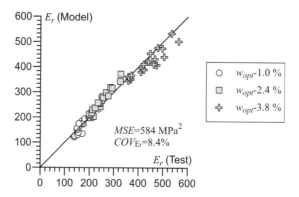

Figure 5.21 Comparison of the measured and predicted Young's moduli obtained for different water contents using an effective stress approach.

- $\text{COV}_{E_r} = 10.7\%$ for the model that uses two state variables: vertical total stress and suction.

- $\text{COV}_{E_r} = 8.4\%$ for the model that uses the mean effective stress.

It is clear that the models that use the results of the three samples, *i.e.*, the model with two state variables and the model in effective stress, lead to more reliable results. From a practical point of view, it is also clear that testing three samples for the resilient Young's moduli and measuring the water retention curve increase the laboriousness of laboratory work; however, the substantial increase in the quality of the prediction largely justifies this complexity.

5.3.3 Fatigue lifespan of the bituminous layer depending on the water content of the granular layer

The last point of this example, and the most important one from the practical point of view, is the evaluation of the fatigue lifespan of the bituminous layer depending on the water content of the granular layer. This section uses the effective stress approach due to its better predictive performance.

In this example, the fatigue life is evaluated for five different water contents from $w_{\text{opt}} - 3.0$ to $w_{\text{opt}} + 1.0$. Therefore, the first step consists in evaluating the degree of saturation corresponding to each water content, using the phase's relationships, and then computing the matric suction using Equation 5.22; Table 5.14 shows the results of this computation.

Considering the stresses of this example, the mean effective stress p' is made up of the sum of the total geostatic mean stress p_0, the matrix suction s and the mean total stress produced by the load on the road structure, Δp, as follows:

$$p' = p_0 + s + \Delta p.$$

Starting with the mean geostatic stress, it involves the vertical and horizontal components of the total stress; the latter can be calculated by multiplying the total vertical stress by the coefficient of rest K_0 as follows:

$$p_0 = \frac{\sigma_v + 2\sigma_h}{3} \quad \rightarrow \quad p_0 = \frac{1 + 2K_0}{3}\sigma_v.$$

Table 5.14 Degree of saturation and matric suction obtained using the van Genuchten water retention equation for the five water contents of the example

w%		S_r	s (kPa)
$w_{opt} + 1.0$	8.0	0.6336	44.4
w_{opt}	7.0	0.5544	64.9
$w_{opt} - 1.0$	6.0	0.4752	92.1
$w_{opt} - 2.0$	5.0	0.3960	129.4
$w_{opt} - 3.0$	4.0	0.3168	182.8

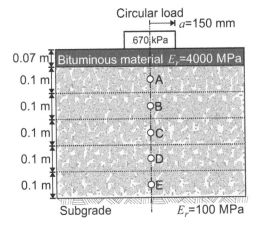

Figure 5.22 Road structure divided in sublayers and points for the analysis of stresses.

This example considers that the water content and the suction pressure are constants throughout the layer, but the total stresses vary with depth. For this reason, it is convenient to divide the granular layer into sublayers, as shown in Figure 5.22, which divides the entire granular layer into five 10-cm-thick sublayers.

Then, the vertical stress at the midpoint of each sublayer becomes

$$\sigma_{v_i} = 0.07\gamma_b + (z - 0.07)\gamma,$$

where γ_b is the unit weight of the bituminous layer calculated as $\gamma_b = \rho_b g$, g is the acceleration of the gravity; and γ is the unit weight of the granular material corresponding to each water content w, it is calculated as $\gamma = \rho_d g(1 + w)$. Table 5.15 presents the geostatic vertical, horizontal, and total and effective mean stress for the different water contents, as well as the mean geostatic effective stress given by $p_0' = p_0 + s$.

Computing the increase in mean total stress due to the load Δp requires an iterative procedure. In fact, vertical and horizontal stresses depend on the resilient Young's modulus of the sublayers, which in turn depend on those stresses. The iterative procedure carried out in this example is as follows:

1   Calculate a first value of the resilient Young's modulus of each sublayer, assuming $\Delta p = 0$.

2   Using this first set of resilient Young's moduli, compute the vertical and horizontal stress using Burmister's method.

3   Calculate a new set of resilient Young's moduli and repeat the process until convergence.

4   Once the convergence achieved, compute the tensile strain at the base of the bituminous layer.

Table 5.15 Geostatic vertical, horizontal, and total and effective mean stress for the different water contents

w = 8%

Point	ρ_d (kg/m³)	γ (kN/m³)	z (m)	σ_v (kPa)	σ_h (kPa)	p_0 (kPa)	s (kPa)	p_0' (kPa)
A	1980	21.0	0.12	2.6	1.8	2.0	44.4	46.4
B	1980	21.0	0.22	4.7	3.3	3.7	44.4	48.1
C	1980	21.0	0.32	6.7	4.7	5.4	44.4	49.8
D	1980	21.0	0.42	8.8	6.2	7.1	44.4	51.5
E	1980	21.0	0.52	10.9	7.7	8.8	44.4	53.2

w = 7%

Point	ρ_d (kg/m³)	γ (kN/m³)	z (m)	σ_v (kPa)	σ_h (kPa)	p_0 (kPa)	s (kPa)	p_0' (kPa)
A	1980	20.8	0.12	2.5	1.8	2.0	64.9	66.9
B	1980	20.8	0.22	4.6	3.2	3.7	64.9	68.6
C	1980	20.8	0.32	6.7	4.7	5.4	64.9	70.3
D	1980	20.8	0.42	8.8	6.1	7.0	64.9	71.9
E	1980	20.8	0.52	10.9	7.6	8.7	64.9	73.6

w = 6%

Point	ρ_d (kg/m³)	γ (kN/m³)	z (m)	σ_v (kPa)	σ_h (kPa)	p_0 (kPa)	s (kPa)	p_0' (kPa)
A	1980	20.6	0.12	2.5	1.8	2.0	92.1	94.1
B	1980	20.6	0.22	4.6	3.2	3.7	92.1	95.8
C	1980	20.6	0.32	6.7	4.7	5.3	92.1	97.4
D	1980	20.6	0.42	8.7	6.1	7.0	92.1	99.1
E	1980	20.6	0.52	10.8	7.5	8.6	92.1	100.7

w = 5%

Point	ρ_d (kg/m³)	γ (kN/m³)	z (m)	σ_v (kPa)	σ_h (kPa)	p_0 (kPa)	s (kPa)	p_0' (kPa)
A	1980	20.4	0.12	2.5	1.8	2.0	129.4	131.4
B	1980	20.4	0.22	4.6	3.2	3.7	129.4	133.1
C	1980	20.4	0.32	6.6	4.6	5.3	129.4	134.7
D	1980	20.4	0.42	8.6	6.0	6.9	129.4	136.3
E	1980	20.4	0.52	10.7	7.5	8.5	129.4	137.9

w = 4%

Point	ρ_d (kg/m³)	γ (kN/m³)	z (m)	σ_v (kPa)	σ_h (kPa)	p_0 (kPa)	s (kPa)	p_0' (kPa)
A	1980	20.2	0.12	2.5	1.8	2.0	182.8	184.8
B	1980	20.2	0.22	4.5	3.2	3.6	182.8	186.4
C	1980	20.2	0.32	6.6	4.6	5.2	182.8	188.0
D	1980	20.2	0.42	8.6	6.0	6.9	182.8	189.7
E	1980	20.2	0.52	10.6	7.4	8.5	182.8	191.3

This iterative procedure converges with a low number of iterations, five in the case of this example. Table 5.16 shows the results of the first and the fifth iterations carried out for the case when the water content of the granular layer is $w = w_{opt} - 3\%$; while Tables 5.17–5.20 show the results of the fifth iteration corresponding to the other values of water content.

Table 5.21 summarizes the values of tensile strain at the base of the bituminous layer computed for each value of water content, these results are also drawn in Figure 5.23a

Table 5.16 Iterative computation of stresses and resilient Young's moduli for $w = w_{opt} - 3\%$

Iteration 1

Layer	Point	p'_0/p_a	E_i (MPa) (kPa)	σ_z (kPa)	σ_h (kPa)	Δp	p'/p_a (MPa)	E_{i+1}
Top				382	56			
2	A	1.85	222.4	306	30	122.00	3.07	412.8
Bottom				230	4			
Top				230	3			
3	B	1.86	224.7	184.5	−2.5	59.83	2.46	315.7
Bottom				139	−8			
Top				139	−9			
4	C	1.88	227.1	112.5	−11	30.17	2.18	272.3
Bottom				86	−13			
Top				86	−14			
5	D	1.90	229.5	70.5	−17	12.17	2.02	247.6
Bottom				55	−20			
Top				55	−20			
6	E	1.91	231.9	46	−26	−2.00	1.89	228.9
Bottom				37	−32			

⋮

Iteration 5

Layer	Point	p'_0/p_a	E_i (MPa)	σ_z (kPa)	σ_h (kPa)	Δp (kPa)	p'/p_a	E_{i+1} (MPa)
Top				432	49			
2	A	1.85	394.0	333.5	−1	110.50	2.95	394.0
Bottom				235	−51			
Top				235	−14			
3	B	1.86	296.5	183.5	−21	47.17	2.34	296.0
Bottom				132	−28			
Top				132	−18			
4	C	1.88	259.7	105.5	−20	21.83	2.10	259.7
Bottom				79	−22			
Top				79	−18			
5	D	1.90	240.9	64	−20.5	7.67	1.97	240.9
Bottom				49	−23			
Top				49	−20			
6	E	1.91	227.5	41	−24.5	−2.67	1.89	227.9
Bottom				33	−29			

| ε_t | 241.7 | μdef |

Table 5.17 Results of the fifth iteration for $w = w_{opt} - 2\%$

				Iteration 5				
Layer	Point	p'_0/p_a	E_i (MPa)	σ_z (kPa)	σ_h (kPa)	Δ_p (kPa)	p'/p_a	E_{i+1} (MPa)
Top				394	41			
2	A	1.31	291.1	305.5	−5	98.50	2.30	290.3
Bottom				217	−51			
Top				217	−12			
3	B	1.33	212.2	171.5	−18.5	44.83	1.78	212.2
Bottom				126	−25			
Top				126	−14			
4	C	1.35	184.4	102.5	−15.5	23.83	1.59	184.4
Bottom				79	−17			
Top				79	−13			
5	D	1.36	170.7	65.5	−14	12.50	1.49	170.7
Bottom				52	−15			
Top				52	−13			
6	E	1.38	161.8	44.5	−15.5	4.50	1.42	161.8
Bottom				37	−18			

ε_t	307	μdef

Table 5.18 Results of the fifth iteration for $w = w_{opt} - 1\%$

				Iteration 5				
Layer	Point	p'_0/p_a	E_i (MPa)	σ_z (kPa)	σ_h (kPa)	Δ_p (kPa)	p'/p_a	E_{i+1} (MPa)
Top				358	36			
2	A	0.94	220.4	279.5	−6.5	88.83	1.83	219.7
Bottom				201	−49			
Top				201	−10			
3	B	0.96	157.2	161	−16	43.00	1.39	156.8
Bottom				121	−22			
Top				121	−11			
4	C	0.97	135.4	99.5	−12	25.17	1.23	134.7
Bottom				78	−13			
Top				78	−9			
5	D	0.99	125.2	66	−9	16.00	1.15	124.7
Bottom				54	−9			
Top				54	−8			
6	E	1.01	119.2	47	−8.5	10.00	1.11	119.0
Bottom				40	−9			

ε_t	373.1	μdef

showing the considerable effect of the water content on the tensile strain. In fact, the tensile strain for the highest water content is more than two times the tensile strain that results from the lower water content. This difference in strain has a significant effect on the fatigue lifespan of the bituminous layer.

Table 5.19 Results of the fifth iteration for $w = w_{opt}\%$

Layer	Point	p'_0/p_a	E_i (MPa)	σ_z (kPa)	σ_h (kPa)	Δ_p (kPa)	p'/p_a	E_{i+1} (MPa)
Top				324	33			
2	A	0.67	169.2	255	−6	81.00	1.48	169.5
Bottom				186	−45			
Top				186	−8			
3	B	0.69	118.8	150.5	−13.5	41.17	1.10	117.7
Bottom				115	−19			
Top				115	−9			
4	C	0.70	100.9	96.5	−9	26.17	0.96	100.5
Bottom				78	−9			
Top				78	−6			
5	D	0.72	93.6	67	−5.5	18.67	0.91	93.1
Bottom				56	−5			
Top				56	−3			
6	E	0.74	89.8	49	−2	15.00	0.89	90.6
Bottom				42	−1			

ε_t	440.3	μdef

Table 5.20 Results of the fifth iteration for $w = w_{opt} + 1\%$

Layer	Point	p'_0/p_a	E_i (MPa)	σ_z (kPa)	σ_h (kPa)	Δ_p (kPa)	p'/p_a	E_{i+1} (MPa)
Top				293	30			
2	A	0.46	131.6	232.5	−5.5	73.83	1.20	131.6
Bottom				172	−41			
Top				172	−6			
3	B	0.48	90.7	141	−10.5	40.00	0.88	90.1
Bottom				110	−15			
Top				110	−6			
4	C	0.50	76.7	93	−6	27.00	0.77	76.1
Bottom				76	−6			
Top				76	−3			
5	D	0.51	71.3	66.5	−2	20.83	0.72	70.7
Bottom				57	−1			
Top				57	0			
6	E	0.53	69.6	50.5	2	18.17	0.71	69.6
Bottom				44	4			

ε_t	508	μdef

Calculating the effect of the variation of tensile strain on the number of loading cycles that the bituminous layer can sustain is possible by using the fatigue equation of the example which is

Table 5.21 Results of the tensile strain at the bottom of the bituminous layer for different water contents, and ratio between the number of loads that this layer withstands at a specific water content and at the optimum water content

w%		ε_t μdef	N_w/N_{opt}
$w_{opt} + 1.0$	8.0	508.0	0.56
w_{opt}	7.0	440.3	1.00
$w_{opt} - 1.0$	6.0	373.1	1.94
$w_{opt} - 2.0$	5.0	307.0	4.23
$w_{opt} - 3.0$	4.0	241.7	11.01

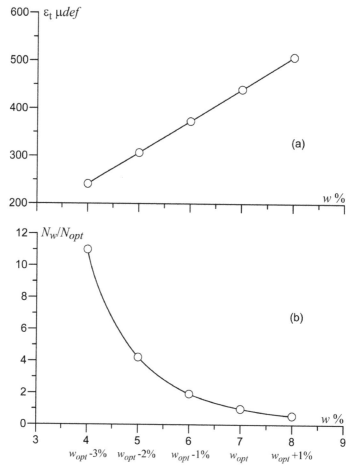

Figure 5.23 (a) Tensile strain at the bottom of the bituminous layer for different water contents and (b) number of loads supported by this layer related to the number of loads supported at the optimum water content.

$$\varepsilon_t = \varepsilon_{ref} N^{-0.25}.$$

According to this equation, the number of cycles up to failure at a particular water content N_w is

$$N_w = \left(\frac{\varepsilon_{ref}}{\varepsilon_{t_w}} \right)^4 .$$

Therefore, the ratio between the number of loading cycles the bituminous layer withstands when the granular layer has particular water content, and the number of cycles when the granular layer is at its optimal water content is

$$\frac{N_w}{N_{opt}} = \frac{\left(\frac{\varepsilon_{ref}}{\varepsilon_{t_w}} \right)^4}{\left(\frac{\varepsilon_{ref}}{\varepsilon_{t_{opt}}} \right)^4} = \left(\frac{\varepsilon_{t_{opt}}}{\varepsilon_{t_w}} \right)^4 .$$

Table 5.21 and Figure 5.23b present the results of the ratio $\frac{N_w}{N_{opt}}$ showing the astonishing effect of the water content on the fatigue life of the bituminous layer. Certainly, when the water content increases only one point above the optimum water content, fatigue life decreases to 56%. On the other hand, decreasing the water content two points below the optimum water content increases fatigue life more than four times, and more than eleven times if the water content decreases three points below the optimum.

5.3.4   Concluding remarks

The results of this example lead to the following important conclusions for practical purposes:

- A first remark is the relevance of a proper characterization of the resilient Young's modulus of the granular material, including measures at different water contents and also the measure in the laboratory of the water retention curve. Despite the increase in the laboriousness of the laboratory work, it is highly compensated by the increase in the accuracy of the computations.
- Secondly, maintaining the water content of the granular layer below the optimum water content has a remarkable importance, it requires well drainage and avoiding as much as possible the infiltration through cracks or through the bituminous layer itself.
- Finally, it is essential to highlight that for this type of low traffic road structure, the role of the granular layers is undoubtedly the most critical component of the performance of the whole system.

Chapter 6

Climate effects

6.1 RELEVANT EQUATIONS

6.1.1 Heat flow in road structures

Evaluating the temperature changes that occur in the structure of a road in the face of changes in weather conditions is crucial because temperature determines the behavior of materials, and particularly it has a significant impact on the behavior of bituminous materials.

A first step in the process of assessing temperature changes is the evaluation of the heat flux that affects the surface of the road, which is called sensible heat q_{sens}. By neglecting interactions with water in the form of rain or evaporation, this heat flux results from the sum of the effects of radiation, q_{rad}, thermal emissions, q_{th}, and convection, q_{conv}, so that

$$q_{sens} = q_{rad} - q_{th} - q_{conv}, \tag{6.1}$$

$$q_{rad} = \alpha I, \tag{6.2}$$

$$q_{th} = \epsilon \sigma (T_s^4 - T_{sky}^4), \tag{6.3}$$

$$T_{sky} = T_a [0.77 + 0.0038(T_d - 273.15)]^{0.25}, \tag{6.4}$$

$$q_{conv} = h_c(T_s - T_a), \tag{6.5}$$

where α is the mean absorptivity coefficient, I is the irradiance, ϵ is the emissivity coefficient of the surface (that, according to Kirchhoff's law, $\epsilon = \alpha$ when reaching thermal equilibrium), σ is the Stefan–Boltzmann constant $\sigma = 5.67 \cdot 10^{-8}$ W/m^{2}K^4, T_s is the temperature at the surface of the road, T_{sky} is a hypothetical temperature above the surface of the road, T_d is the dew point temperature, h_c is the convection coefficient, and T_a is the air temperature.

It is important to note that all temperatures in Equation 6.4 are in K. In addition, the Magnus formula gives the dew point temperature as

$$T_d = T_n \frac{\ln\left(\frac{u_v}{A}\right)}{m - \ln\left(\frac{u_v}{A}\right)}, \tag{6.6}$$

$$u_v = u_{vs} U_w,$$

$$u_{vs} = A e^{\left(\frac{mT_a}{T_n + T_a}\right)} [Pa],$$

where u_v is the vapor pressure, u_{vs} is the saturation vapor pressure, U_w is the relative humidity, and T_a is the air temperature in °C. Moreover, A, m, and T_n are constants whose values are $A = 611.2$, $m = 17.62$, and $T_n = 243.12$.

The sun is the primary source of energy on earth, and road structures are no exception. Therefore, the heat that results from solar radiation is probably the essential factor that affects the temperature change in the structure of a road. The heat flux due to radiation is given in Equation 6.2, where the irradiance I is the factor that quantifies the solar energy reaching the road. This total irradiance I results from the combination of the direct, I_b, and diffuse irradiances, I_d.

Direct irradiance depends on the position of the sun in relation to the surface of the road given by several angles which are depicted in Figure 6.1. These angles are h which is the angle between the center of the disc of the sun and the horizon, the solar zenith angle, θ, which is the angle between a normal to the surface of the road and the direct line joining the sun (note that $h + \theta = 90°$), the solar declination δ is the angle between the Equator and a line joining the center of the earth and the center of the sun, the angle Γ indicates the position of the earth in its orbit for a particular day of the year, and w is the hour angle which provides information about the position of the sun in each elapsed time of the day.

Each one of those angles is calculated using the following equations:

$$\cos\theta = \sin h = \sin\phi \sin\delta + \cos\phi \cos\delta \cos w, \tag{6.7}$$

$$w = 15(12 - T_{sv}), \tag{6.8}$$

$$T_{sv} = T_l - \Delta T_l + D_{hg}/60, \tag{6.9}$$

$$\Gamma = \frac{2\pi(d_n - 1)}{365}, \tag{6.10}$$

$$\delta = \left(\frac{180}{\pi}\right)(0.006918 - 0.399912\cos\Gamma + 0.070257\sin\Gamma - 0.006758\cos 2\Gamma$$
$$+ 0.000907\sin 2\Gamma - 0.002697\cos 3\Gamma + 0.00148\sin 3\Gamma), \tag{6.11}$$

where ϕ is the latitude on the earth of the site, T_{sv} (hours) is the true solar time at the site, T_l is the local time in the zone of the site, ΔT_l is the time difference between local time and standard time for the reference point of that time zone, D_{hg} is the time difference of the site from the time zone's reference point (4 min per meridian degree), and d_n is the Julian day so that $d_n = 1$ for January 1 and $d_n = 365$ for December 31.

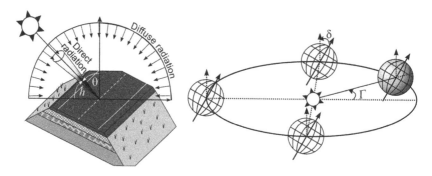

Figure 6.1 Schematic drawing describing the different angles that allow characterizing the position of the sun.

Moreover, a correction factor, E_0, is required for computing direct radiation because of the position of the earth in its orbit varies throughout the year, this correction factor is

$$E_0 = 1.000110 + 0.034221 \cos \Gamma + 0.001280 \sin \Gamma$$
$$+ 0.000719 \cos 2\Gamma + 0.000077 \sin 2\Gamma. \tag{6.12}$$

As a result, the normal component of the direct radiation is

$$I_b(t) \cos \theta = I_0 E_0 \sin h, \tag{6.13}$$

where I_0 is the solar constant, usually $I_0 = 1367$ W/m^2.

After applying an approximate correction factor accounting for the ozone, gas and water attenuations, the direct irradiance becomes

$$I_b(t) \cos \theta = 0.798 I_0 E_0 e^{-0.13/\sin h} \sin h \quad \text{for } \sin h > 0, \text{ and}$$
$$I_b(t) \cos \theta = 0 \quad \text{for } \sin h < 0. \tag{6.14}$$

This value of direct irradiance corresponds to a day with clear skies. Regarding the diffuse irradiance, the following approximation is given in Ref. [28]:

$$I_d(t) = 120 \cdot 0.798 e^{-1/(0.4511 + \sin h)}. \tag{6.15}$$

In fact, some amount of diffuse irradiance remains during the night. Nevertheless, Equation 6.15 should apply only during the day.

Finally, the total irradiance at a particular time of the day, $I(t)$, is

$$I(t) = I_b(t) \cos \theta + I_d(t). \tag{6.16}$$

Regarding convection, which is another essential component of heat transfer, it results from the interaction between the material's molecules on the road surface and the air around it. The heat flux due to convection is given by Equation 6.5, where the convection coefficient is the main element. Most of the approaches that allow obtaining the convection coefficient are empirical and involve the air velocity V_a as summarized in Table 6.1.

Subsequently, due to the sensible heat entering the road, the temperature on its surface changes, generating a heat flux toward deeper layers. Therefore, the net heat flux must consider the interchanges with those more profound layers of the road. Fourier describes this heat flux that results from heat conduction as

$$q_{\text{conduct}} = k_H \frac{dT}{dx}, \tag{6.17}$$

where k_H is the thermal conductivity of the first layer of the road, T is the temperature, and x is the flux direction.

Therefore, the net heat flow is given by

$$q_{\text{net}} = q_{\text{sens}} - q_{\text{conduct}}. \tag{6.18}$$

Depending on whether this net heat flux is positive or negative, which in turn depends on the time of day, the road surface heats up or cools down, as it is described in detail in Example 19.

Table 6.1 Empirical relationships for calculating the convection coefficient h_c [36]

Equation	Model
$h_c = 698.24\left[0.00144T_{avg}^{0.3}V_a^{0.7}+0.00097(T_s - T_a)^{0.3}\right]$ $T_{avg} = (T_s + T_a)/2$ $0.8< V_a <8.5$ m/s, $6.7°C< T_s <27°C$	Vehrencamp
$h_c = 7.55 + 4.35V_a$	Nicol
$h_c = 5.8 + 4.1V_a$	Jurges
$h_c = 1.824 + 6.22V_a$	Kimura
$h_c = 18.6V_a^{0.605}$	ASHRAE
$h_c = 5.7 + 6.0V_a$	Sturrock
$h_c = 16.15V_a^{0.4}$	Loveday

h_c is the convection coefficient in W/m^{2}K,
T_a and T_s are the air and surface temperatures in K,
and V_a is the air velocity in m/s.

6.1.2 Flow of water through a drainage layer

Chapter 2 of this book describes a numerical method for computing water flow in porous materials and road structures. Even though this analysis is rigorous, its numerical solution requires the use of small time steps, which makes it computationally costly. This problem is particularly evident when computing the flow of water through granular materials that have an abrupt change between the saturated and the unsaturated states. Precisely, for those cases, the approximate method that uses Dupuit's approach gives a straightforward solution. This approach concentrates the flow of water in a saturated layer having a free surface on top. As a result, water coming from the infiltration due to rain acts as a source that supplies water directly to the free surface.

Dupuit's solution gives the horizontal velocity, V_w, of water in a drainage layer resting over an impervious layer (see Figure 6.2). The three following solutions are possible for this particular case:

$$\text{for } \frac{4q_{inf}}{k_w\xi^2} > 1: \ x = L\left(\frac{q_{inf}}{k_w}\right)^{1/2} r^{-1/2}e^{\frac{1}{m}\left(\arctan(-1/m)-\arctan\frac{2u-\xi}{\xi m}\right)} \tag{6.19}$$

where $m^2 = \dfrac{4q_{inf}}{k_w\xi^2} - 1$, or

$$\text{for } \frac{4q_{inf}}{k_w\xi^2} = 1: \ x = L\left(\frac{q_{inf}}{k_w}\right)^{1/2} \frac{2}{2u - \xi}e^{\left(\frac{2u}{2u-\xi}\right)}, \ \text{or} \tag{6.20}$$

$$\text{for } \frac{4q_{inf}}{k_w\xi^2} < 1: \ x = L\left(\frac{q_{inf}}{k_w}\right)^{1/2} r^{-1/2}\left(\frac{1+n}{1-n}\cdot\frac{1-n-2u/\xi}{1+n-2u/\xi}\right)^{\frac{1}{2n}} \tag{6.21}$$

where $n^2 = 1 - \dfrac{4q_{inf}}{k_w\xi^2}$, and

k_w is the hydraulic conductivity, $\xi > 0$ is the slope of the base of the layer, $h(x)$ is the height of the free surface, L is the length of the drainage layer, q_{inf} is the uniform infiltration, and u and r are ancillary variables given by $h = ux$ and $r = u^2 - \xi u + q_{inf}/k_w$. When the base is permeable, q_{inf} must be substituted by the net flow of water given by $q_{net} = q_{inf} - k_{w-s}$, where k_{w-s} is the water infiltrating through the base of the layer.

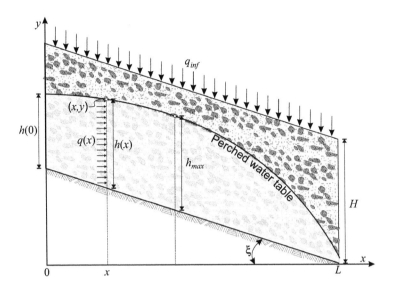

Figure 6.2 Schematic drawing of a drainage layer.

Another useful information that is obtained using this solution is the maximum height of the free surface, h_{max}, which determines the thickness of the drainage layer. This maximum height is given by

$$h_{\text{max}} = L \left(\frac{q_{\text{inf}}}{k_w} \right)^{1/2} F, \text{ where}$$

(6.22)

$$F = e^{\frac{1}{m} \left(\arctan(-1/m) - \arctan \frac{m^2-1}{2m} \right)}.$$

Also, this solution provides the maximum uniform infiltration that the layer can evacuate, q_{max}, which is

$$q_{\text{max}} = \frac{k_w H^2}{L^2 F^2}.$$

(6.23)

6.2 EXAMPLE 19: EVOLUTION OF THE TEMPERATURE IN A ROAD STRUCTURE DEPENDING ON THE ENVIRONMENTAL VARIABLES

The purpose of this example is to calculate the evolution of the temperature within the structure of a road on a particular day and at a specific place in the world.

The road structure has four layers: first, a layer of bituminous material, then granular base and sub-base layers, and finally, the subgrade. Regarding the thermal properties of each layer, which are the thermal conductivity and the volumetric heat capacity, they are the same as in Example 11, which are summarized here in Table 6.2.

Additional information regarding the particular location and the day of the year for the computations are

- Location, Bogotá, Colombia, South America, whose latitude is $\phi = 4.57°$.
- Day of the year, January 15.
- Difference between the local and true solar time for Bogotá, -6 min.
- This example assumes that the weather conditions for this particular day are sunny and with clear sky. Moreover, for this particular day, the air temperature, T_a, the relative humidity, U_w, and the wind velocity, V_a, are given in Table 6.3.

This example describes the procedure to calculate the following elements:

1   The evolution along the day of the components of sensible heat.
2   The evolution of the temperature throughout the structure along the day. This point allows to trace the change of temperature along the day at the interfaces between layers, the temperature in the vertical profile of the road for four particular hours of the day (i.e., 6, 12, 18, and 24 h), and the temperature map for different depths and hours of the day.

Table 6.2 Thermal properties of the materials

Layer	Thickness (m)	k_H (W/m°C)	c_{H_y} (J/m³K)	Absorptivity α (-)
Bituminous layer	0.1	1.500	$2.000 \cdot 10^6$	0.85
Granular base	0.2	2.159	$2.578 \cdot 10^6$	
Granular sub-base	0.3	1.793	$2.706 \cdot 10^6$	
Subgrade	∞	1.506	$3.053 \cdot 10^6$	

Table 6.3 Climatic variables along the day

Time (h)	T_a (°C)	U_w (%)	V_a (km/h)	Time (h)	T_a (°C)	U_w (%)	V_a (km/h)
0 : 1	6	80	10	12 : 13	20	50	5
1 : 2	4	80	10	13 : 14	22	50	5
2 : 3	2	85	10	14 : 15	22	50	5
3 : 4	2	85	10	15 : 16	22	50	5
4 : 5	4	80	10	16 : 17	20	50	5
5 : 6	6	80	5	17 : 18	18	50	5
6 : 7	8	80	5	18 : 19	16	60	5
7 : 8	12	60	5	19 : 20	16	60	5
8 : 9	14	60	5	20 : 21	16	60	10
9 : 10	16	60	5	21 : 22	14	70	10
10 : 11	18	50	5	22 : 23	10	70	10
11 : 12	20	50	5	23 : 24	8	80	10

The Finite Difference Method, using an explicit scheme, allows solving this problem numerically, and the solution requires the following steps:

1   Assess the environmental variables which are required for computing the heat flux (*i.e.*, the dew point temperature, T_d, and the temperature above the road, T_{sky}).
2   Evaluate the heat flux produced by the direct and diffuse irradiances.
3   Discretize the problem in space.
4   Discretize the problem in time.
5   Develop an equation accounting for the continuity of heat flow between layers.
6   Develop equations that allow considering the top and the bottom boundary conditions.
7   Choose the proper time step for the computation.
8   Perform a numerical solution using the equations developed in the previous steps.

6.2.1   Environmental variables

As shown in Equation 6.3, the loss of heat through the road surface due to thermal emission, q_{th}, requires the evaluation of the temperature surrounding the road T_{sky}, which, in turn, requires the dew point temperature T_d.

Using Equations 6.6, the expression that gives the dew point temperature, which depends on the temperature of the air T_a, and the relative humidity U_w is

$$T_d = \frac{T_n \left[\frac{mT_a}{T_n + T_a} + \ln(U_w) \right]}{m - \left[\frac{mT_a}{T_n + T_a} + \ln(U_w) \right]}.$$

Usually, this equation uses the following constants: $m = 17.62$ and $T_n = 243.12$, leading to

$$T_d = \frac{243.12 \left[\frac{17.62 T_a}{243.12 + T_a} + \ln(U_w) \right]}{17.62 - \left[\frac{17.62 T_a}{243.12 + T_a} + \ln(U_w) \right]}. \tag{6.24}$$

Note that in the previous equation, T_d and T_a are temperatures in °C and U_w is in decimals.

Once the dew point temperature has been calculated, Equation 6.4 allows obtaining the temperature above the road, T_{sky}, as

$$T_{sky} = (T_a + 273.15)(0.77 + 0.0038 T_d)^{0.25} \ \text{[K]}.$$

Regarding the convection coefficient h_c, any equation shown in Table 6.1 could be used. However, this example adopts the recommendation given by Hall et al. [36] who recommend using the Jurges equation which is

$$h_c = 5.8 + 4.1 V_a \ \ \text{[W/m}^2\text{K]}. \tag{6.25}$$

Note that in the previous equation, the air velocity V_a is in m/s.

Likewise, according to Ref. [36], the radiation heat transfer coefficient, h_{rad}, is

$$h_{rad} = \epsilon \sigma (T_0 + T_{sky})(T_0^2 + T_{sky}^2) \ \ \text{[W/m}^2\text{K]}, \tag{6.26}$$

where ϵ is the emissivity of the material of the surface of the road, σ is the Stefan–Boltzmann constant $\sigma = 5.67 \cdot 10^{-8}$ W/m²K⁴, and T_0 is the absolute temperature of the surface in K.

Table 6.4 shows the results of T_d, T_{sky}, h_c, and h_{rad} obtained using the previous equations. Also, Figure 6.3 shows the evolution of these variables along the day. It is important to remark that, for evaluating h_{rad}, the temperature of the surface is assumed to have a maximum value of 50°C (i.e., $T_0 = 323.15$ K). This approximate calculation gives the highest value of h_{rad}, which, in turn, is useful for the assessment of the maximum time step allowed to avoid instabilities in the computations, as it is described in this example in the section that evaluates the proper time step.

6.2.2 Heat flow due to solar radiation

Calculating the irradiation in a particular day in a specific location requires the evaluation of various angles. The first angle is Γ, which relates the position of the earth in its orbit for a particular day; it is evaluated using Equation 6.10. For January 15, which is the day given for this example, this angle is

$$\Gamma = \frac{2 \cdot 180(15 - 1)}{365} = 13.808°.$$

Table 6.4 Results of T_d, T_{sky}, h_c, and h_{rad} along the day

Time (h)	T_a (°C)	U_w (%)	V_a (km/h)	T_d (°C)	T_{sky} (K)	h_c (W/m²K)	h_{rad}
0 : 1	6	80	10	2.81	262.39	17.19	4.89
1 : 2	4	80	10	0.86	259.89	17.19	4.83
2 : 3	2	85	10	−0.26	257.66	17.19	4.78
3 : 4	2	85	10	−0.26	257.66	17.19	4.78
4 : 5	4	80	10	0.86	259.89	17.19	4.83
5 : 6	6	80	5	2.81	262.39	11.49	4.89
6 : 7	8	80	5	4.76	264.90	11.49	4.95
7 : 8	12	60	5	4.47	268.57	11.49	5.03
8 : 9	14	60	5	6.35	271.07	11.49	5.09
9 : 10	16	60	5	8.23	273.57	11.49	5.15
10 : 11	18	50	5	7.41	275.20	11.49	5.19
11 : 12	20	50	5	9.26	277.69	11.49	5.26
12 : 13	20	50	5	9.26	277.69	11.49	5.26
13 : 14	22	50	5	11.09	280.19	11.49	5.32
14 : 15	22	50	5	11.09	280.19	11.49	5.32
15 : 16	22	50	5	11.09	280.19	11.49	5.32
16 : 17	20	50	5	9.26	277.69	11.49	5.26
17 : 18	18	50	5	7.41	275.20	11.49	5.19
18 : 19	16	60	5	8.23	273.57	11.49	5.15
19 : 20	16	60	5	8.23	273.57	11.49	5.15
20 : 21	16	60	10	8.23	273.57	17.19	5.15
21 : 22	14	70	10	8.61	271.80	17.19	5.11
22 : 23	10	70	10	4.78	266.79	17.19	4.99
23 : 24	8	80	10	4.76	264.90	17.19	4.95

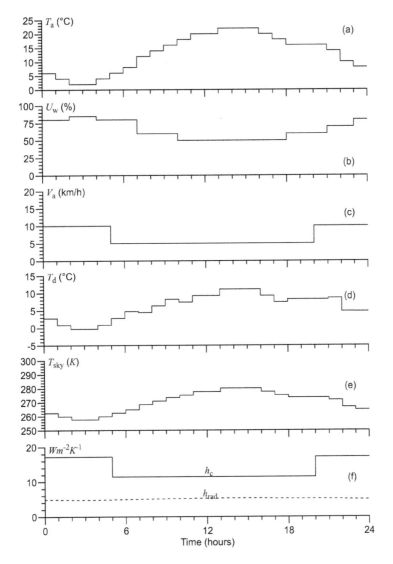

Figure 6.3 Evolution of the atmospheric variables along the day, (a) air temperature, (b) relative humidity, (c) air velocity, (d) dew point temperature, (e) sky temperature, and (f) convection and radiation heat transfer coefficients.

This value corresponds to the beginning of the day, but, as the earth goes through its orbit continuously, the angle Γ changes at the end of January 15 to $14.795°$, which corresponds to an increase of around one degree each day.

Another angle required to evaluate the irradiance is the declination δ that represents the tilting angle of the earth with respect to the Equatorial plane. The declination δ is calculated using Equation 6.11 as

$$\delta = \left(\frac{180}{\pi}\right)[0.006918 - 0.399912\cos(13.808) + 0.070257\sin(13.808)$$

$$- 0.006758 \cos(2 \cdot 13.808) + 0.000907 \sin(2 \cdot 13.808)$$
$$- 0.002697 \cos(3 \cdot 13.808) + 0.00148 \sin(3 \cdot 13.808)] = -21.273°.$$

Likewise, the declination also evolves continuously during the day and, at the end of January 15, it changes to $-21.09°$. Figure 6.4a and b shows the evolution of the declination angle throughout the year and its value at the beginning of January 15.

In addition, the evaluation of solar radiation must consider the effect of the distance between the sun and the earth throughout the year using a correction factor, which is given by Equation 6.12. This equation applies to any day of the year and, for January 15, which corresponds to an angle $\Gamma = 13.808°$, the correction factor E_0 is

$$E_0 = 1.000110 + 0.034221 \cos(13.808) + 0.001280 \sin(13.808)$$
$$+ 0.000719 \cos(2 \cdot 13.808) + 0.000077 \sin(2 \cdot 13.808) = 1.0343$$

As shown in Figure 6.4c, the maximum value of the correction factor in a year is $E_0 = 1.035$ while its minimum value is $E_0 = 0.9666$, indicating a low variation along the year. Consequently, the change of this correction factor during the day is minimal (*i.e.*, $E_0 = 1.0341$ at the end of January 15).

Since the difference between the local time and the actual solar time for Bogotá is -6 min, then the hour angle of the sun, w, is

$$w = 15[12 - (T_l - 6/60)].$$

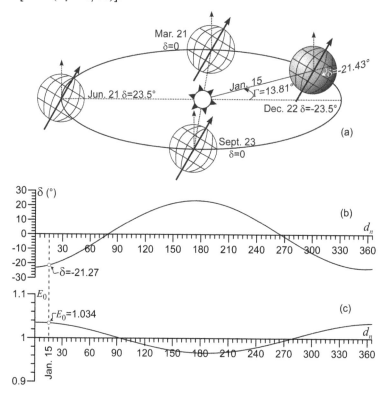

Figure 6.4 (a and b) Evolution of the angles describing the position of the earth in its orbit, and (c) evolution throughout the year of the correction factor depending on the distance sun–earth.

After calculating the declination and the hour angle, and considering that the latitude of the city of Bogotá is $\phi = 4.57°$, Equation 6.7, which allows computing the elevation angle of the sun during the day depending on the local time T_l (see Figure 6.5a), leads to the following equation for the cosine of θ:

$$\cos\theta = \sin h = \sin(4.57)\sin(-21.273) + \cos(4.57)\cos(-21.273)$$
$$\times \cos[180 - 15(T_l - 0.1)].$$

Note that, as depicted in Figure 6.5a, $\cos\theta = \sin h$.

Subsequently, Equations 6.14 and 6.15 permit to calculate the direct and diffuse irradiances along the day as follows:

$$I_b(t)\cos\theta = 0.798 \cdot 1367 E_0 e^{-0.13/\sin h}\sin h \quad \text{direct, and}$$
$$I_d(t) = 120 \cdot 0.798 e^{-1/(0.4511+\sin h)} \qquad \text{diffuse.}$$

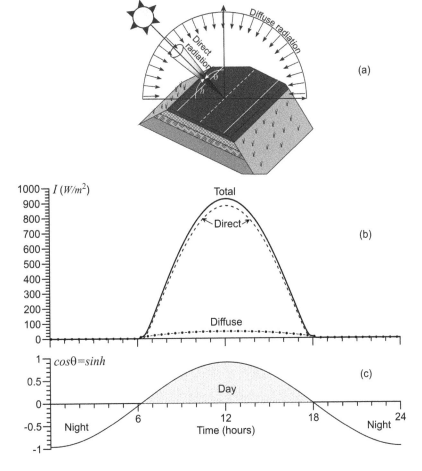

Figure 6.5 Results of the direct, diffuse, and total irradiances for January 15 in the city of Bogotá, (a) angles describing the position of the sun along the day, (b) direct and diffuse irradiances, and (c) evolution of the angles describing the position of the sun during the day January 15.

Figure 6.5b shows the results of the direct, diffuse, and total irradiances for January 15 in the city of Bogotá. These values of irradiances are essential as input data to calculate the evolution of the temperature within the road structure.

6.2.3 Discretization in space

The following diffusion equation permits to calculate the change in temperature within a road structure due to heat conduction:

$$c_{H_v}\frac{\partial T}{\partial t} = -\nabla \cdot q_H, \tag{6.27}$$

where T is the temperature, and c_{H_v} and k_H are the volumetric heat capacity and the thermal conductivity of the material, respectively.

The space discretization aims to transform the right hand of the differential Equation 6.27 into a system of linear equations. Since in this example, heat flows only in the vertical direction, therefore the discretization involves only one dimension of the space. In other words, when considering only the vertical dimension, the equation can be expanded considering a unit length in the horizontal directions, as shown in Figure 6.6.

The following procedure permits expanding Equation 6.27 in a discrete form.

First, it is possible to consider a node i located in a homogeneous layer (*i.e.*, excluding the boundaries and interfaces between layers). For this node, heat fluxes up and down the node are

- Heat flow through the face BC of Figure 6.6b is: $Q_{H-BC} = q_H^{i-1/2}$,
- Heat flow through the face AD of Figure 6.6b is: $Q_{H-AD} = q_H^{i+1/2}$.

Then, according to Fourier's law, heat fluxes q_H are

$$q_H^{i-1/2} = k_H^{i-1/2}\frac{T_{i-1} - T_i}{\Delta z}, \text{ and}$$

$$q_H^{i+1/2} = k_H^{i+1/2}\frac{T_i - T_{i+1}}{\Delta z},$$

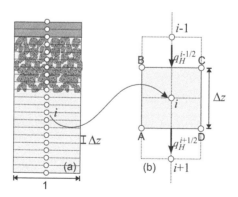

Figure 6.6 Discretization in space of Equation 6.27, (a) column describing the road structure, and (b) Close-up view of a node in the road structure.

where $k_H^{i-1/2}$ and $k_H^{i+1/2}$ are the heat conductivities of the materials located above and below the node i, respectively. However, since node i is located within a homogeneous material $k_H^{i-1/2} = k_H^{i+1/2}$, then the heat fluxes become

$$q_H^{i-1/2} = k_H \frac{T_{i-1} - T_i}{\Delta z}, \text{ and} \tag{6.28}$$

$$q_H^{i+1/2} = k_H \frac{T_i - T_{i+1}}{\Delta z}. \tag{6.29}$$

Moreover, the balance of heat flux per unit volume, $\nabla \cdot q_H$, in a discrete form is

$$-\nabla \cdot q_H \approx \frac{Q_{H-AD} - Q_{H-BC}}{\Delta z}. \tag{6.30}$$

Note that, because this analysis considers unit lengths and unit area in the horizontal plane, in the previous equations $Q_H = q_H$, and the unit volume of material is Δz.

Finally, from Equations 6.28, 6.29, and 6.30, the heat balance given by $\nabla \cdot q_H$ becomes

$$-\nabla \cdot q_H \approx \frac{k_H}{\Delta z^2} T_{i-1} + \frac{k_H}{\Delta z^2} T_{i+1} - 2T_i \frac{k_H}{\Delta z^2}. \tag{6.31}$$

Equation 6.31 is the discretized form of the right hand of Equation 6.27.

6.2.4 Discretization in time

In the same way as the previous analysis, it is possible to discretize the left hand of Equation 6.27 that represents the evolution of temperature over time. In a discrete form, this side of the equation becomes

$$c_{H_v} \frac{\partial T}{\partial t} \approx c_{H_v} \frac{T_i^{t+\Delta t} - T_i^t}{\Delta t}, \tag{6.32}$$

where c_{H_v} is the volumetric specific heat of the material located in the node i (because the material is homogeneous around the node), and Δt is the time step used for the discretization in time.

The combination of Equations 6.31 and 6.32 allows computing the evolution of the temperature in the road structure using linear equations instead of a differential equation.

However, before combining these equations, it is essential to decide whether the temperatures in Equation 6.31 are considered in the time t, or the time $t + \Delta t$. Both solutions are possible, and both have advantages and disadvantages. The first option corresponds to an explicit solution, which is a straightforward solution, but which could eventually generate instabilities in the calculation. In contrast, the second option corresponds to an implicit solution that is unconditionally stable, but it is more costly computationally.

This example uses the explicit solution, and then, temperatures in Equation 6.31 are considered in time t. Therefore, this equation together with Equation 6.32 leads to the following recursive expression that allows computing the temperature in the time $t + \Delta t$ when the temperatures in time t are known

$$T_i^{t+\Delta t} = \frac{k_H \Delta t}{c_{H_v} \Delta z^2} \left(T_{i-1}^t + T_{i+1}^t - 2T_i^t \right) + T_i^t. \tag{6.33}$$

Equation 6.33 should be used for computing the change in temperature of all internal nodes for which the thermal properties are homogeneous (*i.e.,* all the internal nodes excepting the interfaces between materials).

Since the explicit solution could produce instabilities of the numerical solution, Section 6.2.7 of this example describes the analysis of the proper time step allowed to avoid such unpredictability.

6.2.5 Continuity equation between layers

Because the interfaces between layers are discontinuities in both thermal conductivity and specific heat, the assumption of homogeneous material is invalid. Therefore, the equations must involve the properties of the materials above and below such interfaces.

As a result, the balance of heat flux above and below the node i, which is located in an interface, is

$$q_H^{i-1/2} - q_H^{i+1/2} = k_H^{i-1/2} \frac{T_{i-1}^t - T_i^t}{\Delta z} - k_H^{i+1/2} \frac{T_i^t - T_{i+1}^t}{\Delta z}.$$

Moreover, as shown in Figure 6.7, the increase in heat over time should be analyzed for a volume that involves half the separation between nodes (*i.e.,* $\Delta z/2$) as

$$c_{H_v} \frac{\partial T}{\partial t} \approx \frac{c_{H_v}^{i-1/2} \Delta z}{2} \frac{T_i^{t+\Delta t} - T_i^t}{\Delta t} + \frac{c_{H_v}^{i+1/2} \Delta z}{2} \frac{T_i^{t+\Delta t} - T_i^t}{\Delta t}.$$

Therefore, the equation for the interfaces becomes

$$k_H^{i-1/2} \frac{T_{i-1}^t - T_i^t}{\Delta z} - k_H^{i+1/2} \frac{T_i^t - T_{i+1}^t}{\Delta z}$$
$$= \frac{c_{H_v}^{i-1/2} \Delta z}{2} \left(\frac{T_i^{t+\Delta t} - T_i^t}{\Delta t} \right) + \frac{c_{H_v}^{i+1/2} \Delta z}{2} \left(\frac{T_i^{t+\Delta t} - T_i^t}{\Delta t} \right).$$

leading to

$$T_i^{t+\Delta t} = \frac{\frac{k_H^{i-1/2}}{\Delta z} T_{i-1}^t + \frac{k_H^{i+1/2}}{\Delta z} T_{i+1}^t - \left(\frac{k_H^{i-1/2}}{\Delta z} + \frac{k_H^{i+1/2}}{\Delta z} - \frac{c_{H_v}^{i-1/2} \Delta z}{2\Delta t} - \frac{c_{H_v}^{i+1/2} \Delta z}{2\Delta t} \right) T_i^t}{\frac{c_{H_v}^{i-1/2} \Delta z}{2\Delta t} + \frac{c_{H_v}^{i+1/2} \Delta z}{2\Delta t}}. \tag{6.34}$$

A simplified form of Equation 6.34 is

$$T_i^{t+\Delta t} = A_{\text{int}} T_{i-1}^t + B_{\text{int}} T_i^t + C_{\text{int}} T_{i+1}^t, \tag{6.35}$$

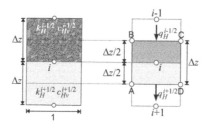

Figure 6.7 Discretization in space in the interfaces between layers.

where

$$A_{\text{int}} = \frac{2\Delta t}{\Delta z^2}\left(\frac{k_H^{i-1/2}}{c_{H_v}^{i-1/2} + c_{H_v}^{i+1/2}}\right),$$

$$B_{\text{int}} = 1 - \frac{2\Delta t}{\Delta z^2}\left(\frac{k_H^{i-1/2} + k_H^{i+1/2}}{c_{H_v}^{i-1/2} + c_{H_v}^{i-1/2}}\right) = 1 - (A_{\text{int}} + C_{\text{int}}), \text{ and}$$

$$C_{\text{int}} = \frac{2\Delta t}{\Delta z^2}\left(\frac{k_H^{i+1/2}}{c_{H_v}^{i-1/2} + c_{H_v}^{i+1/2}}\right).$$

As a result, Equation 6.35 is the expression to be used to calculate the temperature change at each interface between materials in the road structure.

6.2.6 Analysis of the boundary conditions

One-dimensional models, such as the one of this example, have two boundaries: the boundary on the road surface, in which the temperature is directly controlled by the climatic variables, and the lower boundary that, if chosen far enough from the surface, can be considered as a boundary with zero heat flow.

As already mentioned, heat flux on the road surface, $Q_{H_{\text{surf}}}$, involves heat radiation, convection, thermal emission, and the heat flux entering the road structure, as shown in Figure 6.8. The following equation gives the sum of these fluxes:

$$Q_{H_{\text{surf}}} = \underbrace{\alpha I^t}_{\text{Radiation}} + \underbrace{h_c^t(T_a^t - T_1^t)}_{\text{Convection}} + \underbrace{\epsilon\sigma\left[\left(T_{\text{sky}}^t\right)^4 - \left(T_1^t\right)^4\right]}_{\textit{Thermal emission}} - \underbrace{k_H\frac{T_1^t - T_2^t}{\Delta z}}_{\text{Heat flux entering}},$$

where T_1 is the temperature on the road surface, and T_2 the temperature of the node directly below it.

Heat flux at the surface of the road produces a change of the temperature in a small layer whose thickness is $\Delta z/2$, and then,

$$Q_{H_{\text{surf}}} = c_{H_v}\frac{\Delta z}{2}\frac{T_1^{t+\Delta t} - T_1^t}{\Delta t}.$$

Therefore,

$$c_{H_v}\frac{\Delta z}{2}\frac{T_1^{t+\Delta t} - T_1^t}{\Delta t} = q_{\text{rad}}^t + h_c^t(T_a^t - T_1^t) + \epsilon\sigma\left[\left(T_{\text{sky}}^t\right)^4 - \left(T_1^t\right)^4\right] + k_H\frac{T_2^t - T_1^t}{\Delta z}.$$

Resulting in the following equation that allows calculating the temperature on the road surface depending on the climatic variables:

$$T_1^{t+\Delta t} = \frac{2\Delta t\left[\alpha I^t + h_c^t(T_a^t - T_1^t) + \epsilon\sigma\left(\left(T_{\text{sky}}^t\right)^4 - \left(T_1^t\right)^4\right) + k_H\frac{T_2^t - T_1^t}{\Delta z}\right]}{c_{H_v}\Delta z} + T_1^t.$$

$$(6.36)$$

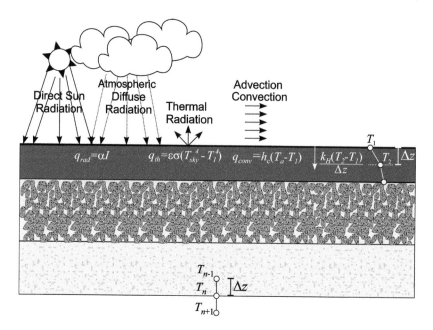

Figure 6.8 Heat fluxes on the surface of the road and the bottom of the model.

On the other hand, the temperature at the bottom of the model, which is T_n, n being the number of the last node, is calculated assuming that the bottom of the model is chosen far enough the surface. Therefore, the heat flux can be considered zero. As described in Figure 6.8, for this type of boundary condition, it is useful to assume an imaginary node outside the model (*i.e.*, node $n+1$). The temperature of this imaginary node is unknown. However, zero heat flux implies zero temperature gradient, leading to equal temperatures above and below the node n, $T_{n+1} = T_{n-1}$. Therefore, the use of Equation 6.33 for node n leads to

$$T_n^{t+\Delta t} = \frac{k_H \Delta t}{c_{H_V} \Delta z^2} \left(2T_{n-1}^t - 2T_n^t \right) + T_n^t. \tag{6.37}$$

As a result, Equation 6.37 permits to compute the temperature of the node n when assuming zero heat flux at the bottom of the model.

In summary, the set of equations required to calculate the change in temperature in the road structure is

- Surface of the road: Equation 6.36.
- Internal nodes: Equation 6.33.
- Interfaces: Equation 6.35.
- Bottom boundary: Equation 6.37.

6.2.7 Analysis of the time step

As already mentioned, the explicit solution is a straightforward method that requires a little computational cost, but it could generate instabilities in the numerical solution.

For this reason, it is crucial to choose a suitable time step Δt. Instabilities can appear in the internal nodes or at the surface of the model. For this reason, the time step must respect the following inequalities:

- The time step to avoid instabilities in the internal nodes must be

$$\Delta t < \frac{1}{2} \frac{c_{H_v} \Delta z^2}{k_H}, \text{ and} \tag{6.38}$$

- The time step to avoid instabilities at the surface of the model must be

$$\Delta t < \frac{1}{2} \frac{c_{H_v} \Delta z^2}{h_{\text{rad}} \Delta z + h_c \Delta z + k_H}. \tag{6.39}$$

The inequality 6.38 must be analyzed for each material of the model, while the inequality 6.39 uses the properties of the surface layer. Of course, the time step used for the calculations should be less than all the evaluations made with the above inequalities.

6.2.8   Numerical solution

This section uses the equations obtained in the previous sections to calculate the evolution of the temperature inside the road structure. For a numerical application, the first step is to choose the size of the model and the length step for the space discretization.

First, the lower limit should be chosen at a depth where the effect of heat flow on the surface is negligible. Since this example intends to calculate the evolution of temperature over a day, choosing a lower limit 2 m below the granular base ensures that low heat flow occurs at this level as a result of temperature changes in the surface. Therefore, the total length of the model for the computations of this example is 2.6 m (i.e., 0.6 m of top layers and 2 m of subgrade). Nevertheless, this depth may be too shallow to calculate the annual evolution of temperature, particularly in regions with extreme seasonal variations in climatic conditions.

Moreover, regarding the length spacing, choosing 0.02 m allows having a proper resolution of the temperature evolution inside each layer of the road structure. Therefore, as shown in Figure 6.9, a 2.6 m deep model with 0.02 m spacing between each node has 131 nodes.

After choosing the length spacing Δz, the next point is the evaluation of the appropriate time step, which is possible using Equations 6.38 and 6.39.

From Equation 6.38, the requirements for the time step to avoid instabilities in the nodes of each layer are

$$\text{Bituminous layer: } \Delta t < \frac{2.0 \cdot 10^6 \cdot 0.02^2}{2 \cdot 1.5} = 266.7 \text{ s.}$$

$$\text{Granular base: } \Delta t < \frac{2.578 \cdot 10^6 \cdot 0.02^2}{2 \cdot 2.159} = 238.8 \text{ s.}$$

$$\text{Granular sub-base : } \Delta t < \frac{2.706 \cdot 10^6 \cdot 0.02^2}{2 \cdot 1.793} = 301.8 \text{ s.}$$

$$\text{Subgrade : } \Delta t < \frac{3.053 \cdot 10^6 \cdot 0.02^2}{2 \cdot 1.506} = 405.4 \text{ s.}$$

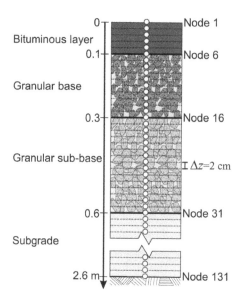

Figure 6.9 Discretization in the space of the entire road structure.

Subsequently, Equation 6.39 provides a time step that avoids instabilities on the model's surface.

$$\Delta t < \frac{0.5 \cdot 2.0 \cdot 10^6 \cdot 0.02^2}{0.02(h_{\text{rad}}^t + h_c^t) + 1.5} = \frac{400}{0.02(h_{\text{rad}}^t + h_c^t) + 1.5} \, s.$$

Since the coefficients h_{rad} and h_c evolves with time, the requirement for the time step also changes. Figure 6.10 shows this evolution resulting in a minimum value of $\Delta t = 205.46$ s.

Based on the previous evaluations, a time step of $\Delta t = 120$ s certainly avoid computational instabilities in both the internal nodes and the surface's node.

Finally, the following recursive equations allow calculating the evolution of the temperature in all the nodes of the model.

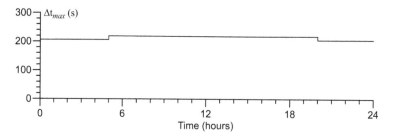

Figure 6.10 Maximum values of the time step required to avoid instabilities of the temperature at the surface of the road.

- **Node 1:** surface of the road,

$$Q_{H_1} = 0.85 I^t + h_c^t (T_a^t - T_1^t) + 0.85 \cdot 5.67 \cdot 10^{-8} \left[\left(T_{sky}^t \right)^4 - \left(T_1^t \right)^4 \right]$$

$$+ 1.5 \frac{T_2^t - T_1^t}{\Delta z},$$

$$T_1^{t+\Delta t} = \frac{2 \cdot 120}{0.02 \cdot 2 \cdot 10^6} Q_{H_1} + T_1^t$$

$$= 6 \cdot 10^{-3} Q_{H_1} + T_1^t. \tag{6.40}$$

- **Nodes 2...5:** internal nodes in the bituminous layer,

$$T_{2...5}^{t+\Delta t} = \frac{1.5 \cdot 120}{0.02^2 \cdot 2 \cdot 10^6} \left(T_{1...4}^t + T_{3...6}^t - 2T_{2...5}^t \right) + T_{2...5}^t$$

$$= 0.225 \left(T_{1...4}^t + T_{3...6}^t - 2T_{2...5}^t \right) + T_{2...5}^t. \tag{6.41}$$

- **Node 6:** interface between the bituminous layer and the granular base,

$$A_{int} = \frac{2 \cdot 120}{0.02^2 \cdot 10^6} \left(\frac{1.5}{2 + 2.578} \right) = 0.197,$$

$$C_{int} = \frac{2 \cdot 120}{0.02^2 \cdot 10^6} \left(\frac{2.159}{2 + 2.578} \right) = 0.283,$$

$$B_{int} = 1 - (0.197 + 0.283) = 0.520,$$

$$T_6^{t+\Delta t} = 0.197 T_5^t + 0.520 T_6^t + 0.283 T_7^t. \tag{6.42}$$

- **Nodes 7...15:** internal nodes in the granular base,

$$T_{7...15}^{t+\Delta t} = \frac{2.159 \cdot 120}{0.02^2 \cdot 2.578 \cdot 10^6} \left(T_{6...14}^t + T_{8...16}^t - 2T_{7...15}^t \right) + T_{7...15}^t$$

$$= 0.251 \left(T_{6...14}^t + T_{8...16}^t - 2T_{7...15}^t \right) + T_{7...15}^t. \tag{6.43}$$

- **Node 16:** interface between the granular base and sub-base,

$$A_{int} = \frac{2 \cdot 120}{0.02^2 \cdot 10^6} \left(\frac{2.159}{2.578 + 2.706} \right) = 0.245,$$

$$C_{int} = \frac{2 \cdot 120}{0.02^2 \cdot 10^6} \left(\frac{1.793}{2.578 + 2.706} \right) = 0.204,$$

$$B_{int} = 1 - (0.245 + 0.204) = 0.551,$$

$$T_{16}^{t+\Delta t} = 0.245 T_{15}^t + 0.551 T_{16}^t + 0.204 T_{17}^t. \tag{6.44}$$

- **Nodes 17...30:** internal nodes in the granular sub-base,

$$T_{17..30}^{t+\Delta t} = \frac{1.793 \cdot 120}{0.02^2 \cdot 2.706 \cdot 10^6} \left(T_{16...29}^t + T_{18...31}^t - 2T_{17...30}^t \right) + T_{17...30}^t$$

$$= 0.199 \left(T_{16...30}^t + T_{18...31}^t - 2T_{17...30}^t \right) + T_{17...30}^t. \tag{6.45}$$

- **Node 31:** interface between the granular sub-base and the subgrade,

$$A_{\text{int}} = \frac{2 \cdot 120}{0.02^2 \cdot 10^6} \left(\frac{1.793}{2.706 + 3.053} \right) = 0.187,$$

$$C_{\text{int}} = \frac{2 \cdot 120}{0.02^2 \cdot 10^6} \left(\frac{1.506}{2.706 + 3.053} \right) = 0.157,$$

$$B_{\text{int}} = 1 - (0.187 + 0.157) = 0.656,$$

$$T_{16}^{t+\Delta t} = 0.187 T_{15}^t + 0.656 T_{16}^t + 0.157 T_{17}^t. \tag{6.46}$$

- **Nodes 32...130:** internal nodes in the subgrade,

$$T_{32..130}^{t+\Delta t} = \frac{1.506 \cdot 120}{0.02^2 \cdot 3.053 \cdot 10^6} \left(T_{31...129}^t + T_{33...131}^t - 2 T_{32...130}^t \right) + T_{32...130}^t$$

$$= 0.148 \left(T_{31...129}^t + T_{33...131}^t - 2 T_{32...130}^t \right) + T_{32...130}^t. \tag{6.47}$$

- **Node 131:** bottom boundary,

$$T_{131}^{t+\Delta t} = \frac{1.506 \cdot 120}{0.02^2 \cdot 3.053 \cdot 10^6} \left(2 T_{130}^t - 2 T_{131}^t \right) + T_{131}^t$$

$$= 0.148 \left(2 T_{130}^t - 2 T_{131}^t \right) + T_{131}^t. \tag{6.48}$$

Equations 6.40–6.48 can be easily implemented into a spreadsheet, as it is depicted in Table 6.5, or in a MATLAB script, such as the one provided with this book.

Besides, solving the set of equations requires to initialize the temperature in all nodes of the model. Any value is valid as initial temperature, and the model progressively reaches an equilibrium state. Nevertheless, it is always necessary to run the model during a certain "heating" time before starting the computation of the temperature for the day of analysis. It is important to remark that, if the initial temperature in the deeper nodes of the model is chosen far from the temperature at the equilibrium state, it could be necessary to run the model for a long time before starting the actual simulation.

In this example, the initial temperature is $T_{\text{init}} = 15°C$, and the "heating" time is 27 days (computing all the days using the same climatic data). This elapsed time of 27 days is adequate to reach an equilibrium state, which can be identified when the temperature at the bottom of the model is approximately constant.

To sum up, Figure 6.11 shows the evolution of the temperature on the surface and the interfaces of the road. In this figure, it is noticeable that during the last 7 days of the simulation, the fluctuations of the temperature in all layers follow a recurrent pattern, confirming the adequate choice of the initial temperature and the *heating* time.

Figure 6.12a focuses on the 28th day of the simulation, showing that the minimum and maximum temperatures on the surface of the road are 6.5°C and 44.2°C that are reached at 4.0 and 13.5 h, respectively. Also, Figure 6.12b–e shows the profiles of temperature in all the road structure at four different times of the day. In these figures, it is noticeable that the more significant changes in temperature occur in the bituminous layer and the granular base. In contrast, the granular sub-base and the subgrade undergo small fluctuations of temperature.

The evolution of the temperature along the day and in each layer can also be identified in the temperature map represented in Figure 6.13.

Table 6.5 Schematic description of a spreadsheet which could be used for computing the evolution of the temperature within the road structure, shaded columns indicate either boundary conditions or interfaces between layers

					Nodes				
	1	2..5	6	7..15	16	17..30	31	32..130	131
	Surface	Internal	Interface	Internal	Interface	Internal	Interface	Internal	Bottom
Time	T_{init}	T_{init}	T_{init}	T_{init}	T_{init}	T_{init}	T_{init}	T_{init}	T_{init}
0	Equation 6.40	Equation 6.41	Equation 6.42	Equation 6.43	Equation 6.44	Equation 6.45	Equation 6.46	Equation 6.47	Equation 6.48
Δt									
$2\Delta t$									
…									

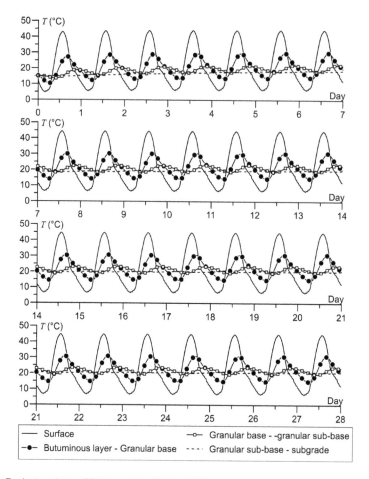

Figure 6.11 Evolution during 28 days of simulation of the temperatures at the surface of the road and the interfaces between the bituminous layer and the granular base, the granular base and sub-base, and the sub-base and the subgrade.

6.3 EXAMPLE 20: ASSESSMENT OF THE LOCAL INFILTRATION THROUGH CRACKS IN THE TOP LAYER OF A ROAD

This example describes a methodology to assess the local infiltration of water through cracks that could exist in the top layer of a road. The methodology assumes that the flow of water occurs in the laminar regime.

After analyzing the usefulness of the Poiseuille solution for the flow of water through cracks, the example evaluates the effect of the following characteristics of a single crack and a net of cracks in the road:

1 *The effect of the width w of a single crack considering cracks in the range of 0.1 mm ≤ w ≤ 1 mm, and temperatures of 10° C, 20° C, and 30° C.*

2 *The infiltration through the surface of the road, assuming a square net of cracks having a separation s between them, and temperature of 20° C.*

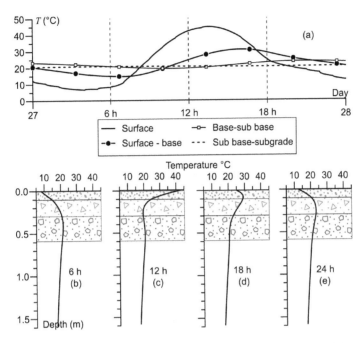

Figure 6.12 (a) Temperature during the 28th day of simulation at the surface of the road and the different interfaces, (b–e) temperature profiles at 6, 12, 18, and 24 h.

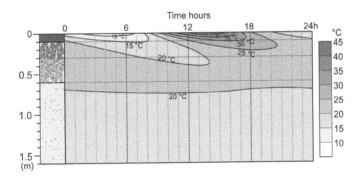

Figure 6.13 Heat map for the 28th day of simulation.

6.3.1 Infiltration through single cracks

The Poiseuille solution permits the assessment of the infiltration of water through cracks [10,63]. When the water flows freely through a crack, there are not overpressures, and therefore the gradient of potential is 1. Under these circumstances, the maximum amount of water flowing into a vertical crack of width w per unit length, denoted as $\bar{\bar{q}}$, is

$$\bar{\bar{q}} = \rho_w g \frac{w^3}{12\mu} \quad \text{for} \quad R_e < 300 \,,$$

$$R_e = \frac{\rho_w^2 g}{6\mu^2} w^3 \,,$$

where ρ_w is the density of water, g is the acceleration of gravity, μ is the dynamic or absolute viscosity of water, and R_e is the Reynolds number. Note that this equation is only valid for Reynolds numbers below 300, and for higher values of this dimensionless number, the exponent that affects w decreases. Nevertheless, high values of the Reynolds numbers correspond to cracks whose width is larger than 0.5 mm, and it is expected that such cracks can be filled with fine particles, and therefore the Poiseuille analysis is invalid.

Dynamic viscosities of water at different temperatures are $\mu = 0.0013076$ *Pas* at $10°C$, $\mu = 0.0010005$ *Pas* at $20°C$, and $\mu = 0.000797$ *Pas* at $30°C$. Therefore, the amounts of water flowing at different temperatures are

$$\bar{\bar{q}} = \frac{1,000 \cdot 9.8}{12 \cdot 0.0013076} w^3 \quad \frac{m^3}{sm} \qquad \text{for } 10°C,$$

$$\bar{\bar{q}} = \frac{1,000 \cdot 9.8}{12 \cdot 0.0010005} w^3 \quad \frac{m^3}{sm} \qquad \text{for } 20°C,$$

$$\bar{\bar{q}} = \frac{1,000 \cdot 9.8}{12 \cdot 0.0007970} w^3 \quad \frac{m^3}{sm} \qquad \text{for } 30°C.$$

Table 6.6 and Figure 6.14 describe how the flow of water grows as the width of the crack increases. Note that the results are given for laminar flow (*i.e.*, $R_e < 300$).

Table 6.6 Infiltration through cracks of different widths

	10°C			20°C			30 °C	
$w\ m$	$\bar{\bar{q}}\ \frac{m^3}{sm}$	R_e	$w\ m$	$\bar{\bar{q}}\ \frac{m^3}{sm}$	R_e	$w\ m$	$\bar{\bar{q}}\ \frac{m^3}{sm}$	R_e
$1.0 \cdot 10^{-4}$	$6.2 \cdot 10^{-7}$	1.0	$1.0 \cdot 10^{-4}$	$8.2 \cdot 10^{-7}$	1.6	$1.0 \cdot 10^{-4}$	$1.0 \cdot 10^{-6}$	2.6
$2.0 \cdot 10^{-4}$	$5.0 \cdot 10^{-6}$	7.6	$2.0 \cdot 10^{-4}$	$6.5 \cdot 10^{-6}$	13.1	$2.0 \cdot 10^{-4}$	$8.2 \cdot 10^{-6}$	20.6
$3.0 \cdot 10^{-4}$	$1.7 \cdot 10^{-5}$	25.8	$3.0 \cdot 10^{-4}$	$2.2 \cdot 10^{-5}$	44.1	$3.0 \cdot 10^{-4}$	$2.8 \cdot 10^{-5}$	69.4
$4.0 \cdot 10^{-4}$	$4.0 \cdot 10^{-5}$	61.1	$4.0 \cdot 10^{-4}$	$5.2 \cdot 10^{-5}$	104.4	$4.0 \cdot 10^{-4}$	$6.6 \cdot 10^{-5}$	164.6
$5.0 \cdot 10^{-4}$	$7.8 \cdot 10^{-5}$	119.4	$5.0 \cdot 10^{-4}$	$1.0 \cdot 10^{-4}$	204.0	$4.9 \cdot 10^{-4}$	$1.2 \cdot 10^{-4}$	302.5
$6.0 \cdot 10^{-4}$	$1.3 \cdot 10^{-4}$	206.3	$5.7 \cdot 10^{-4}$	$1.5 \cdot 10^{-4}$	302.2			
$6.8 \cdot 10^{-4}$	$2.0 \cdot 10^{-4}$	300.4						

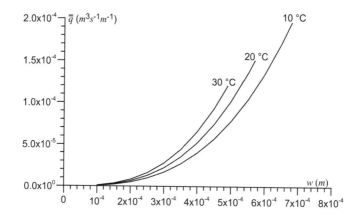

Figure 6.14 Infiltration of water at different temperatures depending on the width of the crack.

6.3.2   Infiltration through a squared net of cracks

The results of infiltration through a single crack allow evaluating the infiltration of water through a network of cracks. A square network of cracks, as in Figure 6.15, is the most straightforward arrangement that allows an evaluation of the flow entering the structure of a road. For such an arrangement, the density of cracks is calculated as the ratio between the length of cracks within the shadowed area in Figure 6.15 and the area of this shadowed zone. Therefore, the flow of water distributed on a unit surface of the road, denoted as q_{inf}^{cracks}, is

$$q_{\text{inf}}^{cracks} = \frac{s+s}{s^2}\bar{\bar{q}} = \frac{2}{s}\bar{\bar{q}} = \bar{\bar{q}}\zeta, \tag{6.49}$$

where s is the distance between cracks, and ζ is the density of cracks on the road, which is $\zeta = 2/s$.

Equation 6.49 allows evaluating the amount of water flowing into the road structure affected by a square network of cracks that have different widths and separations between them. For this purpose, it is possible to use the results of the flow of water through a single crack, carried out in the previous section, and to consider different crack densities. The results of this evaluation are shown in Figure 6.15.

Along with the flow of water through the cracks, evaluating the total infiltration through the road surface requires adding the uniform infiltration that occurs through the pores of the layer, q_{inf}^{pores}. In the case of a bituminous layer, this uniform infiltration depends on the proportion of voids in the mixture. There are several expressions, as described in Ref. [10]. This example uses the following expression:

$$q_{\text{inf}}^{pores} = 10^{-6}10^{6.131 \log V - 4.815} \quad \text{m/s},$$

where V is the proportion of air voids in %.

Figure 6.15 also shows the results of the uniform infiltration depending on the proportion of voids in the bituminous material.

As a result, the total flow of water that can infiltrate into a road's structure is

$$q_{\text{inf}} = q_{\text{inf}}^{cracks} + q_{\text{inf}}^{pores} = \rho_w g \frac{w^3}{12\mu}\frac{2}{s} + 10^{-6}10^{6.131 \log V - 4.815} \quad \text{m/s}.$$

However, the maximum amount of water that can infiltrate the road at a particular time is limited by the precipitation due to rain p. Therefore, to obtain the real amount of water that infiltrates, it is necessary to compare q_{inf} with p in the following way:

$$q_{\text{inf}} = p \text{ for } p < q_{\text{inf}} \quad \text{or} \quad q_{\text{inf}} = q_{\text{inf}} \text{ for } p \geq q_{\text{inf}}.$$

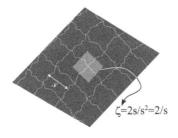

Figure 6.15 Schematic drawing of a road affected by a squared net of cracks with a separation of s between them.

Table 6.7 Categories of infiltration based in Refs. [64,66]

q_{inf} (m/s)	Category	Description
10^{-8} to 10^{-7}	A1	Very low infiltration
10^{-7} to 10^{-6}	A2	Low infiltration
10^{-6} to 10^{-5}	B	Moderate infiltration some water infiltrates under traffic
10^{-5} to 10^{-4}	C	Substantial infiltration under traffic
10^{-4} to 10^{-3}	D	Infiltration freely under traffic or raindrop impact, pumping of fines
10^{-3} to 10^{-2}	E	Free infiltration

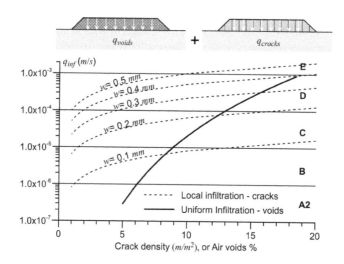

Figure 6.16 Infiltration of water in an asphalt layer with different proportion of air voids and with a square net of cracks with different crack density and width, temperature 20°C.

Finally, Table 6.7, based on Refs. [10,64,66], allows ranking the flow of water into roads. The limits suggested in Table 6.7 are drawn in Figure 6.16. This figure characterizes the road situation concerning infiltration and illustrates the effects of crack density and voids.

6.4   EXAMPLE 2I: DRAINAGE LAYERS IN ROAD STRUCTURES

Example 20 describes a methodology to evaluate infiltration through the top layer of a road due to water flowing through cracks and the voids of the material.

Subsequently, the water entering the road structure must be evacuated through drainage layers, and, eventually, the lower granular layers of the road could fulfill this function. This example describes the methodology for analyzing the flow of water in a drainage layer and calculating the position of the free surface within this layer, which allows estimating the required thickness for this layer.

The following points are evaluated in this example:

1   *The position of the water table in the drainage layer for different relationships q_{inf}/k_w.*
2   *The effect of the slope and the length of the drainage layer on its required thickness.*

As described in Section 6.1.2 of this book, Dupuit's solution permits to compute the position of the water table within a drainage layer (*i.e.,* the free surface); see Figure 6.2.

Although three solutions for the position of the water table are presented in Section 6.1.2, the first solution is the most useful because it corresponds to the higher values of q_{inf}. This solution is as follows:

$$\text{For } \frac{4q_{inf}}{k_w\xi^2} > 1: \; x = L\left(\frac{q_{inf}}{k_w}\right)^{1/2} r^{-1/2} e^{\frac{1}{m}\left(\arctan(-1/m)-\arctan\frac{2u-\xi}{\xi m}\right)}, \tag{6.50}$$

$$\text{where } m^2 = \frac{4q_{inf}}{k_w\xi^2} - 1, \quad h = ux, \quad r = u^2 - \xi u + q_{inf}/k_w,$$

k_w is the water conductivity, $\xi > 0$ is the slope of the base of the layer, $h(x)$ is the height of the free surface, L is the length of the drainage layer, and q_{inf} is the uniform infiltration that acts as a source in the upper part of the free surface. When the base of the layer is permeable, q_{inf} must be replaced by the net flow of water given by $q_{net} = q_{inf} - k_{w-s}$, where k_{w-s} is the water that enters the next layer.

The maximum height of the free surface is h_{max}. This height is also the required thickness of the layer to prevent pressure build-up within it. This maximum height is

$$h_{max} = L\left(\frac{q_{inf}}{k_w}\right)^{1/2} F, \tag{6.51}$$

$$\text{where } F = e^{\frac{1}{m}\left(\arctan(-1/m)-\arctan\frac{m^2-1}{2m}\right)}.$$

Moreover, the profile of the perched water table that flows into the drainage layer is given by its coordinates (x, y). The procedure for calculating this position requires the following steps:

1   Compute the value of the variable m.
2   Choose values of u in the range $(0 \le u < \infty)$.
3   Compute the values of the ancillary variable r.
4   Compute x for the chosen values of u using Equation 6.50.
5   Compute the height of the free surface, $h = ux$.
6   Finally, compute the coordinate $y(x)$ of the free surface as $y(x) = h(x) + \xi(L - x)$.

For example, for a drainage layer of 4 m of length, slope of 2%, and water conductivity $k_w = 10^{-3}$ m/s, which receives an infiltration flow of $q_{inf} = 10^{-5}$ m/s, the position of the perched water table is calculated as follows:

the variable m is

$$\frac{4q_{inf}}{k_w \xi^2} = \frac{4 \cdot 10^{-5}}{10^{-3}0.02^2} = 100, \text{ then } m = \left[\frac{4 \cdot 10^{-5}}{10^{-3}0.02^2} - 1 \right]^{0.5} = 9.9499,$$

now choosing $u = 0.1 \quad \rightarrow \quad r = 0.1^2 - 0.02 \cdot 0.1 + 0.01 = 0.018,$

then, the value of the coordinate x is

$$x = 4 \left(10^{-2}\right)^{1/2} (0.018)^{-1/2} e^{\frac{1}{9.9499} \left(\arctan(-1/9.9499) - \arctan \frac{2 \cdot 0.1 - 0.02}{0.02 \cdot 9.9499} \right)} = 2.741 \ m,$$

the height of the water table is

$h = 0.1 \cdot 2.741 = 0.2741$ m,

and the coordinate y is

$y = 0.2741 + 0.02(4 - 2.741) = 0.299$ m.

Repeating this procedure for different values of u allows calculating the position of the perched water table throughout the drainage layer, as shown in Figure 6.17.

Moreover, Figure 6.18 shows the height of the water table, h, measured from the base of the drainage layer. As mentioned, the maximum value of h corresponds to the required thickness of the layer to avoid the pressure build-up. As observed in Figure 6.18, the maximum value of h is not always located at the extreme of the layer (i.e., at $x = 0$). In fact, its position depends on the relationship q_{inf}/k_w, and the maximum height moves toward $x = 0$ as this relationship increases.

Finally, Equation 6.51 permits to compute the value of h_{max}. Figure 6.19 shows how the required thickness grows as the ratio q_{inf}/k_w increases, and how it also grows as the layer length increases or when its slope decreases.

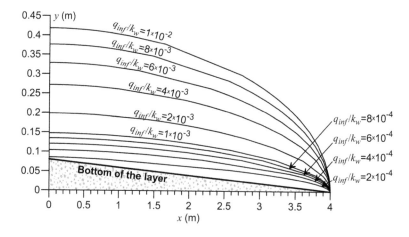

Figure 6.17 Position of the perched water table into a drainage layer for different relationships q_{inf}/k_w, $L = 4$ m, and $\xi = 0.02$.

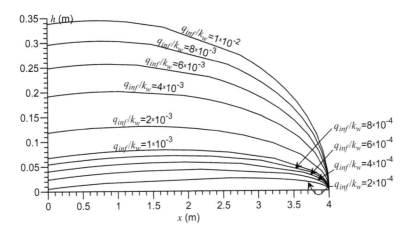

Figure 6.18 Height of the perched water table measured from the base of the drainage layer for different relationships q_{inf}/k_w, $L = 4$ m, and $\xi = 0.02$.

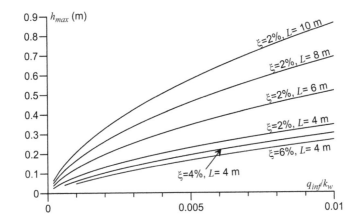

Figure 6.19 Effect of the slope and the length of the drainage layer on its required thickness.

Chapter 7

Nondestructive evaluation and inverse methods

7.1 RELEVANT EQUATIONS

7.1.1 Theoretical analysis of vibratory rollers

The analysis of the movement of vibratory rollers during compaction allows evaluating the performance of the compaction. First, a vibratory roller like the one outlined in Figure 7.1 can be described as a system with two degrees of freedom [1]. Then, the simplifications described in Ref. [10] make it possible to transform the system with two degrees of freedom into a problem with only one degree of freedom, given by the following equation:

$$(m_d + m_e)\ddot{z}_1 = (m_d + m_e + m_f)g + m_e e\omega^2 \sin \omega t - F_s,$$

where m_d is the mass of the drum, m_e is the eccentric mass, m_f is the mass of the frame, e is the eccentricity of the mass, ω is the angular frequency of the rotating mass, F_s is the reaction of the soil, and z_1 is the vertical displacements of the drum.

When the reacting soil remains in the elastic domain of behavior, the reaction of the soil F_s becomes

$$F_s = k_s^e z_s + C_s \dot{z}_s, \tag{7.1}$$

where k_s^e and C_s are the coefficients of the spring–dashpot system representing the soil and z_s its displacement.

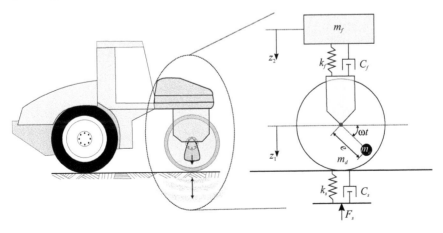

Figure 7.1 Vibratory roller described as a system with two degrees of freedom [1].

Equation 7.1 requires the coefficients for the spring and the dashpot. They can be obtained by combining the Lundberg equation, which provides the depth of penetration of a cylinder, with the contact width resulting from Hertz's theory, as described in Section 7.1.2, and using the cone method, as described in Section 7.1.3.

On the other hand, when the drum moves up, it eventually loses contact with the ground and the soil's reaction returns to zero, which is described as

$$z_s = z_1 \quad \text{for} \quad z_s \geq 0, \text{ or}$$
$$z_s = 0 \quad \text{for} \quad z_1 < 0 \quad \text{and} \quad F_s = 0.$$

Therefore, the dynamic equations that describe the soil–drum interaction become

$$(m_d + m_e)\ddot{z}_1 = (m_d + m_e + m_f)g + m_e e \omega^2 \sin \omega t - k_s^e z_1 - C_s \dot{z}_1 \text{ for } z_s = z_1 \geq 0, \text{ or}$$
$$(m_d + m_e)\ddot{z}_1 = (m_d + m_e + m_f)g + m_e e \omega^2 \sin \omega t \quad \text{for} \quad z_1 < 0. \tag{7.2}$$

7.1.2 Contact between a cylindrical body and an elastic half-space

According to Hertz's theory, half of the contact width a, resulting from a contact between a cylindrical body and an elastic half-space, as shown in Figure 7.2, is

$$a = \left(\frac{4RF}{\pi LE^*}\right)^{\frac{1}{2}}, \tag{7.3}$$

where L is the length of the cylinder, R is its radius, F is the force over it, and E^* is the equivalent Young's modulus given by $E^* = E/(1 - v^2)$.

Also, Hertz's theory provides the maximum pressure, p_0, and the stress distribution on the contact area as

$$p_0 = \left(\frac{E^* F}{\pi LR}\right)^{\frac{1}{2}}, \text{ and} \tag{7.4}$$

$$\sigma_z(x) = p_0 \sqrt{1 - \frac{x^2}{a^2}}, \tag{7.5}$$

where x is the horizontal distance measured from the symmetry plane along the cylinder, and σ_z the vertical stress.

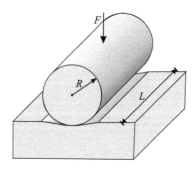

Figure 7.2 Schematic layout of the contact between a cylinder and a half-space.

On the other hand, the Lundberg solution, proposed in Ref. [44], allows obtaining the penetration's depth of the cylinder considering the three-dimensional effect at its edges. This penetration depth d is expressed as

$$d = \frac{2F}{\pi L E^*}\left[1.8864 + \ln\left(\frac{L}{2a}\right)\right].$$ (7.6)

7.1.3 The cone macroelement model

A simplified method for calculating the spring and dashpot constants was proposed by Wolf [68]. This solution assumes that the stresses produced by dynamic loads propagate into the soil within a truncated cone. According to this model, the dashpot coefficient for a rectangularly loaded area is

$$C_s = 4\sqrt{2\rho G \frac{1-\nu}{1-2\nu}} BL \qquad \text{for } \nu \le \frac{1}{3} \text{ or}$$
$$C_s = 8\sqrt{\rho G}BL \qquad \text{for } \frac{1}{3} < \nu \le \frac{1}{2},$$ (7.7)

where B is the width of the contact area (*i.e.*, $B = 2a$), and L is the length of the cylinder.

Depending on Poisson's ratio of the soil, a trapped mass can move together with the loaded area. The value of this trapped mass is

$$\Delta m = \frac{8}{\pi^{0.5}}\mu_m\rho\left(\frac{BL}{4}\right)^{3/2} \qquad \text{where} \quad \mu_m = 2.4\pi\left(\nu - \frac{1}{3}\right).$$

7.1.4 Continuous compaction control (CCC)

As described in *Geotechnics of Roads Fundamentals* [10], there are several parameters to interpret the performance of vibratory drum compactors. This chapter only uses two of these parameters: the Compaction Meter Value (CMV) and the Resonant Meter Value (RMV).

The CMV is a dimensionless parameter that relates the amplitude of the drum movement corresponding to the first harmonic, $A_{2\omega}$, with the amplitude in the fundamental frequency, A_ω, as follows:

$$\text{CMV} = C\frac{A_{2\omega}}{A_\omega},$$ (7.8)

where $C = 300$, according to Ref. [51].

On the other hand, the RMV has been proposed for identifying the occurrence of a double jump of the drum, which reduces the performance of the compaction process. This parameter is computed as

$$\text{RMV} = C\frac{A_{0.5\omega}}{A_\omega}.$$ (7.9)

7.2 EXAMPLE 22: SOIL–DRUM INTERACTION ASSUMING AN ELASTIC SOIL'S RESPONSE

The example solved below describes the mathematical procedure for computing the soil–drum interaction when the soil reacts in its elastic domain of behavior. This analysis permits to compute the parameters for the CCC methodology.

The characteristics of the vibratory drum compactor for this example are

- *drum length $L = 1.6\,m$,*
- *drum radius $R = 0.7\,m$,*
- *mass of the frame $m_f = 3,150\,kg$,*
- *mass of the drum $m_d = 600\,kg$,*
- *eccentric mass, m_e, variable,*
- *eccentricity of the mass $e = 0.3\,m$,*
- *operating frequency, ω, variable.*

The soil characteristics for this example are

- *Young's modulus, E =variable,*
- *Poisson's ratio, $\nu = 0.3$,*
- *density, $\rho = 2,200\,kg/m^3$.*

The procedure to analyze the soil–drum interaction under different operational conditions of the drum and different stiffness of the soil requires the following steps:

1 *Discretize in time the set of Equation 7.2 using the explicit finite differences method.*
2 *Analyze the interaction for different Young's moduli of the soil $E = 5, 20, 50, 100, 500\,MPa$. For this analysis, the maximum dynamic load applied by the drum corresponds to an eccentric mass of 7.4 kg rotating at a frequency of 30 Hz.*
3 *Assuming that Young's modulus of the soil is $E = 50\,MPa$, analyze the effect of the dynamic load obtained using a constant eccentric mass of 7.4 kg that rotates at frequencies of 10, 25, and 50 Hz.*

7.2.1 Discretization in time of the dynamic equation

The Explicit Finite Difference Method permits to discretize the set of Equation 7.2. This discretization is possible linearizing the first and second derivatives in time of the drum's displacement (*i.e.*, $\dot{z}_1$ and $\ddot{z}_1$) as follows:

$$\dot{z}_1 = \frac{z_1^{t+\Delta t} - z_1^t}{\Delta t} \quad \text{and} \quad \ddot{z}_1 = \frac{z_1^{t+\Delta t} - 2z_1^t + z_1^{t-\Delta t}}{\Delta t^2},$$

where Δt is the time-lapse of the discretization.

Therefore, using the explicit scheme, Equation 7.2 becomes

for $z_1 > 0$,

$$(m_d + m_e)\frac{z_1^{t+\Delta t} - 2z_1^t + z_1^{t-\Delta t}}{\Delta t^2} = (m_d + m_e + m_f)g + m_e e\omega^2 \sin \omega t - (k_s^e)^t z_1^t$$

$$- C_s^t \frac{z_1^{t+\Delta t} - z_1^t}{\Delta t},$$

or for $z_s = z_1 \geq 0$,

$$(m_d + m_e)\frac{z_1^{t+\Delta t} - 2z_1^t + z_1^{t-\Delta t}}{\Delta t^2} = (m_d + m_e + m_f)g + m_e e\omega^2 \sin \omega t.$$

This set of equations permits calculating the displacement $z_1^{t+\Delta t}$ when knowing the displacement at times t and $t - \Delta t$ as

for $z_1 > 0$,

$$z_1^{t+\Delta t} = A \left[(m_d + m_e + m_f)g + m_e e\omega^2 \sin \omega t - (k_s^e)^t z_1^t + (m_d + m_e)\frac{2z_1^t - z_1^{t-\Delta t}}{\Delta t^2} \right.$$

$$\left. + C_s^t \frac{z_1^t}{\Delta t} \right],$$

or for $z_1 \leq 0$,

$$z_1^{t+\Delta t} = B \left[(m_d + m_e + m_f)g + m_e e\omega^2 \sin \omega t + (m_d + m_e)\frac{2z_1^t - z_1^{t-\Delta t}}{\Delta t^2} \right], \qquad (7.10)$$

where variables A and B are

$$A = \frac{1}{\frac{m_d + m_e}{\Delta t^2} + \frac{C_s^t}{\Delta t}} \quad \text{and} \quad B = \frac{\Delta t^2}{m_d + m_e}.$$

As the displacement of the soil, z_s, occurs only in compression, it is

$$z_s^{t+\Delta t} = z_1^{t+\Delta t} \qquad \text{for} \quad z_1^{t+\Delta t} \geq 0, \text{ or}$$

$$z_s^{t+\Delta t} = 0 \qquad \text{for} \quad z_1^{t+\Delta t} < 0. \qquad (7.11)$$

Using an explicit scheme, the elastic component of the reaction force on the soil, F_s^e, is

$$\left(F_s^e \right)^{t+\Delta t} = (k_s^e)^t z_s^{t+\Delta t}. \qquad (7.12)$$

It is possible to obtain the spring coefficient representing the elastic response of the ground by combining the Hertz expression, which provides half the contact width of a cylinder resting on an elastic half-space, with the Lundberg equation, which provides the depth of cylinder penetration. In fact, according to Equation 7.3, half the contact width of a cylinder resting on an elastic half-space is

$$a^{t+\Delta t} = \left[\frac{4R \left(F_s^e \right)^{t+\Delta t} (1 - v^2)}{\pi L E} \right]^{\frac{1}{2}}, \qquad (7.13)$$

where L is the length of the cylinder, R its radius, E is the Young modulus of the soil, and v its Poisson ratio.

On the other hand, according to the Lundberg solution given in Equation 7.6, the penetration depth, d, of the cylinder is

$$d^{t+\Delta t} = \frac{2\left(F_s^e\right)^{t+\Delta t}(1-v^2)}{\pi LE}\left[1.8864 + \ln\left(\frac{L}{2a^{t+\Delta t}}\right)\right].$$

Therefore, the spring coefficient at time $t + \Delta t$ becomes

$$(k_s^e)^{t+\Delta t} = \frac{\left(F_s^e\right)^{t+\Delta t}}{d^{t+\Delta t}} = \frac{\pi LE}{2(1-v^2)\left[1.8864 + \ln\left(\frac{L}{2a^{t+\Delta t}}\right)\right]}. \tag{7.14}$$

Besides, the dashpot coefficient at time $t + \Delta t$ becomes

$$C_s^{t+\Delta t} = 8La^{t+\Delta t}\sqrt{2\rho G\frac{1-v}{1-2v}} \quad \text{for} \quad v \le \frac{1}{3}, \text{ or}$$

$$C_s^{t+\Delta t} = 16La^{t+\Delta t}\sqrt{\rho G} \quad \text{for} \quad \frac{1}{3} < v \le \frac{1}{2}. \tag{7.15}$$

Note that in the previous equations, the width, B, of Equation 7.7 is two times the width a given by the Hertz equation (i.e., $B = 2a$).

Given that the spring–dashpot constants are zero when $z_1 \le 0$, Equation 7.10 can be written in a simplified form as the following single equation:

$$z_1^{t+\Delta t} = A\left[F_{fde} + F_D \sin \omega t - \left(F_s^e\right)^t + M_{de}\frac{2z_1^t - z_1^{t-\Delta t}}{\Delta t^2} + C_s^t\frac{z_1^t}{\Delta t}\right], \tag{7.16}$$

where F_{fde} is the whole static load given by $F_{fde} = (m_d + m_e + m_f)g$, F_D is the peak dynamic vibratory load calculated as $F_D = m_e e \omega^2$, F_s^e is the elastic reaction of the soil calculated as $F_s^e = k_s^e z_s$, and M_{de} is the sum of the eccentric and drum masses given by $M_{de} = (m_d + m_e)$.

Equation 7.10 requires knowing the displacement z_1 for times t and $t - \Delta t$. As these displacements are unknown at the beginning of the movement, it is necessary to adopt the following initial values: $z_1(t = 0) = 0$ and $z_1(t = \Delta t) = 0$. Then, the computations begin at $t = 2\Delta t$.

In summary, the following procedure permits to compute the displacement of the drum and the soil:

1   Allocate values of z_1 for $t = 0$ and $t = \Delta t$,
2   Compute $z_1^{t+\Delta t}$ using Equation 7.16,
3   Compute $z_s^{t+\Delta t}$ using Equation 7.11,
4   Compute $\left(F_s^e\right)^{t+\Delta t}$ using Equation 7.12,
5   Compute $a^{t+\Delta t}$ using Equation 7.13,
6   Compute $(k_s^e)^{t+\Delta t}$ using Equation 7.14,
7   Compute $C_s^{t+\Delta t}$ using Equation 7.15,
8   Return to step 2 for computing the displacement of the subsequent time step.

7.2.2 Effect of Young's modulus on the soil–drum interaction

This section describes the procedure for computing the soil–drum interaction for different Young's modulus. For this example, the characteristics of the drum are $L = 1.6$ m, $R = 0.7$ m, $m_f = 3150$ kg, $m_d = 600$ kg, $m_e = 7.4$ kg, $e = 0.3$ m, and the vibratory frequency 30 Hz.

Regarding the soil, the density is $\rho = 2,200$ kg/m^3 and its Poisson's ratio $\nu = 0.3$. Besides, for a detailed explanation, let's consider the case of soil with a Young modulus of $E = 5$ MPa, which corresponds to a shear modulus of $G = \frac{5}{2(1+0.3)} = 1.923$ MPa. Besides, it is possible to adopt a time step of $\Delta t = 10^{-4}$ s.

Static forces and masses, which are constants throughout the computational procedure, are

sum of the static loads

$F_{fde} = (m_d + m_e + m_f)g, = (600 + 7.4 + 3150)9.81 = 36860.1$ N,

angular frequency

$\omega = 2\pi \cdot 30 = 188.496$ rad/s,

peak dynamic force

$F_D = m_e e \omega^2 = 7.4 \cdot 0.3 \cdot 188.496^2 = 78877.9$ N, and

sum of drum and eccentric mass

$M_{de} = (m_d + m_e) = 600 + 7.4 = 607.4$ kg.

As described above, the half of the contact width a is

$$a^{t+\Delta t} = \left[\frac{4 \cdot 0.7 \cdot 0.91 \left(F_s^e\right)^{t+\Delta t}}{\pi \cdot 1.6 \cdot 5 \cdot 10^6} \right]^{\frac{1}{2}} = 3.184 \cdot 10^{-4} \left[(F_s^e)^{t+\Delta t}\right]^{\frac{1}{2}} \text{ m,}$$

and then the spring coefficient becomes

$$(k_s^e)^{t+\Delta t} = \frac{\pi \cdot 1.6 \cdot 5 \cdot 10^6}{2(1-0.3^2)\left[1.8864 + \ln\left(\frac{1.6}{2a^{t+\Delta t}}\right)\right]} = \frac{1.3809 \cdot 10^7}{1.8864 + \ln\left(\frac{1.6}{2a^{t+\Delta t}}\right)} \text{ N/m.}$$

This last evaluation could lead to instabilities when $F_s^e \to 0$, because in this case $a \to 0$. Therefore, it is useful to adopt a minimum value for a when $z_s > 0$; this example uses a value of $a_{min} = 10^{-6}$ m.

As Poisson's ratio of the soil is $\nu < 1/3$, the dashpot coefficient becomes

$$C_s^{t+\Delta t} = 8 \cdot 1.6 \cdot a^{t+\Delta t}\sqrt{2 \cdot 2,200 \cdot 1.923 \cdot 10^6 \frac{1-0.3}{1-2\cdot 0.3}} = 1.558 \cdot 10^6 a^{t+\Delta t} \text{ Ns/m.}$$

The previous numerical values permit to compute the evolution on time of the displacement of the drum as follows:

$$z_1^{t+\Delta t} = \frac{1}{\frac{M_{de}}{\Delta t^2} + \frac{C_s^t}{\Delta t}}\left[F_{fde} + F_D\sin\omega t - (F_s^e)^t + M_{de}\frac{2z_1^t - z_1^{t-\Delta t}}{\Delta t^2} + C_s^t\frac{z_1^t}{\Delta t}\right],$$

using $\Delta t = 10^{-4}$ s.

Table 7.1 Results of the computation of the drum–soil interaction for the ten initial time steps.

t (s)	$F_D \sin(\omega t)$ (N)	z_1 (m)	z_s (m)	F_s^e (N)	a (m)	k_s^e (N/m)	C_s (Ns/m)	F_s (N)
0	0	0	0	0	0	0	0	0
Δt	$1.49{\cdot}10^3$	0	0	0	0	0	0	0
$2\Delta t$	$2.97{\cdot}10^3$	$6.56{\cdot}10^{-7}$	$6.56{\cdot}10^{-7}$	0	$1.00{\cdot}10^{-6}$	$8.92{\cdot}10^5$	1.56	0
$3\Delta t$	$4.46{\cdot}10^3$	$1.99{\cdot}10^{-6}$	$1.99{\cdot}10^{-6}$	1.78	$4.24{\cdot}10^{-4}$	$1.46{\cdot}10^6$	$6.61{\cdot}10^2$	$2.08{\cdot}10^{-2}$
$4\Delta t$	$5.94{\cdot}10^3$	$4.03{\cdot}10^{-6}$	$4.03{\cdot}10^{-6}$	5.91	$7.74{\cdot}10^{-4}$	$1.56{\cdot}10^6$	$1.21{\cdot}10^3$	$1.53{\cdot}10$
$5\Delta t$	$7.42{\cdot}10^3$	$6.80{\cdot}10^{-6}$	$6.80{\cdot}10^{-6}$	10.6	$1.04{\cdot}10^{-3}$	$1.62{\cdot}10^6$	$1.62{\cdot}10^3$	$3.93{\cdot}10$
$6\Delta t$	$8.90{\cdot}10^3$	$1.03{\cdot}10^{-5}$	$1.03{\cdot}10^{-5}$	16.7	$1.30{\cdot}10^{-3}$	$1.66{\cdot}10^6$	$2.03{\cdot}10^3$	$6.76{\cdot}10$
$7\Delta t$	$1.04{\cdot}10^4$	$1.46{\cdot}10^{-5}$	$1.46{\cdot}10^{-5}$	24.3	$1.57{\cdot}10^{-3}$	$1.70{\cdot}10^6$	$2.44{\cdot}10^3$	$1.04{\cdot}10^2$
$8\Delta t$	$1.18{\cdot}10^4$	$1.97{\cdot}10^{-5}$	$1.97{\cdot}10^{-5}$	33.5	$1.84{\cdot}10^{-3}$	$1.73{\cdot}10^6$	$2.87{\cdot}10^3$	$1.49{\cdot}10^2$
$9\Delta t$	$1.33{\cdot}10^4$	$2.56{\cdot}10^{-5}$	$2.56{\cdot}10^{-5}$	44.5	$2.12{\cdot}10^{-3}$	$1.77{\cdot}10^6$	$3.31{\cdot}10^3$	$2.04{\cdot}10^2$
$10\Delta t$	$1.48{\cdot}10^4$	$3.24{\cdot}10^{-5}$	$3.24{\cdot}10^{-5}$	57.2	$2.41{\cdot}10^{-3}$	$1.80{\cdot}10^6$	$3.75{\cdot}10^3$	$2.68{\cdot}10^2$
$\vdots$	$\vdots$	$\vdots$	$\vdots$	$\vdots$	$\vdots$	$\vdots$	$\vdots$	$\vdots$

After computing displacements and forces for the time $t+\Delta t$, it is also useful to compute the whole reaction of the soil, F_s, which includes the elastic and the viscous response. Considering the dashpot coefficient, the viscoelastic reaction of the soil becomes

$$F_s^{t+\Delta t} = (F_s^e)^{t+\Delta t} + C_s^{t+\Delta t}\frac{z_s^{t+\Delta t} - z_s^t}{\Delta t}.$$

The procedure for calculating the soil–drum interaction described in this example can be easily implemented in a spreadsheet, such as in Table 7.1 which shows the results of the first ten time steps of this example.

This procedure can be repeated for different Young's moduli, and the results analyzed in different ways: the transient and frequency responses, as well as the hysteretic cycles that relate forces and displacements.

Figure 7.3 shows the results of the simulations of the interaction between the drum and soils having different Young's moduli. Figure 7.3a–e shows the results in the time domain, while Figure 7.3f–j shows the results in the frequency domain.

These figures allow analyzing the change in soil–drum interaction as soil's stiffness increases as follows:

- First, for soils of low stiffness, such as in Figure 7.3a and f, the drum is in continuous contact with the soil. In the frequency domain, the Fourier transform of this movement is harmonic, and the vibration frequency coincides with the excitation frequency of 30 Hz.

- As the soil's stiffness increases, the movement becomes bumpy, and therefore the drum intermittently lifts off the soil hitting it repeatedly, as shown in Figure 7.3b and c. In the frequency domain, this movement has two characteristic frequencies corresponding to the excitation frequency, ω, and twice this frequency, 2ω, as in Figure 7.3g and h. This mode of operation produces higher loads; thus, it is optimal for compacting soils, but continuous contact is preferred for compacting asphalt materials [8].

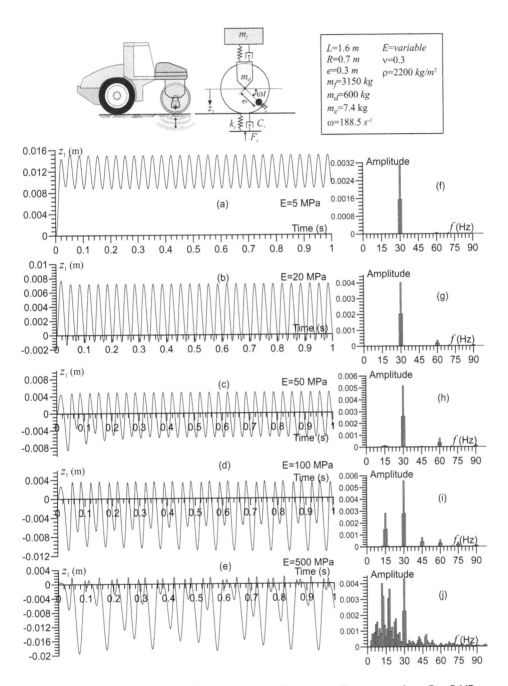

Figure 7.3 Movement of a vibratory drum resting on soils whose stiffness varies from $E = 5$ MPa to $E = 500$ MPa, (a to e) results in the time domain, (f to j) results in the frequency domain. Characteristics of the vibratory drum: $L = 1.6$ m, $R = 0.7$ m, mass of the frame 3,150 kg, mass of the drum 600 kg, eccentric mass 7.4 kg, vibratory frequency 30 Hz. Characteristics of the soil $v = 0.3$ $\rho = 2,200$ kg/m³, elastic response.

- When Young's modulus reaches high values, as shown in Figure 7.3e, the soil's reaction produces a chaotic movement of the drum. This movement is characterized by several harmonics that are proportional to half of the excitation frequency, as shown in Figure 7.3j; such mode of operation may destroy the compacted layer and damage the whole compactor [8].

Linking the hysteretic cycles of dynamic loading graphed against drum displacement, and soil reaction graphed against soil displacement also allow analyzing the soil–drum interaction. Figure 7.4a–e and 7.4f–j shows these hysteretic cycles obtained for the different Young's moduli of this example. These figures were drawn when the movement reaches a periodic response (i.e., $t > 0.5$ s for all cases in the example). It is possible to observe in these figures how the hysteretic cycle, that relates the dynamic load and the displacement of the drum, passes from a cycle with an elliptical shape that corresponds to a viscoelastic response in Figure 7.4a, into a chaotic response in Figure 7.4e.

The CMV and RMV allow evaluating the compaction performance. These parameters require identifying the amplitude of the first harmonic, $A_{2\omega}$, the amplitude of the fundamental frequency, A_ω, and the amplitude of half of the fundamental frequency $A_{0.5\omega}$. All these amplitudes, that come from the Fourier transform of the movement shown in Figure 7.3i and j, are reported in Table 7.2. Then, the CMV and the RMV result from Equations 7.8 and 7.9 as

$$CMV = 300\frac{A_{2\omega}}{A_\omega}, \text{ and}$$

$$RMV = 300\frac{A_{0.5\omega}}{A_\omega}.$$

Figure 7.5 shows the results of the CMV and the RMV for the different Young's moduli. These results suggest a linear relationship between Young's modulus and the CMV, as it has been identified in field research. However, this linear relationship is only valid before the beginning of the double jump behavior of the drum. The RMV makes it possible to identify this limit of validity; in fact, when the RMV grows, it indicates the appearance of a double jump movement.

7.2.3 Effect of the dynamic load

In this section, the effect of the dynamic load is analyzed, remaining constant the eccentric mass m_e and its distance to the axle e, and modifying the vibratory frequency. The methodology for computing the soil–drum interaction for different frequencies follows the same procedure described in the precedent section of this example, and the only changes are the vibratory frequency ω and the peak dynamic force F_D, which are

$$10 \text{ Hz} \longrightarrow \omega = 10 \cdot 2\pi = 62.83 \text{ s}^{-1} \longrightarrow F_D = m_e e \omega^2 = 8764.2 \text{ N},$$

$$25 \text{ Hz} \longrightarrow \omega = 25 \cdot 2\pi = 157.08 \text{ s}^{-1} \longrightarrow F_D = m_e e \omega^2 = 54776.3 \text{ N},$$

$$50 \text{ Hz} \longrightarrow \omega = 50 \cdot 2\pi = 314.16 \text{ s}^{-1} \longrightarrow F_D = m_e e \omega^2 = 219105.2 \text{ N}.$$

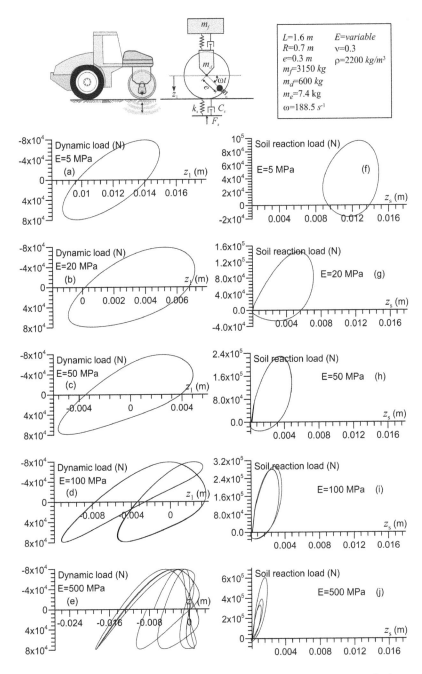

Figure 7.4 Dynamic loading and soil reactions for a vibratory drum resting on a soil whose stiffness varies from $E = 5$ MPa to $E = 100$ MPa, (a to e) dynamic load and drum's displacement, (f to j) soil reaction and displacement. Characteristics of the vibratory drum: $L = 1.6$ m, $R = 0.7$ m, mass of the frame 3,150 kg, mass of the drum 600 kg, eccentric mass 7.4 kg, vibratory frequency 30 Hz. Characteristics of the soil $v = 0.3$ $\rho = 2,200$ kg/m³, elastic response.

Table 7.2 Amplitudes for the frequencies 0.5ω, ω, and 2ω depending on Young's moduli, and CMV and RMV values

Frequency		E (MPa)				
		5	20	50	100	500
0.5ω	15 Hz	6.080·10^{-7}	1.570·10^{-6}	1.330·10^{-4}	2.814·10^{-3}	6.100·10^{-4}
1ω	30 Hz	3.084·10^{-3}	4.015·10^{-3}	5.165·10^{-3}	5.534·10^{-3}	4.312·10^{-3}
2ω	60 Hz	4.460·10^{-5}	3.290·10^{-4}	7.140·10^{-4}	5.630·10^{-4}	2.070·10^{-4}
	CMV	4.34	24.6	41.5	30.5	14.4
	RMV	0.059	0.117	7.73	153	42.4

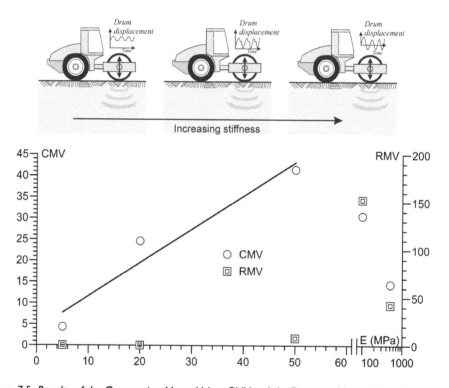

Figure 7.5 Results of the Compaction Meter Value, CMV and the Resonant Meter Value, RMV for the different Young's moduli.

Figures 7.6 and 7.7 shows the result of these calculations in the time and frequency domains as well as the hysteretic cycles. It is essential to keep in mind that chaotic movement can also appear when the dynamic load grows. In fact, smaller dynamic loads produce drum behavior in the continuous contact range, while, as the dynamic load grows, drum behavior progresses toward chaotic motion.

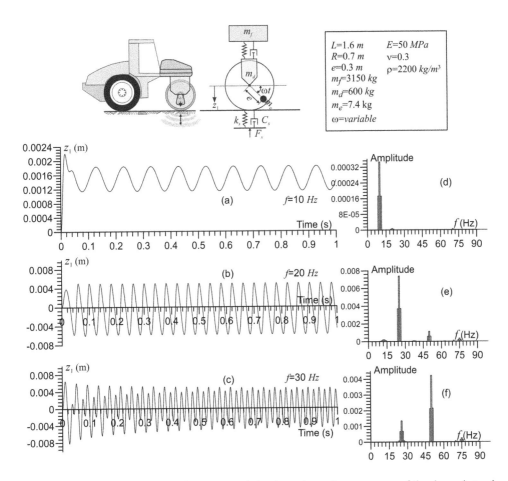

Figure 7.6 Effects of the vibratory frequency and the dynamic on the movement of the drum, (a to c) displacement in the time domain, (d to f) displacement in the frequency domain. Characteristics of the vibratory drum: $L = 1.6$ m, $R = 0.7$ m, mass of the frame 3,150 kg, mass of the drum 600 kg, eccentric mass 7.4 kg, vibratory frequency 10, 25, and 50 Hz. Characteristics of the soil $E = 50$ MPa, $\nu = 0.3$ $\rho = 2,200$ kg/m³, elastic response.

The previous section suggests a possible linear relationship between CMV and Young' modulus. However, this relationship depends not only on the rigidity of the soil but also on the characteristics of the drum. In fact, as shown in Figure 7.8, the CMV and the RMV change even when Young's modulus is kept constant. It is essential to take this evidence into account when performing field calibrations.

In this example, the soil remains in its elastic domain of behavior throughout the simulation. However, the assumption of elastic behavior is contradictory with the intention of the compaction process that is to produce plastic deformations. The following example describes the methodology to take into account when plastic deformations appear.

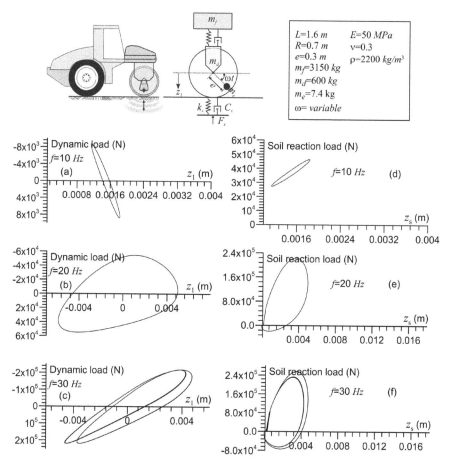

Figure 7.7 Effects of the vibratory frequency and the dynamic load on the dynamic loading and soil reaction, (a to c) hysteretic cycles of dynamic load and drum's displacement, (d to f) hysteretic cycles of soil reaction and displacement. Characteristics of the vibratory drum: $L = 1.6$ m, $R = 0.7$ m, mas of the frame 3,150 kg, mass of the drum 600 kg, eccentric mass 7.4 kg, vibratory frequency 10, 25, and 50 Hz. Characteristics of the soil $E = 50$ MPa, $\nu = 0.3$ $\rho = 2,200$ kg/m^3, elastic response.

7.3 EXAMPLE 23: ANALYSIS OF THE SOIL–DRUM INTERACTION CONSIDERING THE SOIL'S REACTION INTO THE ELASTOPLASTIC DOMAIN OF BEHAVIOR

From a mechanical point of view, the purpose of compaction is to increase the density of the material, which requires producing plastic deformations in the material. This is a requirement that is in contradiction with an elastic behavior of the soil.

The following example develops a method to calculate the soil–drum interaction splitting the soil displacement into its elastic d^e and plastic d^p components, as in the theory of elastoplasticity.

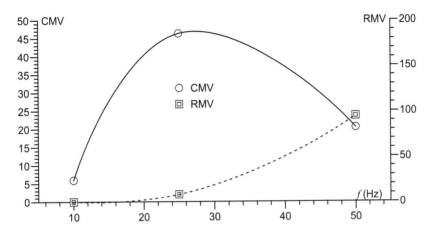

Figure 7.8 Results of the Compaction Meter Value, CMV, and the Resonant Meter Value, RMV for the different dynamic loads and constant Young's modulus.

The characteristics of the drum are similar than in the previous example:

- *drum length $L = 1.6\,m$,*
- *drum radius $R = 0.7\,m$,*
- *mass of the frame $m_f = 3,150\,kg$,*
- *mass of the drum $m_d = 600\,kg$,*
- *eccentric mass, m_e, variable,*
- *eccentricity of the mass $e = 0.3\,m$,*
- *operating frequency 30 Hz.*

While soil's properties are

- *Young's modulus, $E = 100\,MPa$,*
- *Poisson's ratio, $v = 0.3$,*
- *density, $\rho = 2,200\,kg/m^3$,*
- *yield stress P_y, variable and Tresca yielding criterion.*

The analysis is divided into three parts:

1   *Analyze the elastoplastic soil–drum contact under monotonic loading growing up to 10 kN, and compare the results for two yielding stresses: $p_y = 100\,kPa$ and $p_y = 300\,kPa$.*

2   *Using a yielding stress $p_y = 100\,kPa$, analyze the behavior for one loading-unloading cycle up to 10 kN and then unloading to zero.*

3   *Analyze the dynamic soil–drum interaction for three yielding stresses: $p_y=300$, 600 and 800 kPa.*

7.3.1 Contact soil–drum under monotonic loading and elastoplastic behavior

Hertzian contacts produce high stresses along the symmetry axis of the contact area. However, the maximum contact stress is limited by the yield stress. Therefore, when reaching the yield stress, the elliptical shape of the Hertzian diagram of contact stresses turn into a truncated ellipse, as it is shown in Figure 7.9 [15,59]. Besides, as the maximum stress decreases from p_0 to p_y, the contact width required to sustain the external load increases from the elastic contact width B^e, and becomes an elastoplastic contact width of B^{ep}.

The elastoplastic contact width B^{ep} can be calculated by solving the equations that give the area of the stress diagram for a truncated and complete ellipse, as described in Figure 7.10.

Dimensions of the ellipse and the elliptical sector shown in Figure 7.10 in terms of contact width and stresses are $X = B$, $Y = 2p_0$ and $f = p_0 - p_y$, and so the area of the ellipse is

$$A_e = \frac{\pi B p_0}{2}.$$

Therefore, in the elastic domain, the load applied by the cylinder, which is F_s^e, is the product of the area of a half of an ellipse and the cylinder's length given by

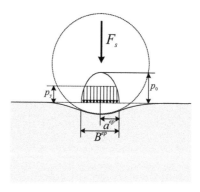

Figure 7.9 Schematic drawing that describes the elastoplastic soil–drum contact.

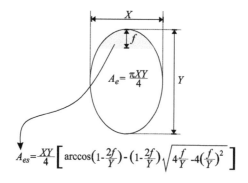

Figure 7.10 Areas of an ellipse and an elliptical sector.

$$F_s^e = \frac{\pi B^e L p_0}{4}.$$

Besides, the area above the yield stress (i.e., the shaded area in Figure 7.10) is

$$A_{es} = \frac{B p_0}{2} \left[\arccos\left(1 - \frac{p_0 - p_y}{p_0}\right) - \left(1 - \frac{p_0 - p_y}{p_0}\right) \sqrt{\frac{2(p_0 - p_y)}{p_0} - \left(\frac{p_0 - p_y}{p_0}\right)^2} \right].$$

Therefore, the load F_s^{ep} applied by the cylinder on the soil when it behaves in its elasto-plastic domain is given by subtracting the area of half of an ellipse minus the shaded area in Figure 7.10 (i.e., $1/2 A_e - A_{es}$) leading to the area of a truncated ellipse, which, multiplied by the cylinder length L gives the following elastoplastic load:

$$F_s^{ep} = L\left(\frac{1}{2} A_e - A_{es}\right),$$

becoming

$$F_s^{ep} = \frac{p_0 B^{ep} L}{2} \left\{ \frac{\pi}{2} - \left[\arccos\left(1 - \frac{p_0 - p_y}{p_0}\right) \right. \right.$$
$$\left. \left. - \left(1 - \frac{p_0 - p_y}{p_0}\right) \sqrt{\frac{2(p_0 - p_y)}{p_0} - \left(\frac{p_0 - p_y}{p_0}\right)^2} \right] \right\}.$$

The change of the contact width, $B^e \rightarrow B^{ep}$, can be calculated by spreading the load given by the elliptical stress distribution, which is the load that the soil would bear in the elastic domain of behavior, into the truncated elliptical area correspond-ing to the elastoplastic response. This computation is possible equating $F_s^e = F_s^{ep}$ as follows:

$$\frac{p_0 B^{ep} L}{2} \left\{ \frac{\pi}{2} - \left[\arccos\left(1 - \frac{p_0 - p_y}{p_0}\right) \right. \right.$$
$$\left. \left. - \left(1 - \frac{p_0 - p_y}{p_0}\right) \sqrt{\frac{2(p_0 - p_y)}{p_0} - \left(\frac{p_0 - p_y}{p_0}\right)^2} \right] \right\} = \frac{\pi p_0 B^e L}{4}.$$

Thus, the ratio between the elastoplastic and elastic contact widths is

$$\frac{B^{ep}}{B^e} = \frac{a^{ep}}{a^e} = \frac{\pi}{2\left\{ \frac{\pi}{2} - \left[\arccos\left(1 - \frac{p_0 - p_y}{p_0}\right) - \left(1 - \frac{p_0 - p_y}{p_0}\right) \sqrt{\frac{2(p_0 - p_y)}{p_0} - \left(\frac{p_0 - p_y}{p_0}\right)^2} \right] \right\}}.$$

For simplicity, this equation can be written as

$$a^{ep} = a^e f\left(\frac{p_0 - p_y}{p_0}\right), \tag{7.17}$$

where a^e is given by Equation 7.3 and $f\left(\frac{p_0 - p_y}{p_0}\right)$ is

$$f\left(\frac{p_0 - p_y}{p_0}\right) = \frac{0.5\pi}{\frac{\pi}{2} - \left[\arccos\left(1 - \frac{p_0 - p_y}{p_0}\right) - \left(1 - \frac{p_0 - p_y}{p_0}\right)\sqrt{\frac{2(p_0 - p_y)}{p_0} - \left(\frac{p_0 - p_y}{p_0}\right)^2}\right]}.$$

(7.18)

According to Equation 7.18, the elastoplastic contact width B^{ep} grows as the ratio $\frac{p_0 - p_y}{p_0}$ increases. Figure 7.11 illustrates the growing of the elastoplastic contact width that grows to infinity when p_y approaches zero.

The elastoplastic distribution of the vertical contact stresses, σ_z, is given by Equation 7.5, but truncated at a maximum value p_y which is the yield stress, as follows:

$$\sigma_z(x) = \min\left(p_0\sqrt{1 - \frac{x^2}{(a^{ep})^2}}, p_y\right),$$

where p_0 is the maximum contact stress given by Equation 7.4, and x is the horizontal distance.

According to the theory of elastoplasticity, the elastoplastic vertical displacement or, in other words, the elastoplastic indentation d^{ep}, is the sum of the elastic and plastic indentations.

$$d^{ep} = d^e + d^p.$$

On the other hand, the Lundberg solution, given in Equation 7.6, allows calculating the elastic indentation, d^e as

$$d^e = \frac{2F_s(1 - v^2)}{\pi LE}\left[1.8864 + \ln\left(\frac{L}{2a^e}\right)\right].$$

(7.19)

The plastic indentation d^p comes from a geometrical construction. In fact, when the cylinder penetrates the soil, both soil and cylinder remain in contact between them.

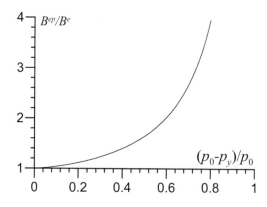

Figure 7.11   Relationship between elastic and elastoplastic contact widths.

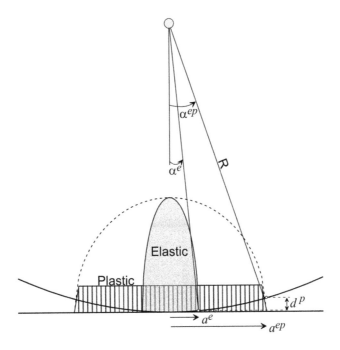

Figure 7.12 Geometrical derivation of the plastic indentation, d^p, resulting from the increase in the contact width.

Then, the contact width grows up to a distance of a^{ep} which, based on the geometrical construction shown in Figure 7.12, becomes

$$d^p = a^{ep} \tan \alpha^{ep} - a^e \tan \alpha^e,$$

where α^e and α^{ep} are angles formed by the vertical axis and the lines that join the edge of the elastic and plastic contact widths with the center of the cylinder.

However, as $\tan \alpha^{ep} \approx a^{ep}/R$ and $\tan \alpha^e \approx a^e/R$, then

$$d^p = \frac{(a^{ep})^2 - (a^e)^2}{R}.$$

And also $a^{ep} = a^e f \left(\frac{p_0 - p_y}{p_0} \right)$, and thus the plastic indentation becomes

$$d^p = \frac{(a^e)^2}{R} \left[f \left(\frac{p_0 - p_y}{p_0} \right)^2 - 1 \right] = \frac{4F_s(1 - v^2)}{\pi LE} \left[f \left(\frac{p_0 - p_y}{p_0} \right)^2 - 1 \right]. \qquad (7.20)$$

Finally, the elastoplastic indentation is

$$d^{ep} = d^e + d^p.$$

Table 7.3 describes the detailed calculations for a load applied over the cylinder of $F_s = 6$ kN, on a soil that could have the two possibilities of yielding stresses given for this example ($p_y = 100$ or $p_y = 300$ kPa).

Table 7.3 Detailed computations of the elastoplastic contact for a load of $F_s = 6$ kN, characteristics of the cylinder ($L = 1.6$ m, $R = 0.7$ m), and soil properties ($E = 100$ MPa, $\nu = 0.3$, $p_y = 100$ or $p_y = 300$ kPa)

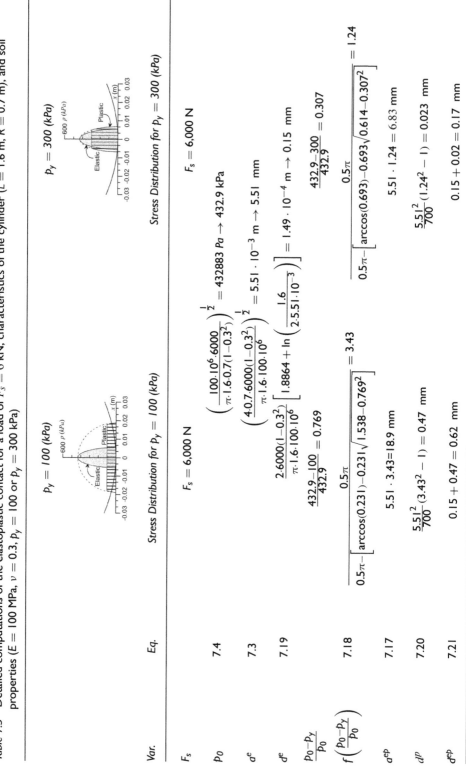

Var.	Eq.	Stress Distribution for $p_y = 100$ (kPa)	Stress Distribution for $p_y = 300$ (kPa)
F_s		$F_s = 6,000$ N	$F_s = 6,000$ N
p_0	7.4	$\left(\dfrac{100 \cdot 10^6 \cdot 6000}{\pi \cdot 1.6 \cdot 0.7(1-0.3^2)}\right)^{\frac{1}{2}} = 432883\ Pa \rightarrow 432.9$ kPa	
σ^e	7.3	$\left(\dfrac{4 \cdot 0.7 \cdot 6000(1-0.3^2)}{\pi \cdot 1.6 \cdot 100 \cdot 10^6}\right)^{\frac{1}{2}} = 5.51 \cdot 10^{-3}\ m \rightarrow 5.51$ mm	
δ^e	7.19	$\dfrac{2 \cdot 6000(1-0.3^2)}{\pi \cdot 1.6 \cdot 100 \cdot 10^6}\left[1.8864 + \ln\left(\dfrac{1.6}{2 \cdot 5.51 \cdot 10^{-3}}\right)\right] = 1.49 \cdot 10^{-4}\ m \rightarrow 0.15$ mm	
$\dfrac{p_0-p_y}{p_0}$		$\dfrac{432.9-100}{432.9} = 0.769$	$\dfrac{432.9-300}{432.9} = 0.307$
$f\left(\dfrac{p_0-p_y}{p_0}\right)$	7.18	$\dfrac{0.5\pi}{0.5\pi-\left[\arccos(0.231)-0.231\sqrt{1.538-0.769^2}\right]} = 3.43$	$\dfrac{0.5\pi}{0.5\pi-\left[\arccos(0.693)-0.693\sqrt{0.614-0.307^2}\right]} = 1.24$
d^{ep}	7.17	$5.51 \cdot 3.43 = 18.9$ mm	$5.51 \cdot 1.24 = 6.83$ mm
d^{lp}	7.20	$\dfrac{5.51^2}{700}(3.43^2 - 1) = 0.47$ mm	$\dfrac{5.51^2}{700}(1.24^2 - 1) = 0.023$ mm
d^{ep}	7.21	$0.15 + 0.47 = 0.62$ mm	$0.15 + 0.02 = 0.17$ mm

On unloading, only the elastic indentation d_e is retrieved. Figure 7.13 shows, for the two yielding stresses ($p_y = 100$ and $p_y = 300$ kPa), the relationship force–displacement when loading up to 10 kN and then unloading. This figure clearly shows the unrecoverable plastic displacement that remains after unloading that are 1.33 mm for the yielding stress of $p_y = 100$ kPa, and 0.1 mm for the yielding stress of $p_y = 300$ kPa. Besides, Figure 7.13b–g shows the elastic and plastic stress distributions on the contact area for three different loads $F_s = 2$, 6, and 10 kN and the two yielding stresses. These figures evidence how the spreading of stresses increases as the yielding stress decreases.

The relationship between force and displacement gives the spring coefficient. Therefore, Equation 7.19 and 7.20, that give the elastic and plastic indentations, permit to obtain the following expression for the elastoplastic spring coefficient:

$$k_s^{ep} = \frac{F_s}{d^e + d^p} = \frac{\pi L E}{2(1 - v^2) \left\{ \left[1.8864 + \ln \left(\frac{L}{2a^e} \right) \right] + 2 \left[f \left(\frac{p_0 - p_y}{p_0} \right)^2 - 1 \right] \right\}}. \qquad (7.21)$$

Note that when the function $f \left(\frac{p_0 - p_y}{p_0} \right) = 1$ that occurs when $p_0 = p_y$, the elastoplastic component in Equation 7.21 is zero, thus leading to the already-known elastic spring coefficient that is

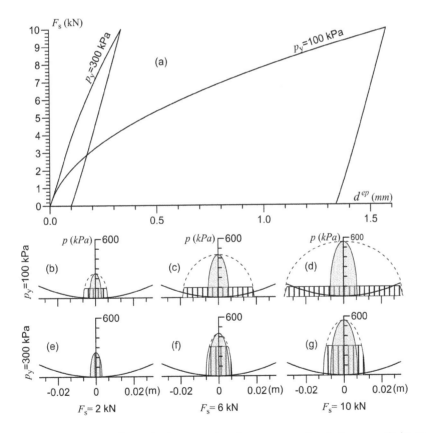

Figure 7.13 (a) Relationships force–displacement for the contact soil–cylinder on soils having two yielding stresses ($p_y = 100$ and $p_y = 300$ kPa), (b–d) elastic and plastic stress distributions for $p_y = 100$ kPa, and (e–g) elastic and plastic stress distributions for $p_y = 300$ kPa.

$$k_s^e = \frac{\pi LE}{2(1 - \nu^2)\left[1.8864 + \ln\left(\frac{L}{2a^e}\right)\right]}.$$

7.3.2 Cyclic loading with elastoplastic soil's response

It is important to note that the expressions for k_s^{ep} and k_s^{ep} are tangent formulations of the spring coefficients starting at zero load and displacement. This consideration requires special treatment when a cyclic load is applied. As shown in Figure 7.14, it is important to record the accumulated plastic displacement d^{pc} which is the plastic displacement when the unloading phase starts.

Taking Figure 7.14 as an example of cyclic loading, the accumulated plastic displacement increases as follows:

- The accumulated plastic displacement begins at zero when the load begins in cycle 1, $d^{pc} = 0$.
- At the end of the loading phase of cycle 1, and when unloading starts, the accumulated plastic displacement shifts to $d^{pc} = d^{pc1}$.
- The value of the accumulated plastic displacement remains constant, equal to d^{pc1}, during the unloading phase of cycle 1 and the reloading phase of cycle 2.
- Then, the accumulated plastic displacement shifts again when beginning the unloading of cycle 2, $d^{pc} = d^{pc2}$.
- This behavior continues for the subsequent cycles of loading and unloading.

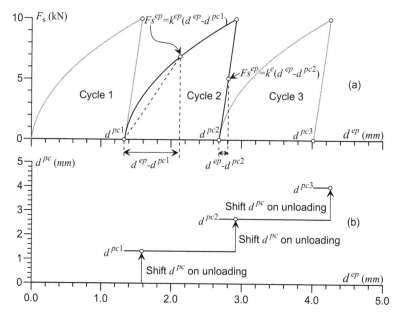

Figure 7.14 (a) Loading-unloading cycles for a soil having a yield stress of $p_y = 100$ kPa and (b) evolution of the accumulated plastic displacement d^{pc} on loading and unloading.

Therefore, it is possible to compute the elastoplastic reaction of the soil, F_s, considering the tangent formulations for k^{ep} and k^e, as follows:

$$F_s = k_s^{ep}(d^{ep} - d^{pc}) \text{ for loading, or}$$
$$F_s = k_s^e(d^{ep} - d^{pc}) \text{ for unloading.}$$

Equation 7.19 and 7.20 are also useful for obtaining the relationship between plastic and elastoplastic indentations as

$$f_d = \frac{d^p}{d^{ep}} = \frac{2\left[f\left(\frac{p_0 - p_y}{p_0}\right)^2 - 1\right]}{\left[1.8864 + \ln\left(\frac{L}{2a^e}\right)\right] + 2\left[f\left(\frac{p_0 - p_y}{p_0}\right)^2 - 1\right]}. \tag{7.22}$$

Equation 7.22 is also a tangent formulation. Therefore, the expression for computing the plastic displacement is

$$d^p = f_d(d^{ep} - d^{pc}).$$

7.3.3 Dynamic soil–drum interaction considering the elastoplastic contact

The discretization in time of Equation 7.2, which leads to Equation 7.10, remains valid for the elastoplastic response. However, it is important to keep in mind that the reaction force of the soil changes from the elastic reaction into the elastoplastic reaction.

The procedure for the numerical computation of the soil–drum interaction, considering the elastoplastic response, is described below.

1   Allocate values of $z_1 = 0$ for $t = 0$ and $t = \Delta t$.
2   Calculate the displacement $z_1^{t+\Delta t}$ as

$$z_1^{t+\Delta t} = A\left[F_{fde} + F_D \sin \omega t - \left(F_s^{ep}\right)^t + M_{de}\frac{2z_1^t - z_1^{t-\Delta t}}{\Delta t^2} + C_s^t \frac{z_1^t}{\Delta t}\right].$$

3   Calculate the soil's displacement $z_s^{t+\Delta t}$ as

$$z_s^{t+\Delta t} = z_1^{t+\Delta t} \quad \text{for} \quad z_1^{t+\Delta t} \geq \left(d^{pc}\right)^t \text{ or}$$
$$z_s^{t+\Delta t} = \left(d^{pc}\right)^t \quad \text{for} \quad z_1^{t+\Delta t} < \left(d^{pc}\right)^t.$$

4   Calculate the plastic displacement $(d^p)^{t+\Delta t}$ as

$$\left(d^p\right)^{t+\Delta t} = \left(\frac{d^p}{d^{ep}}\right)^t \left[z_s^{t+\Delta t} - \left(d^{pc}\right)^t\right].$$

5   Calculate the accumulated plastic displacement $(d^{pc})^{t+\Delta t}$ as follows:

on loading $\rightarrow$ $z_1^{t+\Delta t} > z_1^t$ then
$$\left(d^{pc}\right)^{t+\Delta t} = \left(d^{pc}\right)^t,$$
on starting unloading $\rightarrow$ $z_1^{t+\Delta t} < z_1^t$, and $z_1^{t-\Delta t} < z_1^t$ then
$$\left(d^{pc}\right)^{t+\Delta t} = \left(d^{pc}\right)^t + \left(d^p\right)^{t+\Delta t}.$$

6   Calculate the elastoplastic soil's reaction $\left(F_s^{ep}\right)^{t+\Delta t}$ as follows:

on loading → $z_1^{t+\Delta t} > z_1^t$ and $z_1^{t+\Delta t} > \left(d^{pc}\right)^t$ let

$\left(F_s^{ep}\right)^{t+\Delta t} = \left(k_s^{ep}\right)^t \left[\left(z_s^{ep}\right)^{t+\Delta t} - \left(d^{pc}\right)^t\right],$

on unloading → $z_1^{t+\Delta t} < z_1^t$ and $z_1^{t+\Delta t} > \left(d^{pc}\right)^t$ let

$\left(F_s^{ep}\right)^{t+\Delta t} = \left(k_s^e\right)^t \left[\left(z_s^{ep}\right)^{t+\Delta t} - \left(d^{pc}\right)^t\right]$

for no contact → $z_1^{t+\Delta t} < \left(d^{pc}\right)^t$ let

$\left(F_s^{ep}\right)^{t+\Delta t} = 0.$

7   Calculate the elastic contact width $(a^e)^{t+\Delta t}$ as

$$(a^e)^{t+\Delta t} = \min\left\{\left[\frac{4R\left(F_s^{ep}\right)^{t+\Delta t}(1-v^2)}{\pi LE}\right]^{\frac{1}{2}}, a_{\min}\right\}.$$

8   Calculate the elastic contact stress $p_0^{t+\Delta t}$ as

$$p_0^{t+\Delta t} = \left(\frac{E\left(F_s^{ep}\right)^{t+\Delta t}}{\pi LR(1-v^2)}\right)^{\frac{1}{2}}.$$

9   Calculate the function $f\left(\frac{p_0-p_y}{p_0}\right)^{t+\Delta t}$ as follows:

for $p_0^{t+\Delta t} < p_y$ or $p_0^{t+\Delta t} < p_0^t$ let

$f\left(\frac{p_0-p_y}{p_0}\right)^{t+\Delta t} = 1,$ or

for $p_0^{t+\Delta t} \geq p_y$ and $p_0^{t+\Delta t} \geq p_0^t$ let

$$f\left(\frac{p_0-p_y}{p_0}\right)^{t+\Delta t} =$$

$$\frac{0.5\pi}{\frac{\pi}{2} - \left[\arccos\left(1 - \frac{p_0^{t+\Delta t}-p_y}{p_0^{t+\Delta t}}\right) - \left(1 - \frac{p_0^{t+\Delta t}-p_y}{p_0^{t+\Delta t}}\right)\sqrt{\frac{2(p_0^{t+\Delta t}-p_y)}{p_0^{t+\Delta t}} - \left(\frac{p_0^{t+\Delta t}-p_y}{p_0^{t+\Delta t}}\right)^2}\right]}.$$

10   Calculate the plastic multiplier $\left(\frac{dp}{dep}\right)^{t+\Delta t}$ as

$$\left(\frac{dp}{dep}\right)^{t+\Delta t} = \frac{2\left[\left(f\left(\frac{p_0-p_y}{p_0}\right)^{t+\Delta t}\right)^2 - 1\right]}{\left[1.8864 + \ln\left(\frac{L}{2(a^e)^{t+\Delta t}}\right)\right] + 2\left[\left(f\left(\frac{p_0-p_y}{p_0}\right)^{t+\Delta t}\right)^2 - 1\right]}.$$

11   Calculate the elastic and elastoplastic spring coefficients $\left(k_s^e\right)^{t+\Delta t}$ and $\left(k_s^{ep}\right)^{t+\Delta t}$ as follows:

$$\left(k_s^e\right)^{t+\Delta t} = \frac{\pi L E}{2(1-v^2)\left[1.8864 + \ln\left(\frac{L}{2(a^e)^{t+\Delta t}}\right)\right]}, \text{ and}$$

$$\left(k_s^{ep}\right)^{t+\Delta t} = \frac{\pi L E}{2(1-v^2)\left\{\left[1.8864 + \ln\left(\frac{L}{2(a^e)^{t+\Delta t}}\right)\right] + 2\left[\left(f\left(\frac{p_0-p_y}{p_0}\right)^{t+\Delta t}\right)^2 - 1\right]\right\}}.$$

12   Calculate the dashpot coefficient $C_s^{t+\Delta t}$ as follows:

for contact:

$$C_s^{t+\Delta t} = 8L\left(a^e\right)^{t+\Delta t}\sqrt{2\rho G\frac{1-v}{1-2v}} \quad \text{for} \quad v \le \frac{1}{3}, \text{ or}$$

$$C_s^{t+\Delta t} = 16L\left(a^e\right)^{t+\Delta t}\sqrt{\rho G} \quad \text{for} \quad \frac{1}{3} < v \le \frac{1}{2};$$

for no contact:

$$C_s^{t+\Delta t} = 0.$$

13   Return to step 2 for computing the displacement in the subsequent time step.

Finally, the whole reaction of the soil, F_s, which includes the elastic and viscous response, is calculated as

$$F_s^{t+\Delta t} = \left(F_s^{ep}\right)^{t+\Delta t} + C_s^{t+\Delta t}\frac{z_s^{t+\Delta t} - z_s^t}{\Delta t}.$$

Figure 7.15 shows the results of the hysteretic cycles that relate the dynamic load applied by the cylinder with its displacement; and the soil reaction, considering the viscous component, with the soil displacement. The results clearly show how the plastic displacement of the soil increases as the yielding stress of the soil diminishes. However, for each loading cycle, the irrecoverable displacement is constant because of the assumption of a Tresca yield criterion. More realistic criteria, such as the Mohr–Coulomb criterion, must consider the soil's hardening that results from plastic strains. Therefore, when adopting this kind of criterion, the accumulation of plastic displacement must decrease as the number of cycles increases.

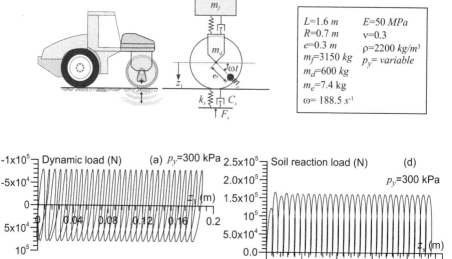

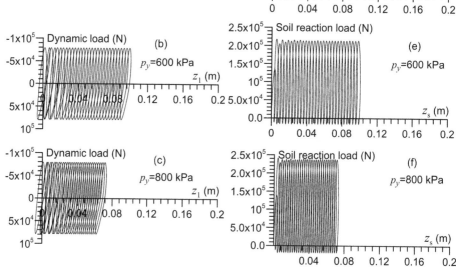

Figure 7.15 Hysteretic cycles relating the dynamic load and the soil reaction with the displacements of the drums and the soils for a vibratory drum resting on soils that react in the elastoplastic domain. Characteristics of the vibratory drum: $L = 1.6$ m, $R = 0.7$ m, mass of the frame 3,150 kg, mass of the drum 600 kg, eccentric mass 7.4 kg, vibratory frequency 30 Hz. Characteristics of the soil $E = 100$ MPa, $v = 0.3$ $\rho = 2,200$ kg/m³, and the yielding stress p_y varies from 300 to 800 kPa.

Bibliography

1. Dietmar Adam and Friedrich Kopf. *Theoretical analysis of dynamically loaded soil*. na, 2000.

2. Eduardo E Alonso, Jean-Michel Pereira, Jean Vaunat, and Sebastia Olivella. A microstructurally based effective stress for unsaturated soils. *Géotechnique*, 60(12):913–925, 2010.

3. EE Alonso, A Josa, and A Gens. Modelling the behaviour of compacted soil upon wetting. *Raúl Marsal Volume, SMMS, México*, pages 207–223, 1992.

4. EE Alonso, NM Pinyol, and A Gens. Compacted soil behaviour: initial state, structure and constitutive modelling. *Géotechnique*, 63(6):463, 2013.

5. Ins ARA. Guide for mechanistic-empirical design of new and rehabilitated pavement structures, 2004.

6. J Boussinesq. Équilibre d'élasticité d'un sol isotrope sans pesanteur, supportant différents poids. *CR Math. Acad. Sci. Paris*, 86(86):1260–1263, 1878.

7. JR Boyce. A non linear model for the elastic behaviour of granular materials under repeated loading. In *Proc. International symposium on soils under cyclic and transient loading, Swansea*, 1980.

8. Jean-Louis Briaud and Jeongbok Seo. Intelligent compaction: overview and research needs. *Report to the Federal Highway Administration*, 2003.

9. Royal Harvard Brooks and Arthur Thomas Corey. Hydraulic properties of porous media and their relation to drainage design. *Transactions of the ASAE*, 7(1):26–0028, 1964.

10. Bernardo Caicedo. *Geotechnics of Roads: Fundamentals*. CRC Press, 2018.

11. Bernardo Caicedo, Octavio Coronado, Jean Marie Fleureau, and A Gomes Correia. Resilient behaviour of non standard unbound granular materials. *Road Materials and Pavement Design*, 10(2):287–312, 2009.

12. Bernardo Caicedo, Manuel Ocampo, and Luis Vallejo. Modelling comminution of granular materials using a linear packing model and Markovian processes. *Computers and Geotechnics*, 2016.

13. Bernardo Caicedo, Manuel Ocampo, Luis Vallejo, and Julieth Monroy. Hollow cylinder apparatus for testing unbound granular materials of pavements. *Road Materials and Pavement Design*, 13(3):455–479, 2012.

14. Bernardo Caicedo, Julián Tristancho, Luc Thorel, and Serge Leroueil. Experimental and analytical framework for modelling soil compaction. *Engineering Geology*, 175:22–34, 2014.

15. Bernardo Caicedo and Luis E Vallejo. Experimental study of the strength and crushing of unsaturated spherical particles. *Unsaturated Soils: Research and Applications*, pages 425–430, 2012.

16. Valentino Cerruti. *Ricerche intorno all'equilibrio de'corpi elastici isotropi: memoria del Valentino Cerruti*. Salviucci, 1882.

17. Octavio Coronado, Bernardo Caicedo, Said Taibi, Antonio Gomes Correia, and Jean-Marie Fleureau. A macro geomechanical approach to rank non-standard unbound granular materials for pavements. *Engineering Geology*, 119(1):64–73, 2011.

18. Octavio Coronado Garcia. *Etude du comportement mécanique de matériaux granulaires compactés non saturés sous chargements cycliques*. PhD thesis, Châtenay-Malabry, Ecole Centrale Paris, 2005.

19. Jean Côté and Jean-Marie Konrad. A generalized thermal conductivity model for soils and construction materials. *Canadian Geotechnical Journal*, 42(2):443–458, 2005.

20. Jean Côté and Jean-Marie Konrad. Thermal conductivity of base-course materials. *Canadian Geotechnical Journal*, 42(1):61–78, 2005.

21. F De Larrard. Compacité et homogénéité des mélanges granulaires. *LC d. P. e. Chaussées (ed) Structures Granulaires et Formulation des Betons, 1st edn. LCPC*, 2000.

22. Francois De Larrard. *Concrete mixture proportioning: a scientific approach*. CRC Press, 1999.

23. François de Larrard, Vincent Ledee, Thierry Sedran, Frédérick Brochu, and Jean-Bernard Ducassou. Nouvel essai de mesure de compacité des fractions granulaires à la table à chocs. *Bulletin des laboratoires des ponts et chaussées*, 1(246-47):101–115, 2003.

24. Daniel A De Vries. Thermal properties of soils. *Physics of plant environment*, 1963.

25. Chandrashekhar S Dharankar, Mahesh Kumar Hada, and Sunil Chandel. Numerical generation of road profile through spectral description for simulation of vehicle suspension. *Journal of the Brazilian Society of Mechanical Sciences and Engineering*, 39(6):1957–1967, 2017.

26. CJ Dodds and JD Robson. The description of road surface roughness. *Journal of sound and vibration*, 31(2):175–183, 1973.

27. Wayne A Dunlap. *A report on a mathematical model describing the deformation characteristics of granular materials*. Texas A&M University, Texas Transportation Institute, 1963.

28. Y El Mghouchi, A El Bouardi, Z Choulli, and T Ajzoul. New model to estimate and evaluate the solar radiation. *International Journal of Sustainable Built Environment*, 3(2):225–234, 2014.

29. International Organization for Standardization, Technical Committee ISO/TC, Mechanical Vibration, Shock. Subcommittee SC2 Measurement, Evaluation of Mechanical Vibration, and Shock as Applied to Machines. *Mechanical Vibration–Road Surface Profiles–Reporting of Measured Data*. International Organization for Standardization, 1995.

30. Delwyn G Fredlund and Anqing Xing. Equations for the soil-water characteristic curve. *Canadian geotechnical journal*, 31(4):521–532, 1994.

31. DG Fredlund, Anqing Xing, and Shangyan Huang. Predicting the permeability function for unsaturated soils using the soil-water characteristic curve. *Canadian Geotechnical Journal*, 31(4):533–546, 1994.

32. NA Fröhlich. *Druckverteilung im Baugrunde: mit besonderer Berücksichtigung der plastischen Erscheinungen*. Springer-Verlag, 2013.

33. Jean-Jacques Fry. *Contribution à l'étude et à la pratique du compactage*. PhD thesis, Ecole Centrale de Paris, 1977.

34. D Gallipoli, SJ Wheeler, and M Karstunen. Modelling the variation of degree of saturation in a deformable unsaturated soil. *Géotechnique.*, 53(1):105–112, 2003.

35. A Gomes Correia. Small strain stiffness under different isotropic and anisotropic stress conditions of two granular granite materials. *Advanced Laboratory Stress-Strain Testing of Geomaterials*, pages 209–216, 2001.

36. Matthew R Hall, Pejman Keikhaei Dehdezi, Andrew R Dawson, James Grenfell, and Riccardo Isola. Influence of the thermophysical properties of pavement materials on the evolution of temperature depth profiles in different climatic regions. *Journal of Materials in Civil Engineering*, 24(1):32–47, 2011.

37. Eqramul Hoque and Fumio Tatsuoka. Anisotropy in elastic deformation of granular materials. *Soils and Foundations*, 38(1):163–179, 1998.

38. P Hornych, A Kazai, and JM Piau. Study of the resilient behaviour of unbound granular materials. *Proc. BCRA*, 98:1277–1287, 1998.

39. Alfreds R Jumikis. *Soil mechanics*. Van Nostrand, 1968.

40. Thomas Keller. A model for the prediction of the contact area and the distribution of vertical stress below agricultural tyres from readily available tyre parameters. *Biosystems engineering*, 92(1):85–96, 2005.

41. Thomas Keller, Pauline Défossez, Peter Weisskopf, Johan Arvidsson, and Guy Richard. Soilflex: A model for prediction of soil stresses and soil compaction due to agricultural field traffic including a synthesis of analytical approaches. *Soil and Tillage Research*, 93(2):391–411, 2007.

42. N Khalili and MH Khabbaz. A unique relationship of chi for the determination of the shear strength of unsaturated soils. *Geotechnique*, 48(5), 1998.

43. William Arthur Lewis. Investigation of the performance of pneumatic tyred rollers in the compaction of soil. Road research technical paper no 45, Department of Scientific and Industrial Research HMSO London, 1959.

44. G Lundberg. Elastische beruehrung zweier halbraeume. *Forschung auf dem Gebiet des Ingenieurwesens A*, 10(5):201–211, 1939.

45. J Mandel and J Salençon. The bearing capacity of soils on a rock foundation/in french. In *Soil Mech & Fdn Eng Conf Proc/Mexico/*, 1969.

46. Melvin Mooney. The viscosity of a concentrated suspension of spherical particles. *Journal of colloid science*, 6(2):162–170, 1951.

47. YY Perera, CE Zapata, WN Houston, and SL Houston. Prediction of the soil-water characteristic curve based on grain-size-distribution and index properties. In *Advances in Pavement Engineering*, pages 1–12. ASCE, 2005.

48. Nadia Pouliot, Thierry Sedran, François de LARRARD, and Jacques Marchand. Prediction of the compactness of roller-compacted concrete using a granular stacking model. *Bulletin des Laboratoires des Ponts et Chaussées*, 1(233):23–36, 2001.

49. Eduardo J Rueda, Silvia Caro, and Bernardo Caicedo. Influence of relative humidity and saturation degree in the mechanical properties of hot mix asphalt (hma) materials. *Construction and Building Materials*, 153:807–815, 2017.

50. Eduardo J Rueda, Silvia Caro, and Bernardo Caicedo. Mechanical response of asphalt mixtures under partial saturation conditions. *Road Materials and Pavement Design*, 20(6):1291–1305, 2019.

51. AJ Sandström and CB Pettersson. Intelligent systems for qa/qc in soil compaction. In *Proc., 83rd Annual Transportation Research Board Meeting*, pages 11–14, 2004.

52. Masanobu Shinozuka. Monte Carlo solution of structural dynamics. *Computers & Structures*, 2(5-6):855–874, 1972.

53. Masanobu Shinozuka and George Deodatis. Simulation of stochastic processes by spectral representation. *Applied Mechanics Reviews*, 44(4):191–204, 1991.

54. T Stovall, F De Larrard, and M Buil. Linear packing density model of grain mixtures. *Powder Technology*, 48(1):1–12, 1986.

55. Alessandro Tarantino and E De Col. Compaction behaviour of clay. *Géotechnique*, 58(3):199–213, 2008.

56. F Tatsuoka, M Ishihara, T Uchimura, and A Gomes-Correia. Non-linear resilient behaviour of unbound granular materials predicted by the cross-anisotropic hypo-quasi-elasticity model. In *Unbound Granular Materials, Laboratory Testing I-Situ Testing and Modelling*, pages 197–204. Balkema, 2999.

57. F Tatsuoka, RJ Jardine, D Lo Presti, H Di Benedetto, and T Kodaka. Characterising the pre-failure deformation properties of geomaterials. In *Fourteenth International Conference on Soil Mechanics and Foundation Engineering. ProceedingsInternational Society for Soil Mechanics and Foundation Engineering*, volume 4, 1999.

58. Fumio Tatsuoka and Yukihiro Kohata. Stiffness of hard soils and soft rocks in engineering applications. In *Pre-Failure Deformation of Geomaterials. Proceedings of The International Symposium, 12-14 September 1994, Sapporo, Japan. 2 Vols*, 1995.

59. C Thornton. Coefficient of restitution for collinear collisions of elastic-perfectly plastic spheres. *Journal of Applied Mechanics*, 64(2):383–386, 1997.

60. Takao Ueda, Takashi Matsushima, and Yasuo Yamada. Effect of particle size ratio and volume fraction on shear strength of binary granular mixture. *Granular Matter*, 13(6):731–742, 2011.

61. Jacob Uzan. Characterization of granular material. *Transportation research record*, 1022(1):52–59, 1985.

62. Luis E Vallejo and Roger Mawby. Porosity influence on the shear strength of granular material–clay mixtures. *Engineering Geology*, 58(2):125–136, 2000.

63. R Van Ganse. Les infiltrations dans les chaussées: évaluations prévisionnelles. In *Symposium on road drainage*, pages 22–24, 1978.

64. PJ Vardanega and TJ Waters. Analysis of asphalt concrete permeability data using representative pore size. *Journal of Materials in Civil Engineering*, 23(2):169–176, 2011.

65. G Verros, S Natsiavas, and C Papadimitriou. Design optimization of quarter-car models with passive and semi-active suspensions under random road excitation. *Modal Analysis*, 11(5): 581–606, 2005.

66. TH Waters. A study of water infiltration through asphalt road surface materials. In *International symposium on subdrainage in roadway pavements and subgrades*, volume 1, pages 311–317, 1998.

67. Peter J Williams. *The surface of the earth: an introduction to geotechnical science*. Addison-Wesley Longman Ltd, 1982.

68. John P Wolf. *Foundation vibration analysis using simple physical models*. Pearson Education, 1994.

69. W-J Xu, R-L Hu, and R-J Tan. Some geomechanical properties of soil–rock mixtures in the hutiao gorge area, china. *Geotechnique*, 57(3):255–264, 2007.

70. TL Youd. Factors controlling maximum and minimum densities of sands. In *Evaluation of relative density and its role in geotechnical projects involving cohesionless soils*. ASTM International, 1973.

71. Nan Zhang and Zhaoyu Wang. Review of soil thermal conductivity and predictive models. *International Journal of Thermal Sciences*, 117:172–183, 2017.

Index